S.M.A.R.T. Environments

The book series will cover a broad spectrum of data-driven multivalence in the built environment. The individual volumes will set the stage for understanding multiple ways in which the exponential escalation of digital ubiquity in the contemporary environment is being absorbed, modulated, processed and actively used for enhancing the performance of our built environment. S.M.A.R.T. is used as an acronym for Systems & Materials in Architectural & Urban Research and Technology, with a specific focus on investigating the intricate relationship between information systems and associative material formations at variable scales within the built environment. This is deeply rooted in exploring multi-domain research and design strategies involving nonlinear processes for developing data-driven generative meta-design systems. This book series is relevant to shaping our built environment and will explore novel design and research paradigms.

Sushobhan Majumdar • Vinay Kandpal
Leonidas G. Anthopoulos
Editors

Smart Cities

Innovations, Challenges and Future Perspectives

Springer

Editors
Sushobhan Majumdar
Pandit Raghunath Murmu Smriti Mahavidyalaya (Affiliated to Bankura University)
Bankura, West Bengal, India

Vinay Kandpal
Department of Management Studies
Graphic Era (Deemed to be University)
Dehradun, Uttarakhand, India

Leonidas G. Anthopoulos
Department of Business Administration
University of Thessaly
Larissa, Greece

ISSN 2523-8469 ISSN 2523-8477 (electronic)
S.M.A.R.T. Environments
ISBN 978-3-031-59845-6 ISBN 978-3-031-59846-3 (eBook)
https://doi.org/10.1007/978-3-031-59846-3

This Springer imprint is published by the registered company Springer Nature Switzerland AG
The registered company address is: Gewerbestrasse 11, 6330 Cham, Switzerland

Foreword

I extend this letter to the consideration of the editorial committee to support the development of the book entitled *Smart Cities: Innovations, Challenges and Future Perspectives*, whose publication is planned for Springer Nature, which, from my experience, represents a great opportunity to establish strategic and visionary challenges and opportunities for the implementation of the concept of smart cities. The book, edited by Dr. Sushobhan Majumdar, Dr. Vinay Kandpal, and Dr. Leonidas Anthopoulos, brings together contributions from leading experts in the field of smart cities and aims to provide a comprehensive overview of the innovations, challenges, and future perspectives in its evolution. The diverse perspectives and insights presented in this publication make it a valuable resource for academics, researchers, and practitioners interested in advancing smart city initiatives.

This book is divided into four parts. Part I—Defining the Smart City and Path to Smart Urbanization—contains four chapters which are oriented towards conceptual and theoretical exploration of the ecosystem of smart cities from data science and the implementation of digital tools aimed at the visualization of information and bibliometric analysis of data from research. Part II—Digital Transformations: Harnessing Technology for Smarter Cities—contains four chapters that encompass new visions around the Internet of Things, metaverse, artificial intelligence and cybersecurity, digital twins, and recent trends that will undoubtedly set a benchmark for innovation and technology in the development of new industries. Part III—Sustainable Urban Ecosystems: Integrating the Circular Economy and Planning in Smart Cities—contains four chapters oriented towards an approach to circularity, which implies the recognition of processes associated with environmental awareness in the management of energy, actions aimed at the regeneration of ecosystems, and initiatives identified as references for new actions within the framework of smart cities. The final Part IV—The Synergy of Tomorrow: Smart Cities and Sustainable Futures—guides new scenarios and perspectives towards the smart cities of the future, from various latitudes and in developing countries, where concepts such as e-governance and innovative urban environments, in addition to integrating smart infrastructure and buildings with minimal environmental impact. Thus, throughout four parts, a great proposal can be analysed based on new perspectives and visions towards the cities of the future, where the smart city not only integrates the use of emerging technologies but also considers the impact of all sectors like social, economic, cultural, environmental, and so on. Given the above, I acknowledge and offer great congratulations to the editors for such a valuable project as well as to all the collaborators, which will undoubtedly be an important reference for all sectors in the international level.

Sincerely

PhD. Arch. Mariana Flores García
Full-Time Researcher Universidad
La Salle Bajío, Mex /
Director National Institute of Architecture
and Urbanism INAU - FCARM
Doctorate in Architecture, Housing
and Contemporary Habitat
UAT, Tamaulipas, Mexico. - EIDUS University
of Seville, Spain
Master MREEB_US Seville,
Spain - MDA_ Universidad La Salle Bajío Méx
email: maryfg83@gmail.com, mfloresg@lasallebajio.edu.mx
Móvil: +52 492 113 0982
https://www.linkedin.com/in/arqmarianaflores/,
https://www.researchgate.net/profile/Mariana-Flores-Garcia

Preface

Amidst the rising urbanization of human settlements, the idea of "smart cities" has evolved as a symbol of sustainable, efficient, and technologically sophisticated urban lifestyles. *Smart Cities Innovations, Challenges and Future Perspectives* is a collection of works that examines the complex idea of smart cities, investigating the interdependence between advanced technology and the growth of urban areas.

This book consolidates a wide range of viewpoints from prominent authorities and academics worldwide. Authored by experts in their respective domains, each chapter offers comprehensive insights into different facets of smart city development. This book explores the fundamental ideas and structures that define smart cities, as well as the complex difficulties related to privacy, security, and governance. It delves into the complicated network of elements that influence the development of smart cities both now and in the future.

The book is organized in a manner that leads the reader through a progression, starting with the fundamental principles that define a smart city. Subsequently, it delves into examining certain uses and advancements, encompassing the impact of the Internet of Things (IoT) in industrial sectors, as well as the consequences of artificial intelligence and the metaverse in urban development. The book emphasizes the crucial element of sustainability, specifically focusing on the contribution of smart cities in promoting a circular economy and sustainable urban ecosystems.

In light of the global challenges posed by growing urbanization, climate change, and the imperative for sustainable development, the ideas and insights provided in this book hold heightened significance. It functions as both a comprehensive scholarly resource and a forward-thinking guide for legislators, urban planners, and technological entrepreneurs who are leading the way in defining the future of our cities.

We commence this expedition with the awareness that the route to intelligent urbanization is fraught with obstacles. However, the combined knowledge and understanding presented in this book shed light on the potential and methods for developing cities that are not only intelligent in their utilization of technology but also astute in their commitment to sustainability, inclusion, and resilience.

Chapter 1 explores the essential components, architecture, applications, and security concerns in smart cities, along with addressing key research questions. Chapter 2 focuses on designing child-friendly urban environments in Indonesia, emphasizing the need for cities that cater to the well-being of children. Chapter 3 discusses the use of graphical analysis and algorithms in understanding and planning smart cities, highlighting the importance of visualization in urban development. Chapter 4 presents a bibliometric analysis of smart city research, providing insights into the evolution and trends in smart city studies. Chapter 5 examines the role of the Internet of Things (IoT) in transforming industrial landscapes within smart cities, showcasing its impact on modern industrial practices. Chapter 6 investigates the challenges and hegemonic aspects of smart city implementation in Indonesia, offering a critical view of urban development strategies. Chapter 7 explores the integration of artificial intelligence and the metaverse in urban planning, addressing cybersecurity challenges in these virtual environments. Chapter 8 discusses the concept of digital twins as transformative tools in smart city planning and management, illustrating their potential in urban development.

Chapter 9 focuses on sustainable city initiatives in Cape Town, highlighting the role of upcycling organizations in promoting a circular economy. Chapter 10 delves into the control towers of energy management systems, demonstrating their importance in optimizing energy flows in smart urban settings. Chapter 11 explores the synergy between smart city concepts and circular economy practices in India, emphasizing innovation in urban sustainability. Chapter 12 examines the case of Visakhapatnam, Andhra Pradesh, to understand how political priorities shape smart city initiatives. Chapter 13 discusses the opportunities and challenges in the journey toward smart and sustainable urban development in Hong Kong. Chapter 14 investigates new urban technologies aimed at energy harvesting and environmental initiatives, crucial for smart city development. Chapter 15 presents perspectives on intuitive e-governance systems in developing nations, focusing on how they contribute to smart city frameworks. Chapter 16 explores the innovations in smart buildings and infrastructure, crucial for intelligent urban living and the future of smart cities. Chapter 17 delves into the transformative role of advanced technologies in shaping urban resilience. It emphasizes the strategic integration of smart technologies as a crucial element in evolving urban paradigms, driving cities toward enhanced sustainability, efficiency, and adaptability in the face of various challenges.

Join us on a journey into the future, where the convergence of technology and urban living presents unprecedented opportunities for the advancement of human society.

Bankura, West Bengal, India — Sushobhan Majumdar

Dehradun, Uttarakhand, India — Vinay Kandpal

Larissa, Greece — Leonidas G. Anthopoulos

The original version of this book was revised. The version supplied here includes the following elements: Foreword, Preface, Acknowledgements, and About the Editors.

Acknowledgements

The enormous amount of effort put forth by urban planners, urban geographers, policymakers, and decision-makers in analyzing the innovations, difficulties, and potential applications of smart cities has served as the inspiration for this book. We owe a debt of thanks to the researchers and working professionals whose substantial efforts have made this book a reality. We are grateful to the anonymous reviewers for their insightful comments, which have raised the book's caliber. Throughout the process of editing this book, we were incredibly grateful for the assistance that came from our instructors, parents, students, colleagues, and partners. Without the unwavering inspiration from our students, the expertise of our teachers, the zeal of our colleagues and collaborators, and the support of our family, none of our efforts would have been possible. Finally, but just as importantly, we would like to thank Springer, the publishing editor, and our publisher for their ongoing support.

Contents

About the Editors

Sushobhan Majumdar is an applied geographer who holds a doctorate in geography from the Department of Geography at the Faculty of Science at Jadavpur University in Kolkata and a postgraduate degree from Presidency University in Kolkata. He is currently employed by the Government of West Bengal as a "Group A Officer" in the "Minority Affairs and Minority Education" Department. He is presently engaged in research on the design and development of cities' geographies. Urban geography, social geography, regional development and planning, remote sensing and GIS, and transport geography are among his main areas of study. Additionally, he is working on different development projects for the Government of West Bengal's Socio-Economic Planning Unit of the Kolkata Metropolitan Development Authority (KMDA). He has more than 50 research papers published in a variety of reputable journals, proceedings, and edited volumes both nationally and internationally. He has already authored two edited volumes for Springer Nature and Routledge Group on urban-related issues. He is now editor of various renowned publishing houses such as Springer, Routledge, Wiley, Elsevier, etc. In excess of 30 national and international seminars and conferences, both in India and overseas, he has presented papers. He gave guest or special talks as part of numerous national and regional programs. He holds memberships in the Geographical Society of India, the Regional Science Association of India (RSAI), the Institute of Indian Geographers (IIG), and the National Association of Geographers of India (NAGI). He has recently been heavily involved in mapping and simulating surface temperatures in Indian metropolises and their effects on the environment.

Vinay Kandpal is a finance expert and educator who teaches and mentors students at Graphic Era University, Dehradun, India, with over 18 years of experience in academia and industry. He is a Professor in the Department of Management Studies, Graphic Era University, Dehradun, India, where he imparts his knowledge and insights on various topics in finance, management, and education. He holds a D.Litt. (Post Doc) and a Ph.D. in Finance from Kumaun University, India, and is also a Postdoctorate Scholar of PG Administracao-Paulista University—UNIP, Brazil.

Dr. Kandpal is also a passionate and prolific researcher and author, who has published more than 62 research papers and 6 books on diverse subjects, such as sustainability, fintech, smart cities, risk management, circular economy, and financial inclusion. He has presented his work at prestigious international conferences and institutions and serves on the editorial and reviewer boards of several well-regarded journals. He is a Full-Time Member of the Association of North America Higher Education International (ANAHEI). Dr. Kandpal's impressive career highlights his commitment to advancing knowledge and fostering innovation in finance.

Leonidas G. Anthopoulos is a Professor in E-Business at the University of Thessaly, Greece. During his current mission and at his previous job positions (as an Expert Counselor at the Greek Ministry of Foreign Affairs, as a Special Advisor of the Mayor of Trikala, Greece, and as an ICT project manager at the Information Society S.A. (Greece)), he planned and supervised several critical projects, worth mentioning the smart city of Trikala. He is the Director of the "Intelligent Production Systems and Smart Cities" Research Institute and the Director of the Postgraduate Program in Agile Management at the University of Thessaly, Greece. He is a member of numerous committees, worth mentioning working for the ITU (Rapporteur of Q13 of the SG5), the head of the Smart and Sustainable Cities' working group for the Greek Standards, the Associate Editor of the IET Smart Cities journal, etc. He is the author and editor of several book collections, which can be located in Cities, Technological Forecasting and Social Change and Government Information Quarterly (Elsevier), ACM Computing Surveys monograph for Springer etc. His research interests concern, among others, smart city, e-Government, Enterprise Architecture, Strategic Management, etc.

Part I
Defining the Smart City and Path to Smart Urbanization

Chapter 1
Shaping the Next Generation Smart City Ecosystem: An Investigation on the Requirements, Applications, Architecture, Security and Privacy, and Open Research Questions

Wasswa Shafik

Abstract The concept of a Smart City Ecosystem has garnered significant attention due to its potential to revolutionize urban living through advanced technologies and data-driven solutions. This chapter explores the multifaceted dimensions of this ecosystem by delving into its requirements, architecture, applications, security considerations, and the open research questions it present. In the realm of requirements, it highlights the intricate balance between technological innovation, social inclusivity, economic sustainability, and environmental preservation that smart cities demand. It navigates these requirements' challenges, shedding light on the trade-offs and synergies with which city planners and stakeholders must grapple. The chapter provides a comprehensive overview of the foundational structure of a smart city ecosystem. It dissects the intricate web of interconnected components, layers, and communication infrastructures that underpin its functionality. The architecture's complexity is unraveled by examining the integration of diverse technologies and data sources. The diverse applications enabled by the smart city ecosystem are categorized across sectors such as transportation, energy, healthcare, and more. These applications are explored in detail, demonstrating their potential to enhance efficiency, sustainability, and citizen well-being. Real-world examples are presented, showcasing the transformative impact of these innovations. Security is critical in this ecosystem, as interconnectivity introduces new vulnerabilities. The chapter discusses the intricate web of security challenges encompassing data privacy, cybersecurity, and potential threats. Strategies and best practices for safeguarding smart city systems are highlighted, underscoring the paramount importance

W. Shafik (✉)
Dig Connectivity Research Laboratory, Kampala, Uganda

School of Digital Science, Universiti Brunei Darussalam, Gadong, Brunei
e-mail: wasswashafik@ieee.org

S. Majumdar et al. (eds.), *Smart Cities*, S.M.A.R.T. Environments,
https://doi.org/10.1007/978-3-031-59846-3_1

of secure design and operation. Then, it raises Open Research Questions that underscore the dynamic nature of the smart city landscape. It identifies areas where further exploration is needed, including the integration of emerging technologies, ethical considerations, and the societal impacts of data-centric urban environments. These questions beckon researchers and practitioners to actively shape the trajectory of smart city development.

Keywords Sustainable cities and communities · Good health and well-being · Response consumption and production · Climate action · Security and privacy · Industry · Innovation and infrastructure · Affordable and clean energy

1.1 Introduction

The ultimate target of the "smart city" concept is to create a sustainable living environment while integrating high-tech information and communication technologies (ICT) approaches (Yang et al., 2021). In an era of rapid urbanization and technological advancement, smart cities have emerged as a beacon of transformative urban development (Tareen et al., 2023). A smart city goes beyond mere urban planning; it is a dynamic ecosystem that harnesses cutting-edge technologies to enhance the quality of life for its inhabitants while fostering sustainable growth (El-Shafai et al., 2023; Wang et al., 2023). As our world becomes increasingly interconnected, the notion of a smart city gains profound significance, promising solutions to the pressing challenges of urban congestion, resource scarcity, and environmental degradation (Gagliardi & Albergo, 2023). These cities leverage data-driven insights to optimize resource allocation, streamline transportation, enhance healthcare accessibility, and elevate the overall urban experience (Mengash et al., 2023). The smart city concept encapsulates a vision of urban living that integrates physical and digital realms, paving the way for innovation across sectors and empowering citizens with unprecedented connectivity and efficiency.

The significance of smart cities becomes all the more pronounced as cities continue to attract a growing majority of the global population (Mengash et al., 2023; Shafik, 2024a). By 2050, around 68% of humanity is projected to reside in urban areas, placing an immense strain on existing infrastructure and necessitating novel solutions for sustainable urban growth. Smart cities offer a roadmap for meeting this challenge by leveraging technology as an accessory and a fundamental driver of progress (Pandya et al., 2023). These cities optimize resource utilization, reduce energy consumption, and minimize environmental impact, aligning urban expansion with the imperative of environmental stewardship.

However, the journey toward smart city realization is not devoid of challenges. The intricate interplay between technological innovation, social equity, regulatory frameworks, and economic sustainability demands a holistic approach to urban planning. Ethical considerations surrounding data privacy, equity of access, and the potential for technology-driven exclusion also need careful navigation (Badr, 2023). This chapter embarks on a comprehensive exploration of the multifaceted smart city ecosystem. It delves into its essential requirements, architectural intricacies, diverse applications, security dimensions, and the intriguing open research questions it bequeaths using generative technologies (Shafik, 2024b; Tripathy et al., 2023). By unpacking the layers of this dynamic paradigm, we strive to offer insights into how smart cities can reshape the urban landscape, elevate citizen well-being, and set a precedent for sustainable urbanization in the twenty-first century and beyond.

Amidst the intricate web of challenges and opportunities, smart cities promise to redefine how we conceive, design, and experience urban life. They epitomize a fusion of innovation and adaptability, where urban infrastructure responds intelligently to its residents' needs, fostering not only an efficient but also inclusive and responsive environment (Berawi, 2023). The convergence of the digital and physical realms empowers citizens with real-time information, enabling them to make informed decisions and actively participate in shaping their surroundings (Mengash et al., 2023).

At the heart of the smart city vision lies the concept of interconnectedness (Fig. 1.1), converging to create an intricate fabric that allows urban systems to communicate, learn, and self-regulate (Shafik, 2023b). Whether optimizing traffic flow to reduce congestion, managing energy consumption to minimize carbon footprint, or providing tailored healthcare services, integrating technology ensures that the city functions seamlessly, efficiently, and sustainably.

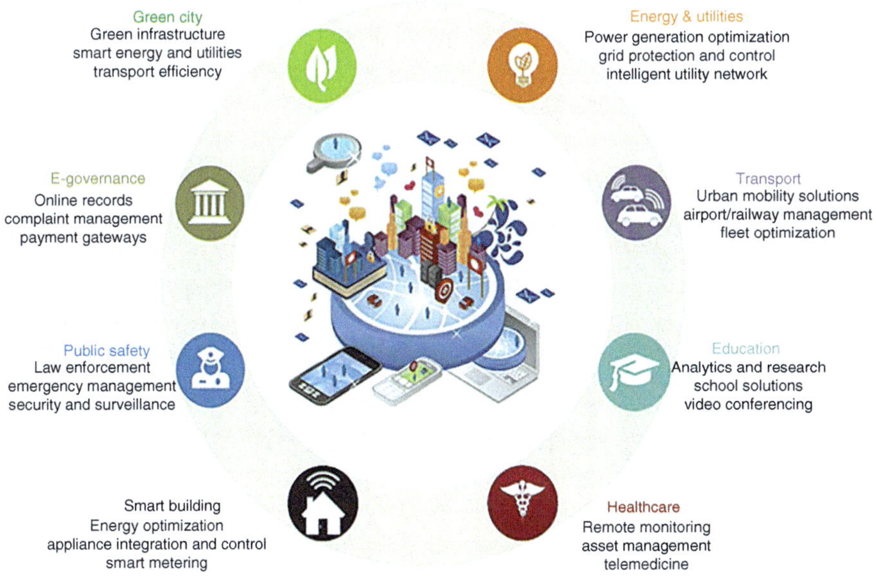

Fig. 1.1 Smart city and related entities

However, the evolution toward smart cities necessitates a collaborative effort between various stakeholders, including governments, businesses, researchers, and citizens. Policymakers are crucial in establishing regulatory frameworks that balance innovation with privacy and security safeguards (Shafik, 2023a; Tareen et al., 2023). Businesses contribute by developing and implementing the technologies that power these cities, while citizens engage through their feedback, demands, and aspirations for a better urban existence (AlZu'bi et al., 2023; Shafik et al., 2023).

As the journey toward smarter urban environments continues, it becomes evident that smart cities are not just a romantic concept but a pragmatic necessity. This chapter embarks on an expedition into the heart of this transformative paradigm, aiming to shed light on the complex interplay of factors that shape smart city ecosystems. From the technical intricacies of architecture to the ethical considerations of data usage, from the potential applications to the security challenges, and from the present-day achievements to the future research horizons, this exploration endeavors to provide a comprehensive roadmap for understanding and contributing to the evolution of smart cities, where technology and humanity harmonize to build the cities of tomorrow (Dutta et al., 2023). Figure 1.2 illustrates the public engagement in smart city acceptance using IoT and non-IoT connections globally.

It is essential to acknowledge that smart cities are not a one-size-fits-all solution. Every city possesses its unique blend of culture, history, and challenges. As such, smart technologies must be adapted to each urban landscape's specific needs and

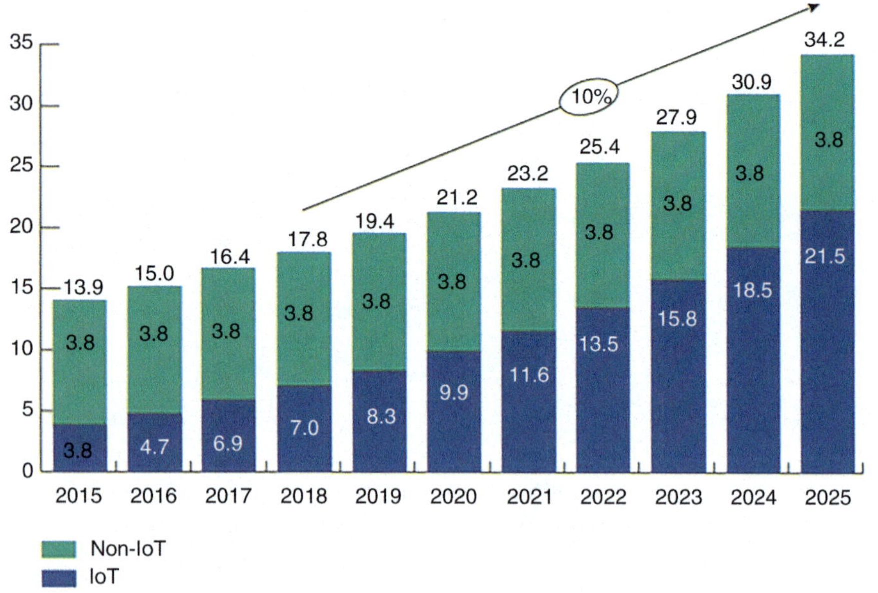

Fig. 1.2 Global Total IoT and Non-IoT active devices: 2015 to 2025 connections

contexts. What works seamlessly in one city might require adjustments and customizations in another (Aldhyani et al., 2023). Moreover, the vision of smart cities extends beyond the boundaries of city limits. Collaboration between neighboring municipalities and even countries is crucial for creating interconnected networks that optimize resource sharing, transportation, and disaster response (Aldossary, 2023). The smart city paradigm extends its reach to create smart regions and interconnected ecosystems, fostering a harmonious coexistence that transcends traditional geopolitical boundaries.

As we embark on this exploration, it is clear that smart cities have the potential to redefine not only urban living but also the very essence of citizenship and community engagement. The influx of data and technology challenges us to redefine privacy norms and develop innovative strategies for ensuring that the benefits of the smart city are distributed equitably to all segments of society (Aldossary, 2023). The journey into the world of smart cities requires holistic thinking, collaboration, and an unwavering commitment to the well-being of citizens and the planet.

The vision is grand, the challenges are intricate, but the rewards are transformative. Figure 1.3 depicts the increase in urbanization from 1990 to 2014 and a projection for 2050. Through this chapter, we aim to unravel the layers of the smart city ecosystem, offering insights, raising questions, and inspiring action to pave the way for cities that are not just smart but truly sustainable, resilient, and human-centric (Alwakeel et al., 2023). As the world continues to evolve, smart cities stand as a

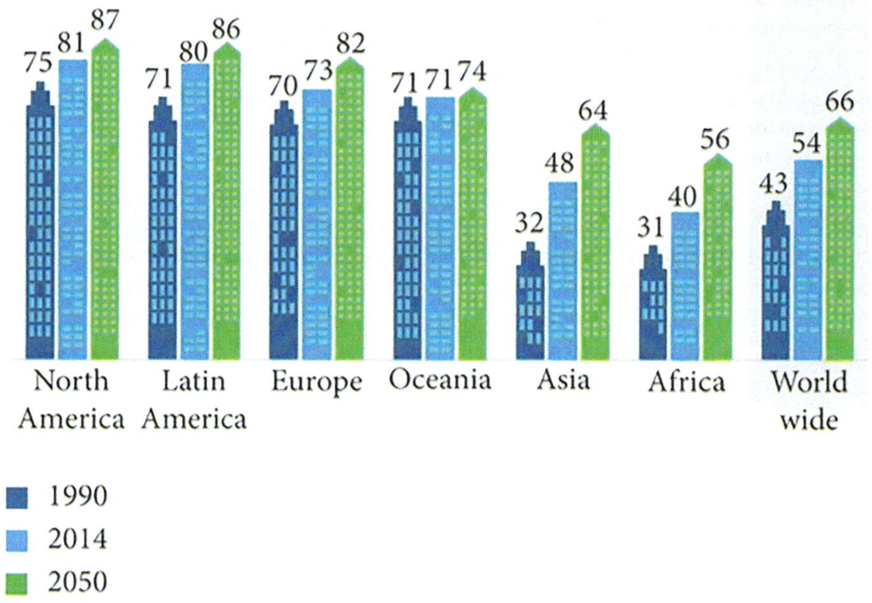

Fig. 1.3 Population increase in cities

testament to human innovation, creativity, and the shared aspiration for a better future.

This study serves as a succinct introduction to the advent of smart cities and the necessity for urbanization. It presents the following contribution as listed below:

- A detailed description and examination of the need for smart cities regarding the necessity, requirements, and architecture. The chapter presents the application of smart cities and their characteristics.
- The chapter presents an overview of the top recognized smart cities (namely, London, New York, Singapore, Busan, Amsterdam, and Sunshine Coast Regions) and their distinctive traits.
- The chapter presents the smart city attributes, such as privacy and security concerns. The attributes leveraged are highlighted, and privacy issues and possible solutions are demonstrated.
- The chapter avails future research motivations for smart city applications, some identified ones entailing blockchain integration, security considerations, collaborative filtering, and infrastructure upgrades.
- The chapter illustrates the open research questions for smart cities that need to be addressed to foster the development, application, and deployment of smart cities.
- Finally, the chapter summarizes the lessons learned that illustrate key points concerning smart cities.

The remainder of this chapter is structured as follows. Section 1.2 presents the smart city environment in three aspects, including smart city necessity (like connectivity, security and privacy, decentralization, sustainability, and connectivity), requirements (for instance, quality of services, sustainable environment, privacy and security components, intelligent services, smart infrastructure, human resource), and architecture (for example, application, data management, transmission, and sensing layer). Section 1.3 presents the application of smarties like smart connectivity infrastructure, intelligent energy and power management systems, smart connectivity, home, and transportation. Section 1.4 presents some selected smart cities globally. Section 1.5 presents the smart attributes, including clould-related security issues, physical attacks on the smart city infrastructure, privacy and security concerns, security analysis, response, and monitoring, updating and patching, secure boot and firmware integrity, security life cycle management, and secure, Mutual Authentication. Section 1.6 demonstrates the security and privacy of smart cities. Section 1.7 presents future research directions. Section 1.8 illustrates the open research questions for smart cities. Section 1.9 demonstrates the lessons learned. Finally, Section 1.10 presents the conclusion.

1.2 Smart City Environment

The requirements of a smart city ecosystem form the bedrock upon which the vision of connected, intelligent urban environments is built. These multifaceted prerequisites encompass technological innovation, social inclusivity, economic sustainability, and environmental preservation, intricately interwoven to shape the trajectory of urban development.

Technological innovation is a pivotal cornerstone, necessitating the integration of advanced technologies such as the Internet of Things (IoT), data analytics, artificial intelligence (AI), and smart infrastructure. These innovations empower cities to collect and analyze vast data streams, enabling real-time decision-making, resource optimization, and improved service delivery (Tripathi et al., 2023). From smart grids that manage energy distribution to sensors that monitor air quality, the technological requirements underpin the functionality of a smart city. However, technology alone is insufficient. Social inclusivity is an equally vital requirement, emphasizing that the benefits of smart cities must be accessible to all citizens (El-Fouly et al., 2023). Inclusivity goes beyond technological access, encompassing issues of digital literacy, affordability, and engagement (Kushwaha et al., 2023). Smart cities must be designed to bridge digital divides and ensure that diverse populations enjoy the advantages of urban innovation.

Economic sustainability is another imperative. Figure 1.4 demonstrates the application of AI and the Internet of Things (IoT). Smart city initiatives entail significant investments; to be viable, they must demonstrate straightforward economic returns. Sustainable revenue models, public-private partnerships, and strategies for long-term funding are pivotal in ensuring that smart city projects can be maintained and expanded over time (Zakzouk et al., 2023). Economic viability aligns the goals of urban development with the realities of budgetary constraints. Environmental

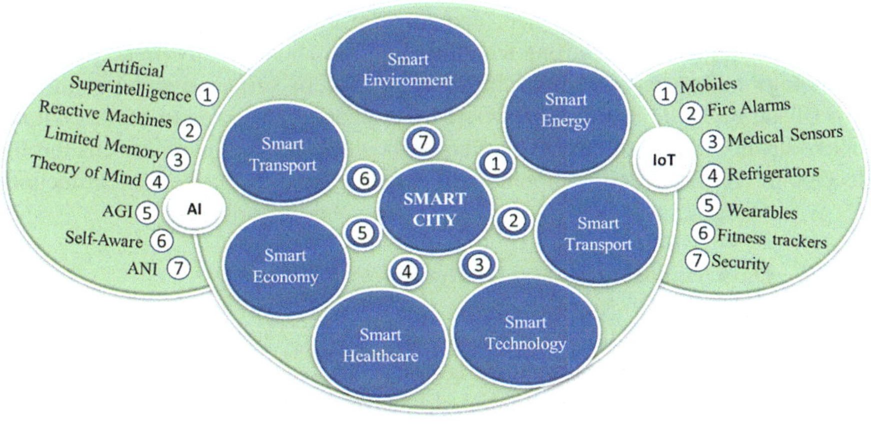

Fig. 1.4 Application Artificial Intelligence and Internet of Things in Smart Cities

preservation is integral to the fabric of smart cities. The essence of these ecosystems revolves around optimizing resource usage and minimizing ecological impact (Zakzouk et al., 2023). From energy-efficient buildings to intelligent waste management systems, the environmental requirements of smart cities address global challenges like climate change and resource depletion, seeking to forge ecologically responsible urban landscapes (Tripathi et al., 2023).

1.2.1 Smart City Necessities

These smart city necessities represent a holistic approach to urban development, where technology, sustainability, citizen well-being, and efficient governance intersect. Addressing these necessities in a balanced and integrated manner is crucial for realizing the full potential of smart cities and creating urban environments that are responsive, adaptable, and vibrant.

1.2.1.1 Technological Infrastructure

The technical facilities criteria for areas are actually basic to the prosperous application of smart metropolitan area projects. An essential criterion is actually a durable and also fast connection basis, such as 5G systems, which develop the structure for interaction between numerous gadgets as well as devices. For example, brilliant visitor traffic monitoring devices greatly rely upon real-time information swaps between sensing units and also streamlined nerve centers to enhance website traffic circulation. In addition, urban areas demand an innovative records structure that has enough storage ability and handling functionalities to deal with the substantial quantity of records created through IoT tools as well as sensing units. This facility permits functions like anticipating routine maintenance for internal solutions, enriching functional performance, and also information usage (El-Fouly et al., 2023). The influence of a strong technical framework is actually great, promoting technology and assisting the combination of modern surfacing technologies like upper-hand computers and expert systems. It encourages areas to boost civil services, enhance source allotment, and boost the general lifestyle of citizens through developing a charity for the smooth performance of assorted brilliant area functions.

1.2.1.2 Data Management and Analytics

Effective information administration and analytics are crucial elements of a smart metropolitan area environment, making it possible for areas to acquire purposeful understandings from the large number of records produced. A vital part is actually the application of strong information administration platforms, making certain information high quality, protected, and observant, along with personal privacy laws. For instance, a smart area's transit device relies upon information analytics to maximize website traffic circulation, minimize blockage, and also boost general movement. The influence appears to have boosted urban planning, lessened ecological impact, and improved performance in information allowance. Advanced analytics devices, like artificial intelligence formulas, allow anticipating upkeep for urban area facilities, lessening recovery time and also improving servicing timetables (Samy El Menchawy et al., 2023). By taking advantage of records properly, urban areas can easily create educated selections, handle difficulties proactively, and produce a structure for ongoing renovation, inevitably boosting the premium of services as well as the total health of locals. Furthermore, data-driven understandings support evidence-based policy-making, cultivating advancement and also lasting progression within the metropolitan garden.

1.2.1.3 Connectivity

Connection is actually a keystone of a smart area involving the smooth assimilation of different gadgets as well as units by means of innovative interaction systems. Fast and dependable connections, including 5G innovation, create the foundation for real-time information swap, making it possible for apps to transport devices. As an example, attached traffic signals and sensing units may interact to enhance website traffic circulation, lowering blockage and enhancing total metropolitan wheelchairs. The effect of a durable connection includes boosted effectiveness in civil services, sustaining functions like distant health care tracking, and smart network administration. Homeowners take advantage of better accessibility to relevant information and solutions, as well as real-time updates, helping in a much more linked and also well-informed area. Furthermore, the connection is actually essential for the World IoT, assisting in the publication of tools varying from intelligent home devices to ecological sensing units (Zakzouk et al., 2023). In rundown, a connection is actually transformative, developing a base for cutting-edge apps that boost city residing, enhance procedures, and lead the way for potential improvements in brilliant urban area ecological communities. A strong and pervasive network ensures citizens and businesses can access information and services without interruption.

1.2.1.4 Sustainability

Ecological durability is actually an essential column of sustainable metropolitan area progression, intending to develop city atmospheres that reduce environmental influence as well as advertising information productivity. Smart urban areas take advantage of innovations to observe as well as minimize power intake, deal with dumps efficiently, and also market environment-friendly strategies. As an example, the application of sustainable frameworks assists in the assimilation of renewable resource resources, enhancing electricity circulation and also intake norms. The influence is actually sizable, triggering a considerable decline in carbon dioxide exhausts, strengthening sky premiums, and improving dependence on tidy power resources (Cabrera et al., 2023). In addition, smart refuse monitoring units use sensing units and also information analytics to improve selection courses, decreasing gas intake as well as minimizing the ecological impact of garbage disposal. By focusing on ecological durability, smart metropolitan areas help in worldwide initiatives to battle environmental improvement, produce much healthier residing health conditions for individuals, and prepare an instance for maintainable city progression methods (Zakari et al., 2024). The assimilation of intelligent modern technologies certainly not only improves source productivity but also encourages a society of ecological understanding, urging residents and organizations to join environment-friendly campaigns definitely.

1.2.1.5 Citizen Engagement

Resident involvement is actually a foundation of brilliant area growth, highlighting the energetic involvement and also cooperation of homeowners in decision-making methods. Digital systems, mobile phone applications, and online forums allow residents to provide comments, document concerns, and definitely design metropolitan plans. An instance of reliable resident interaction is actually the application of participatory budgeting, permitting homeowners to assign a section of the area spending plan to ventures of their option, including playground redesigns or even area occasions. The effect is actually varied, promoting a feeling of permission and also area possession (Joe et al., 2023). Involved consumers end up being companions in city control, bringing about additional reactive as well as comprehensive plans. Also, enhanced clarity and interaction between regional authorizations and individuals enrich rely on communal organizations. Past plans and personal involvement boost social connections, marketing a feeling of being a member, and also discussed tasks. Over time, a metropolitan area that focuses on personal interaction constructs a durable and lively neighborhood where individuals proactively add to the co-creation of a comfortable and maintainable city atmosphere.

1.2.1.6 Security and Privacy

Safety, as well as personal privacy, is actually critical in the progression of brilliant area ecological communities to protect vulnerable information and also make certain residents leave. Durable surveillance solutions, like a shield of encryption and verification methods, are actually vital to shield against unapproved access to cyber hazards. For example, in a secure area's security unit, guaranteeing the stability and discretion of video recording nourishes using security aids to avoid unapproved accessibility or even meddling (Jun et al., 2021). The effect of helpful protection, as well as personal privacy techniques, expands past simple security; it inspires self-confidence in residents to accept brilliant innovations without anxiety about personal privacy violations. Sufficient safety and security procedures additionally secure vital commercial infrastructure, including power frameworks and also transport units, coming from prospective cyber assaults, ensuring the metropolitan area's durability when faced with growing risks (Shafik, 2023c). Personal privacy factors, installed in the layout of brilliant metropolitan area answers, display a devotion to honest information management, creating trust funds as well as nurturing complacency in the individuals. Hitting the best harmony between advancement and safeguarding private personal privacy is actually important for the effective and lasting development of intelligent urban areas (Shafik et al., 2019).

1.2.1.7 Inclusivity and Accessibility

Inclusivity and access are actually important concepts in brilliant area advancement, intending to make certain that technical innovations help all sectors of the populace, featuring those with assorted capabilities and also histories. An instance of ensuring inclusivity is actually the execution of smart mass transit bodies along with functions like real-time info shows, audio statements, and available autos for individuals with impairments (Tareen et al., 2023). The effect of such efforts is actually extensive, promoting social equity and delivering level playing fields for involvement in the city lifestyle. Comprehensive urban area layouts look at the demands of aged consumers and people with impairments. Also, those with differing economic ways generate an atmosphere where every person may effortlessly access civil services and info and also take part in communal tasks (Tripathy et al., 2023). Past bodily ease of access and electronic inclusivity make sure that innovations are actually straightforward and also serve a vast array of customers, promoting a feeling of the neighborhood as well as breaking obstacles that could or else impair equivalent involvement in the advantages of a smart metropolitan area. Inevitably, a definitely brilliant metropolitan area is actually one that focuses on inclusivity, guaranteeing that technical improvements improve the lifestyle of all its individuals.

1.2.1.8 Infrastructure and Mobility

Framework and movement are crucial parts in the advancement of sustainable urban areas, enhancing city gardens to enrich effectiveness, durability, and the general lifestyle of homeowners. An instance of a smart framework is actually the combination of Intelligent Transport Solution (ITS), which consists of smart visitor traffic control, real-time information analytics, and linked mass transit. These units relieve traffic jams, decrease drive time, and reduce ecological influence. The effect of smart framework and flexibility options is actually transformative, triggering strengthened urban planning, enriched connection, and enhanced information use (Mengash et al., 2023).

Smart frameworks, as well as energy-efficient properties, are crucial to maintaining commercial infrastructure, resulting in less electricity usage and ecological durability. Furthermore, smart wheelchair services, including ride-sharing applications as well as self-governing automobiles, reinvent transit, supplying advantages, lessening exhausts, as well as taking care of last-mile connection problems (Tareen et al., 2023). The combination of smart facilities and flexibility certainly not simply enhances day-to-day life but also helps smart spot areas adjust to potential difficulties, producing even tougher and more receptive city settings.

1.2.1.9 Resource Optimization

Information marketing is actually an essential principle in the circumstance of intelligent metropolitan area growth, highlighting effective as well as lasting use of numerous pieces of information to boost metropolitan performance. This features the efficient control of electricity, water, and transport, as well as various other crucial framework parts. In a brilliant area, evolved innovations, including IoT sensing units, records analytics, and computerization, play an essential part in improving information utilization (Tripathy et al., 2023). For example, smart networks make dependable power circulation possible, decreasing misuse and minimizing ecological influence. Water administration units make use of sensing units as well as real-time records to track intake, recognize leakages, and improve preservation attempts. In transport, smart website traffic control bodies maximize visitor traffic circulation, minimizing blockage as well as gas usage (Berawi, 2023). The influence of information marketing is actually multifaceted, including economic performance, ecological durability, and strengthened lifestyles for citizens. Through leveraging modern technology to keep an eye on, assess, as well as react to real-time records, intelligent areas may create knowledgeable choices that trigger additional maintainable information usage, lesser functional expenses, as well as a tough city atmosphere with the ability to adjust to advancing difficulties (Fong et al., 2023).

1.2.1.10 Resilience and Disaster Management

Durability and calamity administration are crucial parts of smart city planning, stressing the ability of city units to endure and recuperate from unforeseen events, including all-natural catastrophes or even emergencies. An instance of durability in a smart area is actually the unification of innovative sensing units as well as observing units to recognize very early indications of prospective catastrophes, including quakes or even floodings (Berawi, 2023). These units allow fast feedback and emptying techniques, reducing the effect on homeowners. In addition, intelligent facilities, including resistant structures and also decentralized power units, boost an area's capability to survive as well as bounce back from catastrophes. The influence of such actions is actually great, lowering the death rate, lessening home harm, and making sure a quicker comeback to normality after a situation (Dutta et al., 2023). By including real-time records, anticipating analytics, as well as successful interaction systems, intelligent urban areas may proactively take care of calamities, allot sources effectively, as well as boost total durability, ensuring the security as well as the health of their residents despite unanticipated difficulties.

1.2.2 Smart City Requirements

By holistically addressing these detailed smart city requirements, cities can create an environment that leverages technology to improve citizens' lives while ensuring sustainability, inclusivity, and economic prosperity (Mohammed & Hasan, 2023). The comprehensive integration of these elements forms the foundation of a successful and thriving smart city ecosystem.

1.2.2.1 Technological Infrastructure

Technical framework is actually a fundamental criterion in smart areas, assisting in the assimilation of state-of-the-art innovations for an enriched city lifestyle. This covers the requirement for fast and also dependable interaction systems, exhibited due to the release of 5G innovation. This structure makes it possible to make a smooth connection between myriad gadgets and units, developing the basis of record swaps in brilliant metropolitan area communities. As an example, intelligent web traffic monitoring depends on real-time interaction between sensing units, traffic signals, and streamlined command bodies, improving visitor traffic circulation and decreasing blockage (Tripathy et al., 2023). The influence of durable technical and commercial infrastructure is actually considerable, encouraging enhanced productivity in civil services, like transit, as well as powers. It prepares for the IoT, enabling the relationship of sensing units, tools, and functions. Therefore, metropolitan areas can easily harness the electrical power of records analytics as well as hands-free operation, causing even more enlightened decision-making, information marketing, as well as, eventually, a smarter, even more reactive metropolitan atmosphere for its residents.

1.2.2.2 Data Management and Analytics

Records administration and analytics are essential needs in brilliant metropolitan areas, and they are crucial for taking advantage of the large number of details produced through a variety of sensing units as well as gadgets. A strong record's structure entails the storage space, handling, and study of information to remove relevant knowledge. As an example, in a brilliant metropolitan area's transport body, records analytics could be utilized to enhance web traffic trends, lessen blockage, and also enrich a general range of motion. The effect of reliable information control is actually significant, adding to evidence-based decision-making, enhanced source appropriation, and efficient city solutions. Also, it makes it possible to anticipate the upkeep of vital commercial infrastructure, making certain the life expectancy as well as the effectiveness of social centers (Tripathy et al., 2023). The application of sophisticated analytics resources, including artificial intelligence formulas, enriches the metropolitan area's capacity to conform to transforming problems and also proactively deal with obstacles. Inevitably, effective records monitoring and analytics steer technology, foster durability, and inspire smart metropolitan areas to supply improved companies and a better lifestyle for their homeowners.

1.2.2.3 Connectivity

Connection is a critical criterion in brilliant spots, important for the smooth assimilation of units, bodies, and also functions that create the city's ecological community. Fast and trusted interaction systems, embodied through innovations like 5G, assist in real-time information swap. For example, in a smart urban area's mass transit device, the connection allows the synchronization of traffic control, sensing units, and transportation automobiles to maximize courses and routines. The influence of sturdy connection is actually transformative, improving total effectiveness as well as cooperation. Homeowners take advantage of strengthened solutions, including well-timed mass transit, real-time web traffic info, and also brilliant electrical administration. Furthermore, the connection is actually a crucial enabler of the IoT, allowing the interconnectedness of several units as well as sensing units (Pandya et al., 2023). This interconnectedness boosts the urban area's capability to pick up as well as study information, delivering knowledge that steers notified decision-making, enhances urban planning, and also brings about the production of an even more receptive as well as smart city atmosphere.

1.2.2.4 Energy Efficiency

Power effectiveness is actually an important need in smart urban areas, concentrating on enhancing electricity usage as well as advertising lasting methods. Smart metropolitan area campaigns focus on the execution of energy-efficient innovations and also facilities to lower ecological influence as well as enrich information durability. For instance, smart networks engage in a critical job by integrating state-of-the-art sensing units and keeping track of bodies to improve power circulation and sense mistakes, as well as ensuring power preservation. The effect of power effectiveness in smart metropolitan areas is actually considerable, minimizing carbon dioxide exhausts, reducing electricity expenses, and strengthening durability to energy disturbances. Furthermore, the assimilation of renewable resource resources, including solar power and wind generators, supports a cleaner and even more maintainable power mix (Gagliardi & Albergo, 2023). By focusing on electricity productivity, intelligent urban areas certainly not only alleviate ecological effects but also lead the way for lasting economic discounts and an even more durable city structure with the ability to fulfill the expanding electricity necessities of a quickly metropolitanizing planet.

1.2.2.5 Sustainable Infrastructure

Electricity performance is actually a vital need in intelligent metropolitan areas, concentrating on improving power intake as well as advertising lasting methods. Smart metropolitan area campaigns focus on the execution of energy-efficient modern technologies as structures to lower ecological effects as well as improve source durability. For instance, smart networks take on a vital task by combining state-of-the-art sensing units and also observing bodies to improve power circulation, discover mistakes, as well as advertise electricity preservation. The influence of electricity performance in intelligent areas is actually significant, causing decreased carbon dioxide exhausts, reduced electricity expenses, and enhanced durability to energy disturbances (Tareen et al., 2023). Furthermore, the combination of renewable resource resources, including photovoltaic panels and wind generators, supports a cleaner and also much more lasting power mix. Through focusing on electricity effectiveness, sustainable urban areas certainly not simply minimize ecological influence but also break the ice for lasting financial cost savings and also an extra tough metropolitan framework efficient in satisfying the developing electricity requirements of a quickly metropolitanizing planet.

1.2.2.6 Citizen Engagement

Person involvement is actually a critical demand in smart metropolitan areas, highlighting the energetic participation of locals in decision-making procedures as well as city control. An instance of personal interaction is actually the execution of electronic systems or even mobile phone apps that enable people to give reviews, document concerns, and take part in neighborhood dialogues. The influence of sturdy consumer interaction is actually extensive, nurturing openness, inclusivity, and also a feeling of area possession. Notifying residents helps in even more receptive and citizen-centric urban planning, triggering the advancement of services and facilities that line up with citizens' demands (Ahmed et al., 2023). Past the prompt effect on decision-making, resident interaction creates a lifestyle of partnership between regional authorizations and also the area, boosting the count on communal organizations. This communication certainly does not simply enable consumers to form their city atmosphere proactively. Likesmart enhances the social material of the urban area, ensuring a mutual feeling of accountability as well as belonging amongst its varied populace.

1.2.2.7 Security, Privacy, and Accessibility

As technology becomes deeply integrated, security and privacy are paramount. Smart city solutions need strong cybersecurity measures to protect against cyber threats. Encryption protocols, secure data transmission, and regular security audits are essential (Dutta et al., 2023). Additionally, robust data privacy policies and transparent consent mechanisms protect citizens' personal information. Smart cities should cater to diverse populations. The requirement for inclusivity involves designing infrastructure and digital interfaces that accommodate people with disabilities or limited mobility (Aldhyani et al., 2023). User-friendly interfaces, tactile elements, and audio cues can enhance accessibility, ensuring that all citizens can access and benefit from smart city services.

1.2.2.8 Efficient Mobility, Governance, and Regulation

Smart transportation systems are a cornerstone of smart cities. This requirement encompasses real-time traffic management systems that optimize traffic flow, reduce congestion, and minimize travel times (Aldossary, 2023). Integrated public transportation networks and digital platforms for route planning and payment enhance mobility. Effective governance is essential for smart city initiatives. The requirement for governance and regulation involves establishing clear guidelines and policies for data usage, sharing, and privacy (Chakraborty et al., 2023). Collaboration between government bodies, private sectors, and technology providers ensures smart city projects align with legal and ethical standards.

1.2.2.9 Resilience and Disaster Management

Smart cities must be prepared for various challenges. Resilience requires designing infrastructure that can withstand natural disasters and emergencies. Implementing early warning systems, disaster response plans, and redundant communication networks helps minimize the impact of unexpected events (Alwakeel et al., 2023). Smart city projects require sustainable funding models. Economic viability requires attracting investment from private sectors and fostering an environment that supports innovation and entrepreneurship (Tripathi et al., 2023). Smart city initiatives can drive economic growth by creating job opportunities, supporting local businesses, and enhancing the city's overall economic competitiveness.

1.2.3 Smart City Architecture

Smart city architecture is a comprehensive framework interconnecting various layers, each critical in creating an intelligent and efficient urban environment. Let us delve into the details of the architecture, focusing on the Application Layer, Data

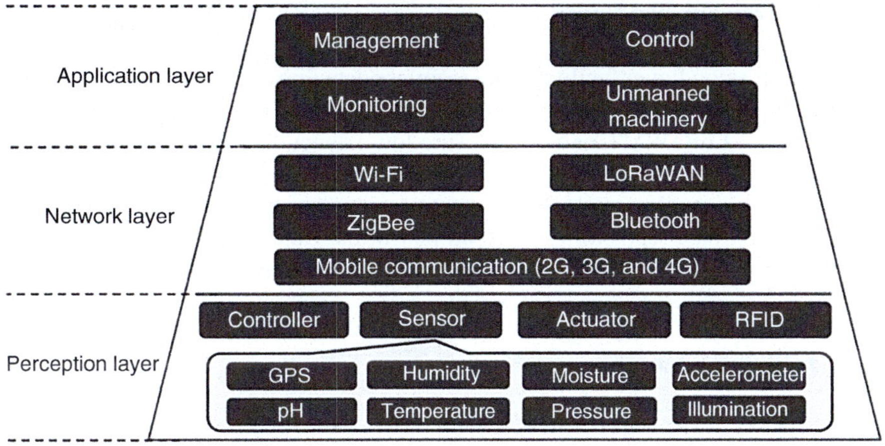

Fig. 1.5 Layers of Smart City

Management Layer, Transmission Layer, and Sensing Layer and how they relate to one another, as demonstrated in Fig. 1.5.

1.2.3.1 Application Layer

At the Application Layer, user-facing interfaces play a crucial role in delivering the benefits of a smart city to its citizens and administrators. These interfaces encompass web and mobile applications that provide real-time access to various services, such as traffic updates, waste management schedules, energy consumption data, and public transportation information. These applications are developed using responsive design principles, ensuring usability across different devices (Sharma et al., 2023). They communicate with underlying layers through Application Programming Interfaces (APIs) to fetch and display data and enable users to input requests or report issues.

1.2.3.2 Data Management Layer

The Data Management Layer handles the lifecycle of data collected from the Sensing Layer. Raw sensor data is ingested into a distributed storage system employing databases and cloud solutions for scalability and redundancy. Data is processed and transformed using Extract, Transform, Load (ETL) pipelines, where cleaning, filtering, and normalization occur (Zakzouk et al., 2023). Advanced analytics tools, including machine learning frameworks, are employed for data exploration and pattern recognition. These tools generate actionable insights, such as predicting traffic congestion, optimizing energy usage, or identifying waste management inefficiencies.

1.2.3.3 Transmission Layer

The Transmission Layer facilitates the seamless data flow between sensors, devices, and data processing or storage components. High-speed communication technologies, like 5G, LoRaWAN, or Wi-Fi, ensure low-latency data transmission. Sensors and IoT devices communicate through these networks, sending data packets to data aggregation points (Joe et al., 2023). As gateways, these points relay the collected data to central data repositories or cloud platforms. Data integrity and security are ensured through encryption protocols and authentication mechanisms.

1.2.3.4 Sensing Layer

The Sensing Layer forms the foundation of the architecture, consisting of an extensive network of sensors and devices distributed throughout the city. These sensors capture various environmental, infrastructure, and social data. Examples include air quality sensors measuring pollutants, traffic cameras tracking vehicle movement, waste level sensors gauging bin fill levels, and smart meters monitoring energy consumption (Tareen et al., 2023). Sensor data is transmitted to aggregation points within the Transmission Layer using various communication protocols, for example, MQTT or CoAP.

1.2.3.5 Relation and Interaction

The relationship among these layers is a dynamic and interconnected process. Sensors within the sensing layer continuously collect data transmitted from the transmission layer to the data management layer. Data is processed, analyzed, and transformed into insights accessible to users through the Application Layer. For instance, an AI algorithm within the Data Management Layer might analyze historical traffic data to predict congestion patterns (El-Shafai et al., 2023). This insight is displayed in real-time on a smart city app in the Application Layer, helping users plan efficient routes.

Conversely, user interactions within the Application Layer trigger actions that traverse the architecture in reverse. A citizen reporting a broken streetlight through the app sends a request through the API in the Application Layer. This request is processed in the Data Management Layer, triggering an alert in the Sensing Layer to dispatch maintenance crews to fix the issue (Gagliardi & Albergo, 2023). This intricate interplay of layers and their components orchestrates a holistic ecosystem that enables a smart city to operate efficiently, respond intelligently to challenges, and enhance the quality of life for its inhabitants through technology-driven insights and services.

Navigating these requirements entails a delicate balancing act. Trade-offs between competing demands arise, such as ensuring technological advancement while safeguarding privacy or promoting economic growth while preserving local culture. Meeting these multifaceted requirements demands a multidisciplinary approach involving urban planners, technologists, policymakers, sociologists, and citizens (El-Shafai et al., 2023). This exploration into the requirements of a Smart City Ecosystem delves into the intricate dance between these four pillars, namely technology, inclusivity, sustainability, and preservation—illuminating the complex challenges, synergies, and possibilities that arise when crafting the blueprint for the cities of the future (Pandya et al., 2023).

1.3 Smart City Application

Smart cities offer many applications across various domains, revolutionizing urban living and governance. These applications leverage technology and data to enhance citizens' efficiency, sustainability, and quality of life.

1.3.1 Transportation and Mobility

This involves real-time data integration from sensors and cameras to optimize traffic flow, reduce congestion, and enhance road safety through intelligent traffic management. Public transportation is elevated with real-time tracking, predictive arrival times, and integrated payment systems that empower commuters to navigate multimodal transit networks efficiently (Gagliardi & Albergo, 2023). Additionally, smart parking solutions employing IoT sensors guide drivers to available spaces, reducing traffic congestion and emissions while maximizing parking efficiency, thereby fostering seamless and sustainable mobility experiences for citizens and visitors alike.

1.3.1.1 Intelligent Traffic Management

Smart website traffic control is actually a transformative app in smart metropolitan areas, making use of state-of-the-art innovations to improve the circulation of auto and also mundane website traffic. Leveraging real-time records coming from sensing units, video cameras, and linked gadgets, smart website traffic control devices dynamically examine and change traffic lights, reroute automobiles, and supply well-timed relevant information to travelers. For instance, flexible traffic lights can easily react to modifying blockage designs, decreasing delay opportunities and also lessening traffic (Tareen et al., 2023). The influence of the use is actually extensive, bringing about boosted website traffic performance, lowered blockage, and reduced exhausts. Smart visitor traffic monitoring certainly not merely boosts day-to-day commute adventures for citizens but also brings about total metropolitan durability through marketing smoother web traffic circulation and minimizing the ecological impact linked with long-term idling. In addition, it constitutes an indispensable aspect of a thorough, smart urban area community, showing the possibility for modern technology to deal with metropolitan obstacles and also boost the lifestyle of people.

1.3.1.2 Public Transportation Enhancement

Public transport improvement is a decisive element of intelligent area advancement, targeting to boost the effectiveness, access, and durability of city flexibility. Leveraging sophisticated innovations, including real-time monitoring, anticipating analytics, and also smart ticketing units, this augmentation improves mass transit companies. For example, real-time monitoring allows travelers to gain access to correct details regarding buses or even learn arrival times, assisting in far better excursion preparation and also minimizing delay opportunities. Smart ticketing devices improve charge remittances, enriching the general customer expertise. The effect is actually complex, advertising a modal switch in the direction of public transport, lessening traffic jams and also decreasing discharges (Mengash et al., 2023). Enriched mass transit certainly delivers handy and trustworthy traveling choices for individuals, but likesmart supports an extra-lasting metropolitan atmosphere. Through combining smart remedies, like path marketing as well as demand-responsive solutions, metropolitan areas can easily make a strong as well as connected public transport system that straightens along with the developing necessities of their neighborhoods, cultivating an even more effective and also green metropolitan movement garden.

1.3.1.3 Smart Parking

Smart auto parking is actually a transformative function within the arena of smart area efforts, taking advantage of innovation to improve garages, boost customer adventure, and also ease metropolitan blockage. Utilizing sensing units, video cameras, and real-time record analytics, intelligent auto parking bodies offer details concerning on-call garages, helping chauffeurs to open up locations as well as lessening the amount of time invested in seeking vehicle parking. As an example, mobile phone applications may deliver real-time updates on vehicle parking supply and also make it possible for individuals to book or even spend for garages from another location. The influence of smart vehicle parking is actually substantial, bringing about lowered traffic jams, lessened energy intake, as well as lesser greenhouse fuel discharges (Tareen et al., 2023). Effective auto parking administration supports enhanced urban planning as well as lowered ecological effects by inhibiting unneeded web traffic blood circulation. By leveraging intelligent vehicle parking remedies, urban areas may boost the general range of motion, lessen ecological impacts, and enhance the advantage and complete satisfaction of both citizens and site visitors getting through metropolitan rooms.

1.3.2 *Energy and Utilities*

Energy and Utilities encompass a transformative approach to managing energy consumption, distribution, and resource utilization. Through cutting-edge technologies and data-driven strategies, smart cities optimize energy efficiency, promote sustainability, and enhance overall utility management. Smart grids utilize real-time data to dynamically manage energy distribution, balance loads, and integrate renewable energy sources, reducing energy wastage and enhancing grid reliability (El-Shafai et al., 2023; Gagliardi & Albergo, 2023). Smart meters and home automation systems empower consumers with real-time insights into energy usage, enabling informed decisions for conservation and cost reduction. Integration of renewable energy, such as solar panels and wind turbines, bolsters the transition to cleaner energy sources, lowering carbon emissions and contributing to a greener urban landscape. By intelligently managing energy and utilities, smart cities pave the way for a more sustainable and resilient urban future.

1.3.2.1 Smart Grids

Smart grids monitor energy distribution in real-time, optimizing energy flow and minimizing wastage. Two-way communication enables load balancing and integration of renewable energy sources. Smart grids play a pivotal role by incorporating real-time data from various sources, allowing utilities to balance energy supply and demand. This results in reduced wastage, enhanced grid stability, and increased integration of renewable energy sources (El-Shafai et al., 2023). Demand-response programs leverage data to adjust energy usage during peak demand periods, lowering costs and preventing overloads.

1.3.2.2 Energy Management

Smart meters provide accurate, real-time data on energy consumption, empowering consumers to make informed decisions to reduce energy usage and costs. Energy management also extends to the individual level, where smart meters and home automation systems empower consumers to track their energy usage in real time (Tareen et al., 2023). By gaining insights into energy consumption patterns, individuals can make informed decisions to optimize their usage, reduce costs, and contribute to overall energy efficiency.

1.3.2.3 Renewable Energy Integration

Solar panels and wind turbines generate clean energy integrated into the grid to reduce reliance on fossil fuels, lower emissions, and promote sustainability. Energy-efficient technologies like LED lighting and smart appliances are integrated into city infrastructure to minimize energy consumption (Mengash et al., 2023). Renewable energy sources like solar panels and wind turbines are harnessed to generate clean power and decrease reliance on fossil fuels.

1.3.3 Waste Management

Waste Management in smart cities involves implementing innovative technologies and data-driven approaches to optimize waste collection, disposal, recycling, and resource management (Pandya et al., 2023). By integrating IoT sensors and data analytics, smart waste management systems enable real-time monitoring of waste bin fill levels. This allows waste collection routes to be dynamically optimized based on actual demand.

1.3.3.1 Smart Bins

IoT-equipped waste bins monitor fill levels and communicate with waste management teams to optimize collection routes. This minimizes unnecessary trips, reduces operational costs, and promotes efficient waste management. These systems help reduce unnecessary collection trips, minimize fuel consumption, and decrease traffic congestion (Badr, 2023). Smart sorting technologies enhance recycling efforts by automating the separation of recyclable materials from general waste, improving recycling rates, and reducing landfill waste.

1.3.3.2 Recycling Initiatives

Smart waste sorting systems use technology to identify and separate recyclable materials, increasing recycling rates and reducing landfill waste. Citizen engagement is encouraged through mobile apps that enable residents to report overflowing bins or request bulk waste collection services (Tripathy et al., 2023). Public awareness campaigns and educational initiatives about waste reduction and proper disposal are also facilitated through digital platforms.

1.3.4 *Environmental Monitoring*

Environmental monitoring within smart cities involves using technology to observe, analyze, and manage the urban environment's ecological health and sustainability. By deploying sensors, data analytics, and real-time monitoring, smart cities can gather precise information about air quality, water quality, noise levels, and other environmental parameters (Berawi, 2023). This data is collected from various points across the city, accurately understanding pollution levels, resource consumption, and potential ecological hazards.

1.3.4.1 Air Quality Monitoring

In real-time, distributed air quality monitors measure pollutants like nitrogen dioxide and particulate matter. Quick action by the authorities can enhance air quality, safeguard public health, and respond to pollution emergencies (Mohammed & Hasan, 2023). Authorities can respond quickly to pollution sources, boost air and water quality, and improve environmental conditions due to environmental monitoring systems.

1.3.4.2 Water Quality Management

IoT devices monitor water body factors like pH and dissolved oxygen levels. These data sustain aquatic ecosystems, assure clean water supplies, and swiftly resolve pollution problems. For instance, air quality sensors can assist in locating areas of high pollution, enabling targeted emission reduction measures (Fan et al., 2023). Monitoring noise helps control noise pollution in highly populated regions, whereas monitoring water quality helps assure the safety of water sources and aquatic ecosystems.

1.3.5 *Governance and Civic Engagement*

Online and mobile platforms make it easier for citizens to engage in civic affairs by letting them report problems, offer opinions, and actively participate in local affairs. By establishing a direct line of communication between individuals and the government, these instruments encourage a sense of shared responsibility and ownership for urban development (Shafik et al., 2023). Initiatives for electronic involvement, such as online polls and virtual town hall meetings, allow people to weigh in on policy debates and urban planning procedures. In order to help authorities make decisions that align with community demands, smart cities also use data analytics to acquire insights from citizen interactions and sentiment analysis. Administrative procedures are streamlined by digital governance technologies, which also improve service quality and resource allocation.

1.3.5.1 Digital Services

Government services are accessible to the public through online platforms, improving transparency and easing administrative hassles. Citizens can engage with government offices, get permits, and pay bills from their homes. Platforms for digital services give citizens easy online access to government services, lowering administrative barriers and enhancing the delivery of public services (Shafik et al., 2023). Open data initiatives make Government information available, empowering citizens to hold officials responsible and make educated decisions.

1.3.5.2 Citizen Feedback

Through mobile apps and social media outlets, citizens can report problems like potholes, broken lamps, and waste overflow. Through direct communication, issues are resolved more quickly, and citizen satisfaction is increased. Civic participation is promoted through social media and mobile apps, allowing citizens to report problems, give opinions, and participate in decision-making (El-Shafai et al., 2023). Citizens can participate in online town halls and surveys that support policy and urban planning discussions because of e-participation platforms.

1.3.6 Safety and Security

Advanced cameras, video analytics, and facial recognition are used in smart surveillance systems to monitor public locations, spot anomalies, and take immediate action in the event of a threat. During accidents, disasters, or security, integrated emergency response systems incorporate data from different sources, such as sensors and social media, to deliver prompt notifications and coordinate swift actions (Tareen et al., 2023). In order to improve community engagement and crime prevention, public safety applications allow users to report occurrences, connect with law enforcement, and get alerts. Public places that are well-lit and secure benefit from smart lighting and smart infrastructure.

1.3.6.1 Surveillance and Monitoring

Real-time detection of suspicious activity and potential security concerns is possible with smart cameras equipped with AI-based video analytics. The public is safer, and law enforcement is helped by doing this (Gagliardi & Albergo, 2023). AI-powered cameras are used in smart surveillance systems to monitor public locations, spot irregularities, and spot potential security risks in real-time in order to enable quicker and better-coordinated responses to emergencies and disasters; integrated emergency response systems collect data from numerous sources.

1.3.6.2 Emergency Response

Integrated systems pool information from many sources to speed up emergency reaction times. Through smartphone apps, alerts can be issued to the public with guidance during emergencies, accidents, or natural disasters. Mobile apps provide citizens with notifications, safety information, and emergency contacts, fostering community readiness and resilience (Mengash et al., 2023). Smart lights and sensors in public spaces also increase visibility and discourage criminal activity, making cities safer.

1.3.7 Healthcare and Well-Being

Platforms for telemedicine allow for remote consultations and tests, improving access to healthcare for underprivileged people and delivering prompt medical guidance. Wearable technology and health sensors track exercise levels, monitor vital signs, and provide real-time health insights to support individualized wellness and preventive care (Pandya et al., 2023). By providing secure pedestrian routes, bike lanes, and recreational areas that promote exercise, smart city infrastructure fosters healthy living. Using data analytics and predictive modeling, authorities can more efficiently address health emergencies and disease outbreaks.

1.3.7.1 Telemedicine

Remote medical consultations are made possible by digital platforms, providing healthcare to people who might not othersmart have access because of accessibility barriers like distance or mobility. Vital signs and health parameters are tracked by wearable health devices, which deliver real-time data to healthcare professionals for ongoing remote monitoring and the early identification of health problems (Badr, 2023). Additionally, smart health apps provide individualized wellness guidance, monitor physical objectives, and assist with mental health.

1.3.7.2 Health Monitoring

Health professionals receive real-time data from wearable devices that monitor vital signs and other health variables. This enables ongoing observation, early identification of health problems, and specialized care (Berawi, 2023). These developments enable individuals to proactively manage their well-being, enhance public health outcomes, and contribute to delivering affordable and effective healthcare.

1.3.8 Education and Smart Learning

Remote learning is made possible by digital classrooms and e-learning platforms, which always provide dynamic, multimedia-rich content (Fong et al., 2023). Smart campuses integrate technology to improve administrative effectiveness, security, and interactive learning, giving students resources for personalized and collaborative learning.

1.3.8.1 Digital Classrooms

Digital classes in education and learning, commonly linked with brilliant understanding, entail the combination of modern technology to improve the standard understanding atmosphere. Making use of electronic devices, including active whiteboards, tablet computers, as well as online information, educators can easily make powerful as well as stimulating courses that deal with varied knowledge types. As an example, active academic software programs and online likenesses may promote immersive discovering adventures, making it possible for trainees to understand sophisticated principles in an extra involved and tailored fashion (Fong et al., 2023). The influence is actually great, promoting boosted trainee interaction, cooperation, as well as crucial reasoning skill-sets. Digital classes make it possible for accessibility to an extensive range of informative components, exceeding geographic barricades as well as supplying additional nondiscriminatory discovering possibilities. In addition, real-time evaluations and also records analytics help in individualized finding out roads and resolving specific trainee necessities (Mohammed & Hasan, 2023). The assimilation of modern electronic technologies in class certainly not only preps pupils for a technology-driven future but likesmart enhances the instructional yard through marketing development and also availability.

1.3.8.2 Smart Campus

A smart grounds leverages innovative modern technologies to make a connected and also smart knowing atmosphere. It incorporates IoT gadgets, sensing units, as well as records analytics to improve working productivity, protection, and also durability. Smart universities take advantage of innovation for components including computerized structure administration, energy-efficient units, and also tailored knowing expertises. Along with smart class, reactive commercial infrastructure, as well as data-driven knowledge, these grounds improve source application, enhance security, and also give a scientifically immersive academic expertise for pupils and also professors.

On college campuses, connectivity and automation improve security, speed up administrative procedures, and give students access to digital tools for learning and collaboration. Students actively learn through digital resources and interactive software, fostering critical thinking and problem-solving abilities (Berawi, 2023). Smart cities enable educators and students to go beyond conventional bounds by utilizing technology, providing quality education, lifelong learning, and developing skills for the digital age.

1.3.9 Tourism and Culture

Through interactive views of historical buildings, museums, and landmarks, augmented reality and virtual reality technologies can enhance cultural understanding (Badr, 2023). Additionally, digitizing artifacts and works of art, digital archives, and virtual reproductions ensures the preservation of cultural heritage for future generations while making them available to a global audience.

1.3.9.1 Smart Tourism

Smart tourist entails the assimilation of cutting-edge innovations to improve the general vacationer knowledge, improve place administration, as well as advertise lasting methods. By means of using mobile phone applications, boosted truth, as well as location-based companies, intelligent tourist gives real-time details on tourist attractions, individualized suggestions, as well as involved overviews. Advanced analytics assist place supervisors in recognizing website visitor actions, permitting effective group administration as well as information allotment. For instance, smart tourists could consist of QR codes at archaological sites, causing mixed media web content on website visitors' smart devices. The influence is actually an extra smooth as well as delightful traveling knowledge for travelers, while places take advantage of boosted commercial infrastructure organizing, lessened ecological influence, as well as the potential to adapt offerings to progressing site visitor choices (Gagliardi & Albergo, 2023). Smart tourist helps in the development of maintainable as well as technology-driven tourists, improving economic chances for regional neighborhoods and also encouraging an extra-linked and well-informed international trip garden.

1.3.9.2 Heritage Preservation

Smart tourism entails the assimilation of impressive modern technologies to enrich the general vacationer's knowledge, maximize place administration, as well as market maintainable methods. Via making use of mobile phone applications, increased fact, and also location-based solutions, sustainable tourist delivers real-time relevant information on tourist attractions, customized suggestions, and also involved overviews. Advanced analytics aid place supervisors know website visitor actions, permitting effective group monitoring and also information appropriation. For instance, smart tourists may feature QR codes at archaological sites, causing interactive media material on site visitors' smart devices (El-Shafai et al., 2023). The effect is actually an even smoother and delightful trip adventure for vacationers, while locations take advantage of enhanced structure preparation, lessened ecological effects, and also the capacity to modify offerings to progressing site visitor desires. Smart tourist helps in the development of maintainable and also technology-driven tourists, enriching financial chances for local area areas as well as promoting an extra linked as well as educated worldwide trip yard.

1.3.9.3 Economic Development

It includes campaigns to promote development, develop job opportunities, and also enrich general financial wealth. Tactics for financial growth commonly consist of commercial infrastructure growth, assets in learning as well as labor force instruction, encouraging advancement, enticing services, and also ensuring entrepreneurship. For instance, producing service incubators or even development centers can easily develop a helpful setting for start-ups and technical innovations, resulting in economic diversity. The effect of prosperous economic progression is actually significant, triggering raised job fees, a much higher standard of comfort, and an extra durable local area economic climate (Tareen et al., 2023). Economic development focuses on supporting entrepreneurship, innovation, and sustainable growth through technology-driven initiatives. Smart cities foster favorable conditions for economic growth by utilizing digital infrastructure and data analytics. Technology incubators and innovation hubs support start-ups and businesses, promoting teamwork and accelerating the creation of ground-breaking ideas.

1.3.9.4 Innovation Hubs and Digital Marketplaces

Tech hubs and innovation centers promote teamwork, entrepreneurship, and the production of new technology, which stimulates economic growth and employment creation (Mengash et al., 2023). Smart cities encourage economic diversification, boost productivity, and allow people and businesses to prosper in a digitally empowered urban environment through focused economic policies and technological integration. Also, online platforms link consumers and local businesses, boosting convenience and encouraging sustainable business practices. Digital marketplaces link nearby businesses with customers, fostering trade and helping the neighborhood economy. Improved infrastructure and effective transportation systems are examples of smart city technology that draw investments, create jobs, and boost company competitiveness (Gagliardi & Albergo, 2023).

1.4 Globally Selected Smart Cities

Several cities worldwide have been recognized for their attempts to become smart cities, as of my most recent information update in September 2023. Six smart cities that were chosen globally have become well-known for their creative programs.

1.4.1 Singapore, Singapore

Singapore, a self-governed city-state and also isle nation in Southeast Asia, stands up as an ideal of city technology, economical success, as well as multiculturalism. Recognized for its own exquisite urban planning, Singapore has actually perfectly incorporated intelligent urban area innovations to make a dependable and also lasting city. The city-state has actually originated efforts in locations like smart transport, making use of real-time information for visitor traffic administration as well as effective social transportation units. Its own dedication to ecological durability appears by means of environment-friendly areas, water administration, and also ecological construction. Singapore's Smart Country sight has actually thrust developments in electronic administration, including innovation to boost resident companies as well as connection. The area's sky line is actually a proof to home wonders, while the Harbor Gulf Sands as well as Gardens due to the Gulf exhibit the blend of originality as well as ecological awareness. Past framework, Singapore is actually an international monetary center, cultivating financial development with development, profession, as well as critical assets. The nation's social grandeur is actually commemorated via celebrations, unique cooking knowledge, as well as an enlightening mixture of heritages. Singapore's excellence account exists certainly not simply in its own technical success however additionally in its own potential to preserve a premium quality of everyday life, enticing international ability and also working as a version for intelligent as well as maintainable city residing.

1.4.2 Seoul, South Korea

Seoul, South Korea, stands up as a worldwide standard for smart area efforts, effortlessly incorporating enhanced modern technologies to take care of city obstacles as well as increase the total city take in. Popular for its own cutting-edge method to urban planning, Seoul has actually applied advanced smart transit units, consisting of included social transportation as well as real-time visitor traffic control. The area's devotion to electronic administration appears in its own use innovation to improve resident companies, ensure clarity, as well as help with communal involvement. Seoul's sky line flaunts environmentally friendly design, and also the urban area focuses on durability via eco-friendly efforts as well as energy-efficient methods. The smooth blend of heritage and originality is actually embodied in city eco-friendly areas and also social maintenance attempts. Along with a sturdy focus on development as well as technology-driven remedies, Seoul remains to be actually a powerful brilliant area, using a premium of lifestyle for its own locals while working as an ideas for city facilities worldwide in getting through the problems as well as options offered due to the ever-evolving yard of smart as well as lasting metropolitan advancement.

1.4.3 Barcelona, Spain

Barcelona, Spain, stands up as an ideal of smart metropolitan area technology, effortlessly mixturing its own abundant social culture along with ground-breaking modern technologies to boost city lifestyle. Prominent for its own ideal urban planning, Barcelona has actually carried out an assortment of brilliant area services that redefine metropolitan take ins. Noteworthy projects feature smart lighting fixtures units, reliable rubbish administration with sensing unit systems, and also the assimilation of electronic systems for resident involvement. The area's dedication to durability appears in its own eco-friendly rooms, environment-friendly transport possibilities, and also attempts to lower ecological effect. Barcelona's smart urban area undertakings expand past commercial infrastructure, incorporating social conservation and also the advertising of electronic incorporation. The well-known Sagrada Família, as well as Playground Güell, sympathetically exist side-by-side along with campaigns like totally free social Wi-Fi areas, showing Barcelona's dedication to nurturing an attached, scientifically evolved city setting while commemorating its own historic and also grandeur social. This mixture of heritage and also originality spots Barcelona as a worldwide forerunner in the smart area yard, impressive various other areas to accept development and also produce metropolitan rooms that are actually each smart and also culturally lively.

1.4.4 Copenhagen, Denmark

Copenhagen, Denmark, stands up as a flare of smart metropolitan growth, effortlessly incorporating durability, advancement, and also a premium quality of life. Prominent for its own dedication to environment-friendly projects as well as its ecological mindset, the metropolitan area has actually executed a detailed, intelligent metropolitan area structure. Copenhagen focuses on maintainable transport along with comprehensive bicycling facilities, smart website traffic monitoring units, as well as electrical mass transit. The metropolitan area's initiatives in renewable resources are actually good, along with paying attention to renewable energy and also energy-efficient properties, bringing about its own objective of ending up being carbon-neutral through 2025. Copenhagen uses smart urban area modern technologies for dependable rubbish administration, environmental monitoring, as well as intelligent frameworks. Past commercial infrastructure, the urban area welcomes smart administration processes, leveraging electronic services for consumer involvement as well as effective civil services. The beneficial combination of intelligent modern technologies along with the area's famous appeal, social grandeur, as well as importance on work-life harmony produces Copenhagen an international prototype of a brilliant, maintainable, as well as habitable city atmosphere that focuses on the health of its own homeowners.

1.4.5 Dubai, United Arab Emirates

Dubai, United Arab Emirates, characterizes the marital relationship of luxury and also sophisticated innovation, placing on its own as a leading international smart area. Well-known for its own advanced sky line and also progressive city progression, Dubai has actually utilized smart urban area options to redefine the metropolitan encounter. The metropolitan area's devotion to technical technology appears in legendary designs like the Burj Khalifa, which has gone well with smart projects that cover numerous fields. Dubai's transport structure integrates smart modern technologies for visitor traffic administration, independent motor vehicles, and also the advancement of the Hyperloop for rapid transit. The metropolitan area welcomes smart administration, providing e-services and also smart systems that enrich person involvement and also enhance management methods. Durability is actually a centerpiece, along with campaigns like the Mohammed container Rashid Al Maktoum Solar Playground and also environment-friendly property requirements, bringing about Dubai's outlook of a brilliant, lasting area. The combination of expert systems, blockchain, as well as IoT innovations emphasizes Dubai's resolve to make a flawlessly hooked-up metropolitan atmosphere, using individuals and also website visitors to glance right into the future of smart, sophisticated residing.

1.4.6 Tokyo, Japan

Tokyo's efforts to become a smart city focus on energy efficiency, disaster management, and urban transportation. The city has implemented earthquake early warning systems and disaster response mechanisms to increase resilience. District heating and cooling systems and energy-efficient buildings facilitate energy savings. Through integrated transportation networks, Tokyo also seeks to increase urban mobility while encouraging effective transit options for its citizens. These cities are models for numerous smart city programs focusing on data-driven decision-making, environmentally friendly infrastructure, technology innovation, and public involvement. Their combined efforts show how metropolitan areas worldwide use technology to build more effective, livable, and resilient ecosystems.

1.5 Smart Cities Attributes

Smart cities exhibit a range of key attributes that distinguish them as technologically advanced, efficient, and sustainable urban environments. These attributes collectively contribute to enhancing the quality of life for residents, promoting economic growth, and addressing urban challenges.

1.5.1 Digital Infrastructure

The digital framework is actually a fundamental feature of smart areas, including the technical basis that assists in smooth connection, effective information control, and also impressive city services. Along with fast and sturdy interaction systems featuring 5G innovation, smart areas develop the foundation for the Web of Traits (IoT) as well as real-time information trade. This electronic structure encompasses smart transit devices, where sensing units as well as interaction innovations enhance web traffic circulation, lowering blockage and also enriching the range of motion. The execution of smart frameworks as well as energy-efficient structures showcases electronic facilities' task in maintainable techniques, enabling dependable electricity administration and also source use (El-Shafai et al., 2023). Additionally, electronic control systems allow straightforward and available civil services, nurturing personal interaction and participatory decision-making. The durability of smart areas when faced with problems, including catastrophes or even urgent, is actually reinforced through the electronic framework with very early caution units as well as data-driven approaches. Inevitably, electronic facilities are actually a cornerstone characteristic, deriving the unique elements of brilliant areas, coming from connection and durability to citizen-centric solutions and also general technical development.

1.5.2 Data Analytics and Insights

Information analytics, as well as understanding, make up a crucial feature within intelligent metropolitan areas, embodying the transformative electrical power of leveraging huge datasets for updated decision-making as well as improved city administration. Via the continual selection and evaluation of real-time records created through sensing units, units, and consumers, intelligent areas increase important knowledge in several parts of metropolitan life. This consists of website traffic designs, power usage, civil services use, and also ecological disorders. Advanced analytics allow area representatives to enhance information allotment, boost functional performance, and also proactively deal with surfacing obstacles (Berawi, 2023). As an example, anticipating analytics can easily assist in foretelling traffic jams, allowing preemptive procedures to minimize traffic jams. The application of data-driven ideas in urban planning supports the development of even more receptive as well as maintainable areas. In addition, citizen-centric apps, including customized companies and also mobile phone applications, are actually enhanced by means of information analytics, boosting total individual knowledge as well as involvement (Kushwaha et al., 2023). The potential to harness information for workable knowledge identifies intelligent areas, encouraging development and durability, as well as ongoing development in the direction of city settings that properly satisfy the growing requirements of their individuals.

1.5.3 IoT Integration

IoT assimilation stands up as a keystone feature in the development of brilliant metropolitan areas, militarizing a transformative unity between bodily facilities as well as the electronic arena. In intelligent metropolitan areas, the IoT participates in a core part by relating a myriad of sensing units, gadgets, as well as devices to compile real-time information and also allow smart decision-making. This combination is extended throughout varied fields, including transit, electricity, health care, and civil services. For example, brilliant website traffic administration takes advantage of IoT-connected sensing units to check and maximize visitor traffic circulation, lessening blockage and improving total movement. Brilliant energy-efficient networks utilize modern IoT technologies for real-time tracking and administration, making certain maximized source use (Pandya et al., 2023). The combination of IoT likesmart encompasses social safety and security by means of smart monitoring units as well as environmental monitoring. The wide range of information produced through IoT gadgets offers areas along with knowledge right into personal habits, information consumption styles, and surfacing styles, permitting positive and data-driven urban planning.

1.5.4 Sustainability and Energy Efficiency

Durability as well as electricity performance work with center features in the material of smart metropolitan areas, mirroring a devotion to ecological duty and also resource-conscious city advancement. Smart metropolitan areas incorporate cutting-edge modern technologies to develop an extra maintainable as well as energy-efficient city yard. This consists of the execution of smart frameworks, renewable resource resources, energy-efficient property layouts that lessen carbon dioxide impacts, and market-liability power usage. Maintainable urban planning involving eco-friendly rooms, environmentally friendly transit, and refuse control units shows the commitment to ecological conservation even more (Berawi, 2023). The fostering of smart transport bodies, like power public transport and also brilliant visitor traffic monitoring, helps lower discharges and maximize movement. These efforts certainly do not simply take care of existing ecological difficulties; likesmart, they posture smart metropolitan areas as innovators in the worldwide activity in the direction of greener, much tougher city settings (Sharma et al., 2023). Via a comprehensive technique for durability and power productivity, smart urban areas make every effort to make an enlightening harmony between technical innovation, lifestyle, and ecological preservation for the advantage of current and potential age groups.

1.5.5 Urban Mobility

Urban range of motion is actually a vital characteristic of smart metropolitan areas, showcasing a standard switch in transit by means of ingenious modern technologies that focus on productivity, durability, and also availability. Smart metropolitan areas set up smart transit bodies that incorporate real-time information coming from sensing units as well as brilliant units to enhance website traffic circulation, decrease blockage, and boost total flexibility. The importance of maintainable city wheelchairs appears via the promo of environmentally friendly transport alternatives, including electricity lorries, bike-sharing systems, and also reliable social transportation devices. Also, smart metropolitan areas utilize electronic systems and mobile phones to supplying relevant information on transport timetables, website traffic problems, and alternative routes in real-time, enabling citizens to create enlightened choices (Mohammed & Hasan, 2023). The smooth combination of brilliant wheelchair remedies certainly not simply handles city blockage as well as ecological problems but likesmart supports making a much more comfortable and also hooked-up city setting. By focusing on a reliable and maintainable city range of motion, intelligent areas aim to enrich the total lifestyle of their citizens while marketing a healthier, extra durable, and technically elevated city garden.

1.5.6 Integrated Public Services

Integrated civil services are actually a foundation characteristic of smart urban areas, characterizing sleek as well as citizen-centric administration. Leveraging electronic systems as well as complementary devices, excellent areas offer locals smooth accessibility to a selection of civil services, varying from health care and education and learning to transit and metropolitan interaction. The combination guarantees that residents experience a comprehensive as well as complementary solution environment where data-driven decision-making boosts productivity. Via central electronic user interfaces, homeowners may get through and also get access to assorted solutions, nurturing an extra reactive as well as uncomplicated city atmosphere that shows the highly sophisticated and citizen-focused attributes of brilliant urban areas (Joe et al., 2023). In the end, centralized platforms improve the citizen-government connection by streamlining contacts, reducing paperwork, and increasing administrative effectiveness.

1.5.7 Smart Governance

Technology is used in smart governance to improve transparency and speed administrative procedures. Digital document management, e-governance portals, and regular task automation increase government efficiency. Citizens are more actively involved in civic processes when they receive real-time updates on government projects and policies (Berawi, 2023). Each of these characteristics helps to turn cities overall into responsive, intelligent and linked urban ecosystems. Smart cities attempt to optimize resource usage, enhance quality of life, and address the complicated issues associated with urbanization by fusing technology, data, and public engagement.

1.6 Security and Privacy for Smart Cities

Security and privacy are paramount considerations in developing smart cities, as integrating technology and data collection can raise concerns about potential vulnerabilities and breaches.

1.6.1 Security

To foster trust and preserve the efficacy of smart city initiatives, it is essential to ensure the security of the systems and the data as well as to protect the privacy of individuals (Sharma et al., 2023).

1.6.1.1 Cybersecurity Frameworks

Smart city cybersecurity guidelines should be adopted in their entirety. Procedures for risk assessment, vulnerability management, intrusion detection, and incident response should be included in these frameworks (Samy El Menchawy et al., 2023). They make it easier for many parties, such as individuals, IT companies, and government agencies, to work cooperatively. These frameworks guarantee the integrity of networked devices, data, and crucial infrastructure by encouraging regular audits, real-time Monitoring, and adaptive responses. The early identification of possible threats and vulnerabilities is made easier by these frameworks.

1.6.1.2 Secure Data Transmission

Data exchanged between devices, sensors, and the central infrastructure should be secured using encrypted communication protocols, like Transport Layer Security (TLS) (Kushwaha et al., 2023; Zakzouk et al., 2023). Encryption encodes the data into an unreadable format, which can only be deciphered by the intended recipient with the proper decryption key. This shields the data from unauthorized access or tampering during transit, ensuring confidentiality and integrity. Also, protocols like SSL/TLS provide secure communication over the internet. This guards against data manipulation and eavesdropping during transmission.

1.6.1.3 Network Segmentation

A cybersecurity technique called network segmentation involves breaking up a bigger network into more isolated, smaller parts or subnetworks. Each section has its security controls and rules and runs independently. By isolating potential breaches or attacks within a particular segment and restricting their lateral migration across the network, this strategy improves security. The overall impact is reduced even if one component is compromised because the other segments are protected (Cabrera et al., 2023). To reduce the attack surface and restrict unauthorized access to sensitive systems, network segmentation helps isolate critical infrastructure from public networks.

1.6.1.4 Regular Audits and Penetration Testing

Audits assess security practices and ensure compliance, while penetration tests simulate real attacks to uncover vulnerabilities. Both practices help organizations stay proactive, identify weaknesses, and patch potential entry points, enhancing overall defense against data breaches and cyber threats. Regular security audits and penetration tests assist in detecting vulnerabilities, evaluating the system's security posture, and remedying possible issues before they become exploitable.

1.6.1.5 Secure Device Management

It encompasses measures such as authentication, authorization, and remote management. Utilizing strong authentication methods and access controls ensures that only authorized personnel can manage devices (Mohammed & Hasan, 2023). Regular software updates and patches safeguard against vulnerabilities. Additionally, remote monitoring and selective access enable efficient troubleshooting without compromising security. Use strong authentication techniques for device registration and management to prevent unwanted access to IoT devices.

1.6.1.6 Firmware and Software Updates

Update and patch IoT device firmware and software frequently to fix known vulnerabilities and keep safe from new dangers. Regular updates patch vulnerabilities, enhancing system resilience against evolving threats (Fong et al., 2023). They address security flaws, improve functionality, and ensure compatibility with the latest defenses. Timely updates mitigate the risk of exploits and breaches by keeping systems fortified. This practice extends across devices, applications, and operating systems, bolstering overall protection against cyberattacks and data compromises.

1.6.1.7 Secure Boot and Chain of Trust

Secure boot checks the validity and integrity of a device's firmware before it starts up, prohibiting the execution of unauthorized or malicious code. This idea is expanded upon by the chain of trust, which makes sure that each system component is validated before the next one loads. This multi-layered strategy creates a safe base, preventing attacks during bootup and fostering confidence throughout system operation. These controls are necessary to protect against firmware-level compromises and ensure the integrity of devices (Fong et al., 2023).

1.6.2 Privacy

Keeping security and privacy in check in smart cities is a continuous effort that calls for cooperation between local government officials, tech companies, residents, and regulatory organizations (Kushwaha et al., 2023). Smart cities may develop an atmosphere that encourages innovation, economic progress, and public trust while preserving individual rights and security by giving priority to these factors.

1.6.2.1 Data Classification

This practice helps organizations prioritize security measures and define access controls. Data is typically classified as public, internal, confidential, and restricted. Each category has specific handling and protection requirements. By categorizing data, organizations can allocate appropriate security resources and implement encryption, access restrictions, and monitoring, ensuring that sensitive information is safeguarded while enabling efficient data management and compliance with privacy regulations (Cabrera et al., 2023). Sort gathered information into categories based on how sensitive it is. Stronger protection measures are required for personal and sensitive data.

1.6.2.2 Explicit Consent

Before obtaining a person's personal information, make sure to get their express, informed consent. Please give them the choice to withdraw their consent and make it clear how to do so (Kushwaha et al., 2023). It ensures that individuals are fully informed about how their data will be used and allows them to make informed decisions About their privacy. Obtaining explicit consent involves providing understandable information, giving individuals the choice to opt in or out, and respecting their right to control their data.

1.6.2.3 Privacy Impact Assessments (PIA)

Personal PIAs are actually an essential characteristic in intelligent urban areas, recognizing the value of securing personal privacy within the arena of innovative innovations. As brilliant urban areas release linked units and data-driven options, PIAs come to be important devices for examining and reducing possible personal privacy threats. These evaluations guarantee that the execution of smart modern technologies is associated with personal privacy rules as well as reliable specifications (Sharma et al., 2023). Through administering PIAs, brilliant urban areas may proactively resolve personal privacy worries, foster openness, and also create social count. The combination of sturdy personal privacy amounts additionally focuses on the dedication of smart metropolitan areas to stabilizing technical technology along with the security of private personal privacy civil rights Shafik et al. (2020).

1.6.2.4 User Access Controls

Individuals' access to management is actually an essential feature in smart metropolitan areas, stressing the significance of safety and security as well as controlled accessibility to the extensive selection of connected bodies and information. As intelligent urban areas combine state-of-the-art innovations for administration and also solutions, durable access to managers makes certain that only licensed people or even facilities possess proper gain access to advantages. This safety step certainly not only buffers vulnerable info but also relieves the danger of unapproved use of or even harmful tasks (AlZu'bi et al., 2023). Through applying rigid customer access commands, smart urban areas display their dedication to record protection, personal privacy, and the liable control of electronic facilities, nurturing a trustworthy and durable city setting for locals and stakeholders. Access controls can include authentication methods (passwords, biometrics), authorization levels (user roles, permissions), and monitoring to maintain data integrity and confidentiality.

1.6.2.5 Data Portability

It promotes user control and freedom over their information. This practice ensures that users can easily transfer their data while maintaining its structure and usability. Implementing secure mechanisms for data portability, including encryption and authentication, is essential to safeguarding privacy during the transfer process (Dutta et al., 2023). Data portability empowers users to switch services without losing their data, enhancing competition and individual privacy rights. Make it possible for users to transmit their data to other services and give them access to it in a machine-readable format.

1.6.2.6 Data Lifecycle Management

It includes stages like data collection, storage, processing, sharing, and deletion. Effective management ensures data accuracy, accessibility, and security throughout its journey. This encompasses data classification, access controls, encryption, and compliance with regulations (Aldossary, 2023). By carefully managing the data lifecycle, organizations minimize risks, maintain data integrity, and uphold privacy standards while deriving value from data-driven insights. Define clear data retention and deletion policies to ensure that data is stored only for as long as necessary and is securely disposed of when no longer needed.

1.6.2.7 Transparent Data Usage

By providing detailed privacy policies and explicit consent mechanisms, users can make informed decisions about sharing their information. Transparent data practices include informing users about data storage, sharing with third parties, and the purposes for which the data will be used. This fosters trust, empowering users to control their data and promoting ethical data handling, which is vital in today's digital landscape (Tripathi et al., 2023). Provide clear and easily understandable privacy notices that detail how data will be used, who will have access to it, and how long it will be retained.

1.6.2.8 Cross-Border Data Transfer

Ensuring its security involves adhering to data protection laws and international regulations. Organizations must assess the legal and technical aspects of such transfers, employing encryption, secure communication channels, and data anonymization when needed. Adequate contractual agreements with third parties involved in the transfer are also crucial (Aldossary, 2023). If data is transferred across borders, ensure compliance with applicable data protection regulations and consider mechanisms such as standard contractual clauses or binding corporate rules.

1.6.2.9 Privacy Training and Regular Audits

Training covers topics like data classification, secure data handling, and compliance with privacy regulations. By fostering a culture of privacy awareness, organizations empower employees to recognize and mitigate potential risks, reducing the likelihood of data breaches and privacy violations (Aldhyani et al., 2023). Regular privacy training cultivates a workforce that actively contributes to maintaining data confidentiality and integrity. Train employees and stakeholders on privacy best practices, emphasizing protecting personal data. Audits provide insights into potential weaknesses, allowing for timely remediation and improvement of security measures. By routinely examining systems and processes, organizations can proactively enhance their defenses, minimize risks, and maintain a robust cybersecurity posture against emerging threats. Conduct periodic privacy audits to evaluate the organization's compliance with privacy policies and regulations (Ahmed et al., 2023; AlZu'bi et al., 2023).

1.7 Future Research Directions

The following are some of the future research directions of smart cities.

1.7.1 Privacy-Preserving Technologies

Privacy-preserving innovations have become important potential instructions for smart urban areas, dealing with the developing worries bordering information, personal privacy, and safety and security as smart metropolitan areas continue to combine varied records resources and accelerated innovations, making certain the discretion and security of delicate relevant information come to be essential. Privacy-preserving modern technologies, consisting of homomorphic file encryption, differential personal privacy, and decentralized identification options, deliver cutting-edge methods to accumulate and refine information without jeopardizing specific personal privacy (Fong et al., 2023). These modern technologies enable intelligent area communities to learn useful knowledge while anonymizing individual details, blowing a fragile equilibrium between data-driven technology and also the security of a person's privacy. Accepting these privacy-centric strategies certainly does not simply straighten smart areas along with progressing governing yards; however, such as, smart creates a base of leaves between metropolitan area authorizations, individuals, as well as various other stakeholders, forming a future where technical improvements exist together sympathetically along with personnel privacy factors to consider in the city yard.

1.7.2 Resilience and Disaster Management

Durability and calamity administration stand up as crucial potential paths for intelligent areas, recognizing the rising problems posed through environmental adjustment as well as metropolitan susceptibilities. Smart urban areas are actually paying considerable attention to innovative modern technologies to boost their capability to reply to as well as recoup coming from unforeseen events. Executing real-time tracking, anticipating analytics, and also IoT-enabled sensing units may dramatically strengthen very early caution devices, assisting in practical calamity readiness. Including resistant frameworks, smart frameworks, and also flexible properties, metropolitan areas are strengthened even more versus the effect of organic calamities (Zakzouk et al., 2023). Using expert systems in calamity feedback as well as healing functions boosts performance and also source allowance.

1.7.3 AI 5G and Beyond Energy Harvesting and Storage

The research will focus on integrating 5G networks with edge computing to reduce latency and enable real-time processing for critical applications such as autonomous vehicles and remote healthcare (Reddy Yeruva, 2023). Additionally, exploring the potential of beyond-5G technologies, such as terahertz communication, will drive innovation in smart city connectivity. Future studies will look into AI-driven predictive analytics for energy optimization, urban planning, and finding irregularities in municipal operations (Aldossary, 2023). As explainable AI advances, transparent decision-making will become possible, giving city planners and inhabitants easier access to AI-driven insights. Through focusing on strength, intelligent metropolitan areas desire to develop flexible city communities that are efficient in resisting and recuperating fast from surprises, ensuring the safety and security health of their homeowners despite a considerably erratic and powerful city setting.

1.7.4 Circular Economy, Waste Management, Ethics, and Governance

Future research will focus on developing smart waste sorting systems that employ AI and robotics to automate recycling processes (Aldhyani et al., 2023). Circular economy models will be integrated into urban planning, promoting waste reduction and materials reuse through innovative supply chain management. The subject of future research will be the development of moral AI frameworks, dealing with algorithmic prejudice, and assuring justice in automated decision-making processes (Ahmed et al., 2023). Creative governance models will be investigated to incorporate community participation and guarantee the responsible and open use of smart city technologies. Researchers will investigate cutting-edge energy harvesting technologies, including piezoelectric materials and thermoelectric generators, to capture energy from urban surroundings (Alwakeel et al., 2023). The distribution and storage of gathered energy will be optimized for energy storage technologies like solid-state batteries and supercapacitors.

1.7.5 Urban Mobility Solutions, Health, and Well-being

To evaluate the effect of autonomous vehicles on traffic patterns and urban planning, researchers will create simulation models (AlZu'bi et al., 2023). It will take sophisticated routing algorithms and real-time data integration to integrate mobility options into a seamless multi-modal transportation network. Also, future research will focus on wearable health sensors that provide real-time health monitoring and predictive analytics for disease prevention (Aldhyani et al., 2023; Shafik, 2024c). Urban planning strategies will incorporate green spaces, pedestrian-friendly zones, and noise reduction measures to enhance citizens' mental and physical well-being.

1.7.6 Social Inclusion and Equity Augmented Reality and Virtual Reality

Innovative approaches to closing the digital divide, like community technology hubs and user interfaces accessible to a wide range of users, will be studied. Independent of socio-economic level, technology solutions will likely meet all residents' requirements due to participatory design techniques. The research will look at how AR and VR may be used in urban planning to give citizens a picture of potential alterations to the cityscape (Ahmed et al., 2023). Experiences in virtual tourism will give visitors and locals alike an intimate look at culture and history.

1.7.7 Blockchain and Distributed Ledgers and Socio-economic Impact Assessment

Future studies will concentrate on growing blockchain networks while maintaining security and decentralization for smart city applications (Dutta et al., 2023). Blockchain-based digital identity systems will improve security and privacy in everyday citizen interactions, from voting to using public services. To evaluate the long-term effects of smart city projects, researchers will create thorough frameworks that consider economic development, social equality, environmental sustainability, and citizen well-being. These evaluations will serve as a guide for making policy decisions and ensuring the comprehensive growth of smart cities (Aldhyani et al., 2023; Shafik, 2024d). These thorough study guidelines demonstrate how smart cities are multidisciplinary, incorporating technology, social sciences, urban planning, and policy-making.

1.8 Open Research Questions for Smart Cities

Open research questions are pivotal to driving innovation and addressing the evolving challenges of smart cities. These questions spark curiosity and guide researchers, policymakers, and practitioners in shaping the future of urban development. Here are some open research questions for smart cities:

1.8.1 Secure and Private Data Sharing, Resilient IoT Networks

How can smart cities facilitate secure and privacy-preserving data sharing among different entities while ensuring compliance with data protection regulations and maintaining citizens' trust? How can smart cities design and implement IoT networks that are resilient to cyberattacks, natural disasters, and disruptions, ensuring continuous operation and reliability?

1.8.2 Human-Centered AI and Integration of Emerging Technologies

How can AI technologies in smart cities be designed and governed to prioritize human well-being, ethical considerations, and citizens' needs while avoiding bias and unintended consequences? What strategies can smart cities employ to integrate emerging technologies like quantum computing, edge AI, and biotechnology into their infrastructure and services?

1.8.3 Interoperability and Standards and Equitable Access

How can smart cities establish interoperable standards that enable different technologies, systems, and devices to communicate and collaborate, fostering a cohesive urban ecosystem seamlessly? How can smart cities ensure that technology benefits all population segments, including marginalized communities, and address potential disparities in access, digital literacy, and service availability?

1.8.4 Digital Inclusion and Climate Resilience

What innovative approaches can bridge the digital divide, providing equitable access to technology and digital services for all residents, regardless of age, income, or education? How can smart cities integrate climate-resilient design principles, renewable energy solutions, and adaptive infrastructure to mitigate the impacts of climate change and promote sustainability?

1.8.5 Citizen Engagement Platforms, Economic, and Social Impact

What strategies can smart cities employ to create effective digital platforms encouraging meaningful citizen participation, collaboration, and co-creation in urban planning and decision-making? What methodologies can be used to comprehensively measure smart city initiatives' economic and social impact, including their effects on job creation, business innovation, and overall quality of life?

1.8.6 Ethical Use of AI, Privacy-Preserving Analytics, Smart Rural Areas

How can smart cities develop guidelines and frameworks for the ethical deployment of AI in critical areas such as public safety, healthcare, and transportation, ensuring transparency, fairness, and accountability? What innovative techniques can be used to perform advanced data analytics while preserving citizens' privacy, such as federated learning, homomorphic encryption, and decentralized data processing? How can the concepts of smart cities be adapted and extended to smart rural areas, addressing unique challenges such as connectivity, agricultural sustainability, and rural development?

1.8.7 Urban Mobility Integration, Legal and Regulatory Frameworks

How can smart cities seamlessly integrate various modes of transportation, including autonomous vehicles, electric scooters, and public transit, to create efficient, multimodal transportation networks? How can smart cities navigate legal and regulatory challenges related to data ownership, liability, cybersecurity, and technology deployment within urban environments? These open research questions underscore the dynamic and evolving nature of smart cities. Addressing these questions will drive innovation, shape policy decisions, and ultimately lead to the development of more inclusive, sustainable, and technologically advanced urban environments.

1.9 Lessons Learned from the Chapter

The following are the lessons learned from the chapter.

1.9.1 Holistic Integration is Essential

The necessity of complete integration is underlined by research on smart cities. Various technologies, systems, and industries must be smoothly integrated for smart city projects to be successful (Dutta et al., 2023). Inefficiencies, data fragmentation, and missed potential for synergy can all result from siloed methods. Lesson discovered: A comprehensive and linked strategy is essential to utilize the promise of smart cities fully.

1.9.2 Citizen-Centric Design Drives Adoption

The acceptance rates and beneficial effects of smart city programs prioritizing citizen demands and engagement are generally greater (AlZu'bi et al., 2023). By integrating users in the planning and decision-making processes, user-centric design ensures that technological solutions align with practical problems and enhance quality of life. Lesson learned: Involving people in the creation process increases acceptance and leads to more long-lasting results.

1.9.3 Data-Driven Decision-Making Is Transformative

Smart city research emphasizes the revolutionary potential of data-driven decision-making. Making informed and quick decisions is made possible by gathering, analyzing, and interpreting data from numerous sources (Dutta et al., 2023). Urban planners, politicians, and stakeholders are given the tools to improve services, distribute resources effectively, and proactively address urban concerns. Data is a priceless resource that fosters innovation and improves city governance.

1.9.4 Privacy and Security Are Paramount

The research on smart cities emphasizes how crucially important privacy and security issues are. Protecting residents' private information and guaranteeing the stability of the digital infrastructure are important considerations as technology enters every area of urban life. Finding the ideal balance between innovation and data security is a difficult task that needs constant attention (AlZu'bi et al., 2023). Trust-building and ensuring the long-term viability of smart cities depend on strong security and privacy protections.

1.9.5 Adaptability and Flexibility Drive Resilience

Smart cities need to be agile and adaptable to handle fast-evolving technical environments and dynamic urban challenges. Lessons from case studies emphasize the necessity of creating infrastructure and processes that can change and grow through time. Smart cities that are resilient utilize open architectures, interoperability standards, and future-proofing techniques (Ahmed et al., 2023). The development of flexible and scalable technologies ensures the endurance and resilience of smart city efforts.

1.10 Conclusion

Exploiting smart city ecosystems has illuminated a transformative vision for urban development. By delving into the requirements, architecture, applications, security considerations, and open research questions, we have uncovered the intricate web of technology, data, and innovation that underpins smart cities. The holistic integration of digital infrastructure, data-driven decision-making, and citizen engagement is a foundational necessity, fostering efficient resource management and improved urban living. With its layered approach encompassing application, data management, transmission, and sensing, the architecture creates a dynamic framework for real-time information exchange and informed governance across transportation, energy, and healthcare domains. While applications of smart cities promise enhanced mobility, sustainability, and services, challenges like security and privacy demand robust solutions. Exploring emerging technologies, climate resilience, equitable access, and ethical deployment remains pivotal. By assimilating the insights gleaned and adhering to the lessons learned, we can forge intelligent, adaptable, and people-centric urban landscapes that respond adeptly to future challenges and aspirations. The journey toward smart cities is an ongoing pursuit of innovation, collaboration, and sustainable progress, potentially redefining urban living for future generations.

References

Ahmed, S. T., Basha, S. M., Ramachandran, M., Daneshmand, M., & Gandomi, A. H. (2023). An edge-AI enabled autonomous connected ambulance route resource recommendation protocol (ACA-R3) for eHealth in smart cities. *IEEE Internet of Things Journal*. https://doi.org/10.1109/JIOT.2023.3243235

Aldhyani, T. H. H., Khan, M. A., Almaiah, M. A., Alnazzawi, N., Hwaitat, A. K. A., Elhag, A., Shehab, R. T., & Alshebami, A. S. (2023). A secure internet of medical things framework for breast cancer detection in sustainable smart cities. *Electronics (Switzerland), 12*(4). https://doi.org/10.3390/electronics12040858

Aldossary, M. (2023). Multi-layer fog-cloud architecture for optimizing the placement of IoT applications in smart cities. *Computers, Materials and Continua, 75*(1). https://doi.org/10.32604/cmc.2023.035414

Alwakeel, A., Alwakeel, M., Zahra, S. R., Saleem, T. J., Hijji, M., Alwakeel, S. S., Alwakeel, A. M., & Alzorgi, S. (2023). Common mental disorders in smart city settings and use of multimodal medical sensor fusion to detect them. *Diagnostics, 13*(6). https://doi.org/10.3390/diagnostics13061082

AlZu'bi, S., Elbes, M., Mughaid, A., Bdair, N., Abualigah, L., Forestiero, A., & Zitar, R. A. (2023). Diabetes monitoring system in smart health cities based on big data intelligence. *Future Internet, 15*(2). https://doi.org/10.3390/fi15020085

Badr, N. G. (2023). Smart cities for people with IDD—Foundations for digitally inclusive healthcare ecosystems. *ITM Web of Conferences, 51*. https://doi.org/10.1051/itmconf/20235104002

Berawi, M. A. (2023). Smart cities: Accelerating sustainable development agendas. *International Journal of Technology, 14*(1). https://doi.org/10.14716/ijtech.v14i1.6323

Cabrera, C., Svorobej, S., Palade, A., Kazmi, A., & Clarke, S. (2023). MAACO: A dynamic service placement model for smart cities. *IEEE Transactions on Services Computing, 16*(1). https://doi.org/10.1109/TSC.2022.3143029

Chakraborty, A., Banerjee, J. S., Bhadra, R., Dutta, A., Ganguly, S., Das, D., Kundu, S., Mahmud, M., & Saha, G. (2023). A framework of intelligent mental health monitoring in smart cities and societies. *IETE Journal of Research*. https://doi.org/10.1080/03772063.2023.2171918

Dutta, A., Kovid, R. K., Thatha, M., & Gupta, J. (2023). Adoption of IoT-based healthcare devices: An empirical study of end consumers in an emerging economy. *Paladyn, 14*(1). https://doi.org/10.1515/pjbr-2022-0106

El-Fouly, F. H., Kachout, M., Alharbi, Y., Alshudukhi, J. S., Alanazi, A., & Ramadan, R. A. (2023). Environment-aware energy efficient and reliable routing in real-time multi-sink wireless sensor networks for smart cities applications. *Applied Sciences (Switzerland), 13*(1). https://doi.org/10.3390/app13010605

El-Shafai, W., Ali, R., Sedik, A., Taha, T. E. S., Abd-Elnaby, M., & Abd El-Samie, F. E. (2023). Analysis of brain MRI: AI-assisted healthcare framework for the smart cities. *Intelligent Automation and Soft Computing, 35*(2). https://doi.org/10.32604/iasc.2023.019198

Fan, J., Yang, W., Liu, Z., Kang, J., Niyato, D., Lam, K. Y., & Du, H. (2023). Understanding security in smart city domains from the ANT-centric perspective. *IEEE Internet of Things Journal*. https://doi.org/10.1109/JIOT.2023.3252040

Fong, B., Kim, H., & Sai, V. (2023). Special section: Consumer healthcare technologies in smart cities. *IEEE Consumer Electronics Magazine*. https://doi.org/10.1109/MCE.2023.3245303

Gagliardi, A. R., & Albergo, F. (2023). The rise of smart healthcare in smart cities: A bibliometric literature review and avenue for a research agenda. *ITM Web of Conferences, 51*. https://doi.org/10.1051/itmconf/20235103002

Joe, S. S. A., Mary, S. J. J., Cordova, R. S., Kalash, H. S., & Al-Badi, A. (2023). IoT-based smart campus monitoring based on an improved chimp optimization-based deep belief neural network. *Indonesian Journal of Electrical Engineering and Informatics (IJEEI), 11*(1). https://doi.org/10.52549/ijeei.v11i1.4410

Jun, Y., Craig, A., Shafik, W., & Sharif, L. (2021). Artificial intelligence application in cybersecurity and cyberdefense. In *Wireless communications and mobile computing* (Vol. 2021). https://doi.org/10.1155/2021/3329581

Kushwaha, R., Shambharkar, R., Gupta, S., & Malik, M. (2023). Integration of block chain model for energy efficient WSN for IOT application. *International Journal for Research in Applied Science and Engineering Technology, 11*(2). https://doi.org/10.22214/ijraset.2023.48942

Mengash, H. A., Alharbi, L. A., Alotaibi, S. S., AlMuhaideb, S., Nemri, N., Alnfiai, M. M., Marzouk, R., Salama, A. S., & Duhayyim, M. A. (2023). Deep learning enabled intelligent healthcare management system in smart cities environment. *Computers, Materials and Continua, 74*(2). https://doi.org/10.32604/cmc.2023.032588

Mohammed, B. G., & Hasan, D. S. (2023). Smart healthcare monitoring system using IoT. *International Journal of Interactive Mobile Technologies, 17*(1). https://doi.org/10.3991/ijim.v17i01.34675

Pandya, S., Srivastava, G., Jhaveri, R., Babu, M. R., Bhattacharya, S., Maddikunta, P. K. R., Mastorakis, S., Piran, M. J., & Gadekallu, T. R. (2023). Federated learning for smart cities: A comprehensive survey. *Sustainable Energy Technologies and Assessments, 55*. https://doi.org/10.1016/j.seta.2022.102987

Reddy Yeruva, A. (2023). Providing a personalized healthcare service to the patients using AIOPs monitoring. *Eduvest—Journal of Universal Studies, 3*(2). https://doi.org/10.36418/eduvest.v3i2.742

Samy-El Menchawy, A., Hassan Moustafa, H., & Ibrahim Abdel-Hamid, N. (2023). Smart capital cities: Towards a smart new administrative capital (NAC). *F1000Research, 12*. https://doi.org/10.12688/f1000research.130322.1

Shafik, W. (2023a). *A comprehensive cybersecurity framework for present and future global information technology organizations*. https://doi.org/10.4018/978-1-6684-9018-1.ch002

Shafik, W. (2023b). Cyber security perspectives in public spaces: Drone case study. In *Handbook of research on cybersecurity risk in contemporary business systems*. https://doi.org/10.4018/978-1-6684-7207-1.ch004

Shafik, W. (2023c). Making cities smarter: IoT and SDN applications, challenges, and future trends. In *Opportunities and challenges of industrial IoT in 5G and 6G networks* (pp. 73–94). IGI Global. https://doi.org/10.4018/978-1-7998-9266-3.ch004

Shafik, W. (2024a). Blockchain-based internet of things (B-IoT): Challenges, solutions, opportunities, open research questions, and future trends. In *Blockchain-based internet of things* (pp. 35–58).

Shafik, W. (2024b). Introduction to ChatGPT. In *Advanced applications of generative AI and natural language processing models* (pp. 1–25). IGI Global. https://doi.org/10.4018/979-8-3693-0502-7.ch001

Shafik, W. (2024c). Navigating emerging challenges in robotics and artificial intelligence in Africa. In *Examining the rapid advance of digital technology in Africa* (pp. 124–144). IGI Global. https://doi.org/10.4018/978-1-6684-9962-7.ch007

Shafik, W. (2024d). Predicting future cybercrime trends in the metaverse era. In *Forecasting cyber crimes in the age of the metaverse* (pp. 78–113). IGI Global. https://doi.org/10.4018/979-8-3693-0220-0.ch005

Shafik, W., Matinkhah, S. M., & Ghasemazade, M. (2019). Fog-mobile edge performance evaluation and analysis on internet of things. *Journal of Advance Research in Mobile Computing, 1*(3), 1–17. https://doi.org/10.5281/zenodo.3591228

Shafik, W., Matinkhah, S. M., & Ghasemzadeh, M. (2020). Internet of things-based energy management, challenges, and solutions in smart cities. *Journal of Communications Technology, Electronics and Computer Science, 27*, 1–11. https://doi.org/10.22385/jctecs.v27i0.302

Shafik, W., Matinkhah, S. M., & Shokoor, F. (2023). Cybersecurity in unmanned aerial vehicles: A review. *International Journal on Smart Sensing and Intelligent Systems, 16*(1). https://doi.org/10.2478/ijssis-2023-0012

Sharma, O., Rathee, G., Kerrache, C. A., & Herrera-Tapia, J. (2023). Two-stage optimal task scheduling for smart home environment using fog computing infrastructures. *Applied Sciences (Switzerland), 13*(5). https://doi.org/10.3390/app13052939

Tareen, F. N., Alvi, A. N., Malik, A. A., Javed, M. A., Khan, M. B., Saudagar, A. K. J., Alkhathami, M., & Abul Hasanat, M. H. (2023). Efficient load balancing for Blockchain-based healthcare system in smart cities. *Applied Sciences (Switzerland), 13*(4). https://doi.org/10.3390/app13042411

Tripathi, A. K., Akul Krishnan, K., & Pandey, A. C. (2023). A novel blockchain and internet of things-based food traceability system for smart cities. *Wireless Personal Communications*. https://doi.org/10.1007/s11277-023-10230-9

Tripathy, S. S., Imoize, A. L., Rath, M., Tripathy, N., Bebortta, S., Lee, C. C., Chen, T. Y., Ojo, S., Isabona, J., & Pani, S. K. (2023). A novel edge-computing-based framework for an intelligent smart healthcare system in smart cities. *Sustainability (Switzerland), 15*(1). https://doi.org/10.3390/su15010735

Wang, Y., Shafik, W., Seong, J. T., Al Mutairi, A., Mustafa, M. S., & Mouhamed, M. R. (2023). Service delay and optimization of the energy efficiency of a system in fog-enabled smart cities. *Alexandria Engineering Journal, 84*, 112–125. https://doi.org/10.1016/j.aej.2023.10.034

Yang, Z., Jianjun, L., Faqiri, H., Shafik, W., Talal Abdulrahman, A., Yusuf, M., & Sharawy, A. M. (2021). Green internet of things and big data application in smart cities development. *Complexity, 2021*. https://doi.org/10.1155/2021/4922697

Zakari, R. Y., Shafik, W., Kalinaki, K., & Iheaturu, C. J. (2024). Internet of forestry things (IoFT) technologies and applications in Forest management. In *Advanced IoT technologies and applications in the industry 4.0 digital economy* (pp. 275–295). CRC Press. https://doi.org/10.1201/9781003434269-15

Zakzouk, A., El-Sayed, A., & Hemdan, E. E. D. (2023). A blockchain-based electronic medical records management framework in smart healthcare infrastructure. *Multimedia Tools and Applications*. https://doi.org/10.1007/s11042-023-15152-z

Wasswa Shafik received bachelor of Science degree in information technology engineering with a minor in mathematics from Ndejje University, Kampala, Uganda, and a master of engineering degree in information technology engineering from Yazd University, Yazd, Iran, and a PhD degree in computer science with the School of Digital Science, Universiti Brunei Darussalam, Brunei Darussalam. He is also the Founder and a Principal Investigator of the Dig Connectivity Research Laboratory (DCRLab) after serving as a Research Associate at Network Interconnectivity Research Laboratory, Yazd University. Prior to this, he worked as a Community Data Analyst at Population Services International, a community data officer at Programme for Accessible Health Communication, a research assistant at the Socio-Economic Data Centre, Prime Minister's Office, Kampala, Uganda, an Assistant Data Officer at TechnoServe, Kampala, IT Support at Thurayya Islam Media, Uganda, and Asmaah Charity Organization. His research interests include Ecological Informatics, Computer Vision, AI-enabled IoT/IoMTs, Cybersecurity, and privacy.

Chapter 2
Smart City for Kids: Creating a Decent City for Children in Indonesia

Hari Harjanto Setiawan

Abstract A child-friendly city (CFC) is one of the requirements that must be present in a smart city, showing the best city environment for children's lives. The city must show good indicators for child development. Indonesia has set five main standards, including the right to freedom, family environment, fulfilment of basic health and education. This chapter describes the implementation of CFC in Indonesia based on these indicators. Quantitative data is obtained through secondary data. Qualitative data was obtained from interviews with children, families, and stakeholders. The research results show that the achievement of implementing a CFC in 2022 shows fluctuations in values. This indicates the need for more serious efforts by the government in formulating policies and developing various child-centered programs and strategies. Children are essential in a smart city because they represent a country's future civilization. The novelty of this book chapter is that it includes indicators of CFC in smart cities.

Keywords Smart city · Child-friendly city · Child protection · Child right

2.1 Introduction

A smart city is an area designed using data and technology that aims to improve the welfare of its people. Through urban-related technology, they seek to overcome various urban challenges (Monfaredzadeh & Berardi, 2014). Smart cities aim to build better and more sustainable cities where residents have a higher quality of life, a more livable environment, and better economic opportunities. Smart cities are important for children because they can provide many benefits and improve their general well-being.

H. H. Setiawan (✉)
National Research and Innovation Agency (BRIN), Jakarta, Indonesia
e-mail: hari047@brin.go.id

S. Majumdar et al. (eds.), *Smart Cities*, S.M.A.R.T. Environments,
https://doi.org/10.1007/978-3-031-59846-3_2

These cities provide opportunities for children to grow and thrive by improving their access to quality education, healthcare, recreational facilities, and safe environments. Additionally, smart cities prioritize environmental sustainability, which is crucial for children's and the planet's future. Smart cities can create a decent urban environment for children in Indonesia by addressing their specific needs and challenges. Children in Indonesia face various challenges and problems in urban areas. These challenges include limited access to quality education, inadequate healthcare facilities, insufficient recreational spaces, and safety concerns. The smart city concept can address these challenges and create a more child-friendly urban landscape (Chan, 2021).

A child-friendly city is an idea that shows the best city environment for children's lives (Vidal & Castro Seixas, 2022). Children have the opportunity and educational resources to learn and explore their environment in society with applicable laws. Child-friendly is a city that guarantees the rights of every child as a city citizen. The city that children want is a city that respects children's rights. This can be realized by providing access to health services, education, clean water and air without pollution.

The city government also provides special policies and budgets for children. Providing a safe environment allows children to develop abilities and protects them from environmental damage and natural disasters (El-Said & Ezzat, 2019). Give special attention to children who work, are sexually exploited, and live with disabilities or without parental support. There is a forum for children to participate in decision-making that directly impacts children's lives.

The child-friendly city policy aims to create a development framework that guarantees the realization of children's rights and is implemented in a planned manner in the long term. Five children's rights must be implemented to realize CFC, namely institutions, the right to freedom, family environment, and basic rights to health and education. Each child's rights will be included in the indicators to be easily assessed (Supeni et al., 2022). Considerations for achieving a child-friendly city include partnerships, guidelines, roles, and dedication.

City governments need collaboration to ensure the creation of child-friendly cities. Partnerships were present between the public and private sectors, community and traditional leaders, representatives from each local government department or sector, non-governmental organizations, and civil society (Soeharnis & Laksmono, 2018). The next things needed to create a CFC are policies and budgets because the absence of such policies and little investment in child development present major challenges. Government executive agencies, legislatures, the commercial sector, non-governmental organizations, and civil society all play a role in creating child-friendly cities. Outreach activities ensure that all parties involved in creating child-friendly cities understand and use these ideas. Lastly, there is commitment. The concept of a CFC must not only be implemented but must also achieve the targets to be achieved.

The state must ensure that its actions meet the requirements of children's rights. In this regard, the government must create policies, programs, and initiatives supporting enforcement efforts by prioritizing related principles (Perera et al., 2022). The government has developed policies, programs, and activities to fulfil children's rights (Kosher et al., 2016). This is to ensure that the government's efforts run as expected. Child-friendly city policies are also implemented to support efforts to enforce these regulations.

This chapter aims to answer several main problems: how to implement child-friendly cities in Indonesia. In particular, it will describe the indicators of four children's rights: the right to freedom, family environment, basic health and education. The novelty of this book chapter is that it contains indicators of child-friendly cities in smart cities so that it can be used as a lesson for other countries.

2.2 The Rationale of the Study

2.2.1 Understanding Smart Cities

Smart cities have received attention from many countries recently. This city has used data by utilizing technology for the welfare of its residents. The city also simplifies administration and pays attention to sustainability. They are integrating technology into various aspects of urban life, including health and transportation. Smart city governance aims to create efficient and sustainable urban areas (Mishra & Chakraborty, 2020).

Smart cities integrate technology into government to create efficiency and improve public services and the welfare of their citizens (Ristvej et al., 2020). One part of a smart city is a child-friendly city. Children need attention because they are part of the city's population. A smart city is a good performance for a city, supported by an intelligent combination of all activities, studies, discoveries, and awareness of the city's people. Smart cities are expected to positively impact government, social life, transportation, and quality of life.

Indonesia has recognized the importance of creating smart cities that prioritize the well-being of its citizens, including children. As a country with a rapidly growing urban population, Indonesia understands the need to develop cities capable of meeting its residents' diverse needs while maintaining sustainability and inclusivity. Child-friendly urban design is a crucial aspect of Indonesia's vision for smart cities. Creating child-friendly urban spaces in Indonesia is very important for the overall development of children (Mishra & Chakraborty, 2020).

2.2.2 Child-Friendly City Design

The biggest users of outdoor activities are children, which directly impacts their personalities (Dagher et al., 2022). Therefore, incorporating child-friendly urban spaces in smart city planning is crucial for children's healthy growth and development. These spaces should prioritize the safety of children, provide ample space for play and exploration, nurture caregiver-child interactions, and promote independent mobility. By including children in the design and decision-making processes, their voices, needs, priorities, and rights can be considered. Moreover, child-friendly urban spaces contribute to smart cities' livability and inclusivity (Hämäläinen, 2020). One example of a city prioritizing children's needs and perspectives is Gothenburg in Sweden.

Cities that focus on children benefit significantly from thoughtful urban planning. It covers cities' physical layout and planning, including housing and public spaces, politics, economics, and ecology (Agarwal et al., 2021). The concept of "play" in urban design has been widely discussed in architecture and urban design. Play in this context refers to the non-functional use of public spaces, particularly those that meet children's recreation, exploration, and interaction needs. However, creating a child-friendly city is not just about providing playgrounds. Placing a playground in an area with heavy motor vehicle traffic, without considering the surrounding environment or other urban furniture design, does not make a city truly "child-friendly."

2.2.3 Indicators of Child-Friendly Cities in Indonesia

Every Indonesian is responsible for implementing regulations related to child protection. Systems related to child protection include family, school, social environment, and government. Policies on Child-Friendly Cities need to be created to ensure child protection. This city has a development system based on children's rights. Cities also integrate government, community, and business resources into policies, programs, and activities (Hu & Zheng, 2021). The aim is to guarantee children's rights comprehensively and sustainably.

Children's ideas, voices, attitudes, and ambitions must be considered in CFC. Because the formation of this city is best for children, children must be directly involved in implementing all programs and activities so that their rights can be respected. The Convention on the Rights of the Child, which the Indonesian government has ratified, is used to measure CFC implementation (Fig. 2.1).

2.3 Materials and Methods

Child-Friendly City Space: Main Component of a Smart City Child-friendly city space is an important component of a smart city (Alias et al., 2023). These spaces prioritize children's needs and well-being, contributing to their overall development and creating a sense of belonging in the community. The report on global change emphasizes that child-friendly urban spatial planning can help achieve urban development goals and overcome challenges by strengthening.

The information described in this book chapter is child-friendly city institutions, rights to freedom, family environment, basic health, and education. Data was

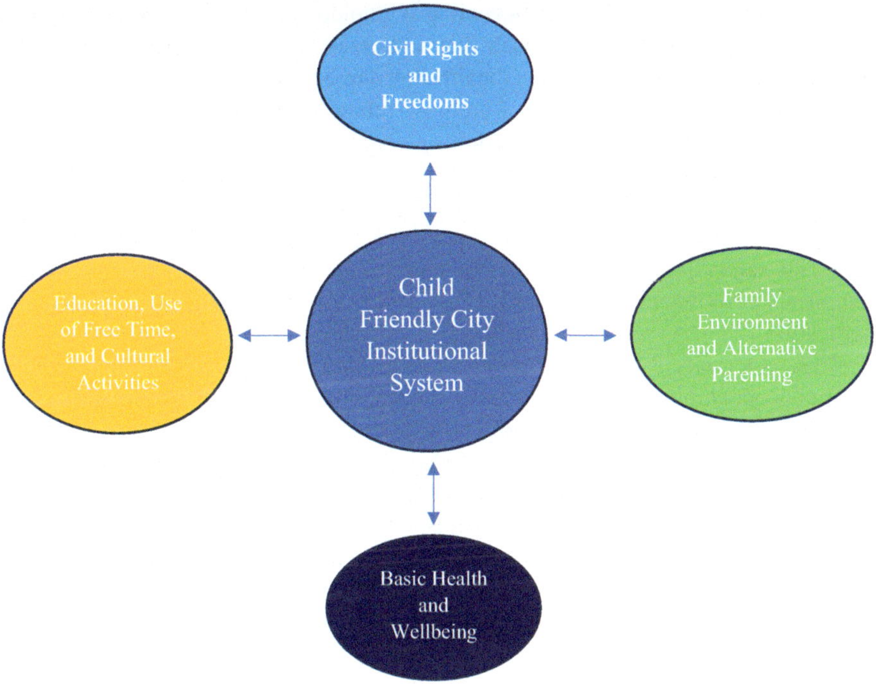

Fig. 2.1 Indicators for child-friendly cities

collected through in-depth interviews (Prentice, 2017) with parties related to children, families, communities, and policymakers. The in-depth interview process took place from January to February 2023. The in-depth interview lasted approximately 45–95 min. Apart from interviews, observation data was also used to strengthen the data (Greatorex, 2014). A documentation study examines quantitative data from annual reports and reputable journals (Jones & McCoy, 2019).

The collected data was analyzed in a thematic way, namely presenting data following predetermined themes (Sundler et al., 2019). The theme was Smart City for Kids: creating a decent city for children in Indonesia. This book chapter has several limitations, including limited time to make observations because the territory of Indonesia is very large. Observations were only carried out in a few cities. All quantitative data was obtained from secondary data.

2.4 Results and Discussion

In Indonesia, several cities have already taken initiatives to become smart cities, focusing on child-friendliness. For instance, Bandung, Malang, Manado, Bogor, Surabaya, and Surakarta/Solo have implemented various innovative city technologies and projects to improve the lives of children. However, it is essential to note that the concept of CFC goes beyond just implementing smart technologies. It requires a comprehensive approach that addresses various aspects of child well-being, including safety, education, healthcare, and accessibility.

The independence of children's mobility refers to their ability to move around their neighborhood or city without adult supervision for play or travel. This concept emphasizes the importance of safe and walkable streets, well-designed public spaces, and amenities that cater to the needs of children and their caregivers. Technological advances must be integrated with policies prioritizing children's welfare and safety (Gerber et al., 2019). One of the first concepts that align with Indonesia's National Policy on CFC is the development of green spaces, parks, and recreation.

CFC is a concept that focuses on building cities that support children's rights. This concept emphasizes creating a safe, healthy environment that stimulating children's development. A child-friendly city seeks to provide easy and safe access to children's educational, health, and recreational facilities and opportunities for them to participate actively in decision-making that impacts city policy. It creates a friendly environment for children and has many positive benefits. A safe and comfortable environment will provide security and comfort for children so they can play and learn better. In addition, a friendly environment for children will also provide opportunities for them to interact with peers and develop social skills that are important for their future lives (Ramioul et al., 2020) (Fig. 2.2).

2.4.1 Child-Friendly Smart City Institutions and Policies in Indonesia

Development is an important aspect that needs to be considered in Indonesia's Child Growth and Development Program (Nugroho, 2021). This is because institutions include regulatory aspects, programs and activities, budgets, and partner support. In measuring child-friendly cities, three indicators fall into the institutional aspect: availability of Regional Regulations and community involvement.

The assessment of institutional aspects is based on the city's answers regarding institutional child-friendly city cluster indicators. Apart from referring to the answer to the child-friendly city indicator question, you must prove the answer with appropriate documents. The average institutional score achieved by Indonesia in 2021 is 58.89. For 1 year, Indonesia's average achievement score increased by 17.68 points, so in 2022, the average achievement score for national institutions will be 76.57. In 2021, there were 14 provinces whose institutional scores were below the national average, which decreased to 11 provinces in 2022.

The province with the highest institutional score in 2021 is Yogyakarta, with a score of 95.43. Meanwhile, West Papua province has the lowest institutional achievement score with a value of 0.08. However, the achievement of the highest institutional score in 2022 has changed. DKI Jakarta Province achieved the best score in Indonesia, with a score of 135.06. Meanwhile, Papua will occupy the

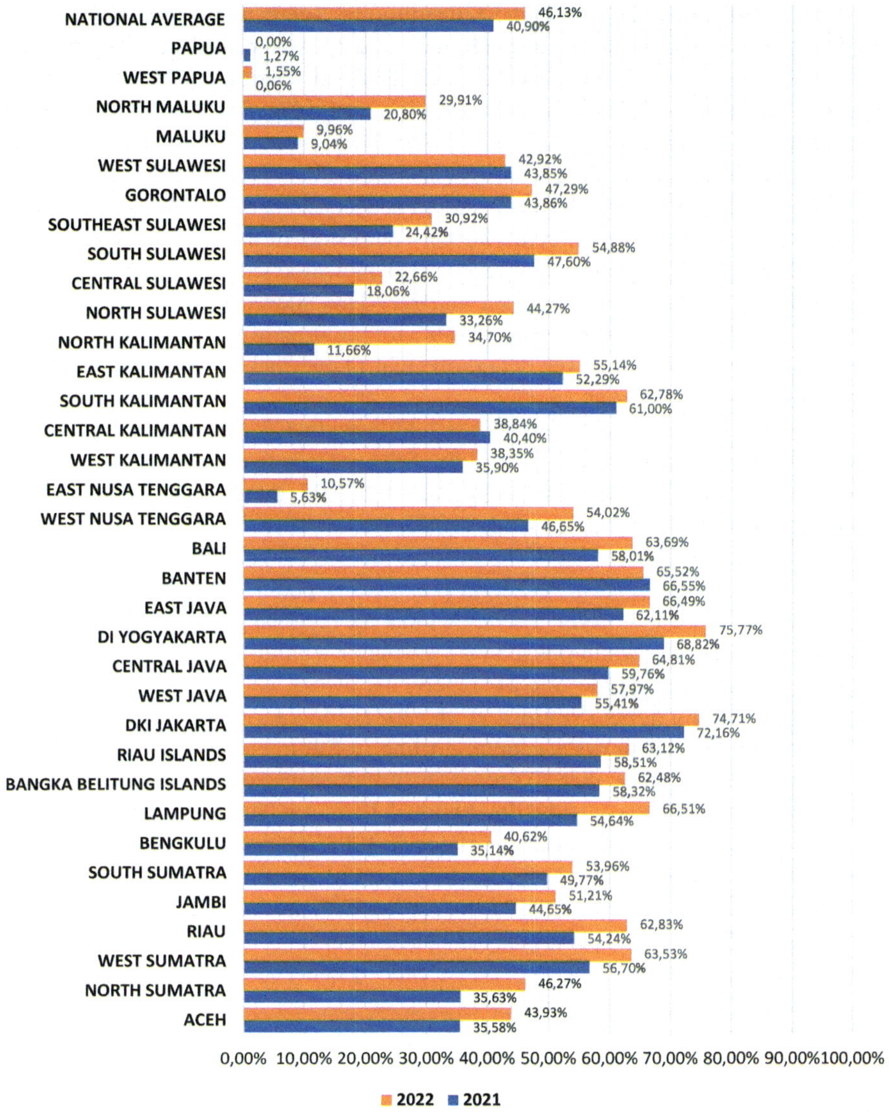

Fig. 2.2 Comparison of evaluation scores for child-friendly cities in 2021 and 2022. (Source: Ministry of Women's Empowerment and Child Protection in 2021–2022)

lowest position in institutional scores in 2022, reaching 0%. This indicates that the institutional aspects of child-friendly cities in Eastern Indonesia, especially in Papua and West Papua Provinces, still need to be improved. Improvements that can be made include issuing regional regulations, forming task forces, establishing action plans and regional profiles, and increasing synergy with various stakeholders who play important roles.

DKI Jakarta received the highest institutional score at the provincial level. Sorong city received the lowest institutional score due to the absence of regulations. There is no institutional strengthening in child-friendly cities, and the role of community institutions, mass media, and the business world in protecting children is not yet running well. Therefore, the Sorong city government must emphasize institutional strengthening by initiating various institutional assessment indicators.

The first indicator from the institutional aspect is the Regional Regulation on Child-Friendly Cities (Brown et al., 2019). The results of the comparative analysis of the 2021 and 2022 evaluations on the first indicator show that there will be a decline in 20 provinces in Indonesia in 2022. This shows that efforts to provide regional regulations regarding cities suitable for children in Indonesia are generally experiencing setbacks. However, data shows that 14 provinces experienced increased efforts to fulfil indicator one.

The second indicator is Strengthening Child-Friendly City Institutions (Dewanti & Anjarani, 2022). Likewise, with the results of the comparative analysis of the evaluation of the fulfilment of indicator one in 2021 and 2022, the data shows that, in general, there has been a decrease in efforts to fulfil the second indicator in the institutional aspect. In 2022, a decrease in efforts to fulfil the second indicator will occur in 26 provinces in Indonesia. However, eight provinces experienced increased efforts to implement the second indicator. This shows that the regional government in the province has strengthened institutions to make the region a child-friendly area by 2022.

The third indicator is the role of society, the media, and the business world in ensuring children's rights (Stoilova et al., 2021). A comparison of the 2021 and 2022 evaluations also shows a general decline in fulfilment of the third indicator. The analysis results show that 30 provinces in Indonesia experienced a decrease in the percentage of efforts to fulfil Institutional Indicator Three in 2022. Only four provinces experienced an increase in efforts to fulfil indicator three. This indicates that the four provinces have succeeded in involving the role of community institutions, media, and the business world in creating CFC in their regions. Meanwhile, for the other 30 provinces, a strategy is needed that involves all stakeholders who play a role in institutional aspects to create child-friendly institutions in their respective regions.

2.4.2 Civil Rights and Freedoms

Civil rights and freedoms are the first determining indicators of CFC (Kron & Lebuhn, 2020). These rights include ownership of a child's birth certificate and identity card and the availability of child-friendly information facilities. The analysis results show that the scores in 2021 and 2022 have the same average, namely 44.89. Even though national achievements have not changed, the data shows several regions have experienced increases and decreases. In 2021, DKI Jakarta Province had the highest national achievement with a score of 79.53. However, in 2022, the Yogyakarta Special Region will be ranked first with a score of 119.00.

The city of East Jakarta received the highest score of 111.5. Malacca Regency achieved the lowest score. Malacca Regency has the lowest score due to the low percentage of children with birth certificates and Identity Cards and the unavailability of Child-Friendly Information Facilities. Therefore, there needs to be a breakthrough, especially from the regional government of Malacca Regency, to increase the percentage related to the realization of CFC.

The fourth indicator is Registered Children and Children with a Birth Certificate Excerpt. The results of the comparative analysis of the evaluation of the achievement of indicator four in 2021 and 2022 show an increase. Nationally, data shows that there will be an increase in efforts to fulfil indicators in 27 provinces in Indonesia in 2022. Meanwhile, seven provinces will experience a decrease in efforts to fulfil indicators.

The fifth indicator is the availability of child-friendly information facilities. A comparison of evaluations on indicator five in 2021 and 2022 shows that there has been an increase in efforts to fulfil the indicators in 18 provinces in Indonesia, which is indicated by an increase in the percentage in 2022. Meanwhile, 16 provinces experienced a decline in efforts to fulfil indicators. This shows that efforts are needed from local governments, especially from the 16 provinces whose achievement scores have decreased, to increase the availability of child-friendly information facilities.

The sixth indicator is the Institutionalization of Children's Participation. A comparison of the evaluation of the indicators for six child participation institutions in 2021 and 2022 shows a decrease in efforts to fulfil the indicators in 18 provinces in Indonesia, as indicated by a decrease in the percentage in 2022. Meanwhile, 16 provinces experienced an increase in efforts to fulfil the indicators.

2.4.3 *Family Environment and Alternative Parenting*

The family environment and alternative care are determining indicators of a CFC. This cluster is defined as fulfilling a child's needs by biological parents, biological father only, biological mother only, or other family, including adoptive parents, adoptive parents, or guardians. The results of the analysis show that the average value of the family environment and alternative care clusters in 2022 has increased compared to the previous year. In 2021, Indonesia's average score was 62.94, increasing in 2022 to 74.12. Fifteen provinces had average achievement scores below the national average in 2021. However, in 2022, there will be a decrease in the number of provinces with scores below the national average to 14.

The data shows that Sumenep Regency obtained the highest achievement score, with a score of 170. Meanwhile, the lowest score was obtained by Malacca Regency. Malaka Regency received the lowest score due to several factors. These factors include (1) prevention of child marriage in the district is less than optimal; (2) strengthening the capacity of consultant institutions providing child care services for families which have not been implemented; (3) holistic and integrative early childhood development has not gone well; (4) standardization of alternative service institutions has not gone well; (5) There is no availability of child-friendly infrastructure in public spaces. This shows the need for special attention for Malaka Regency to participate in realizing CFC in its region by making various improvements to the factors causing Malacca Regency's low achievement scores.

Indicator seven is the Prevention of Child Marriage. Comparative data on the achievements of the seven indicators in 2021 and 2022 shows a decrease in efforts to fulfil indicators in 20 provinces in Indonesia, as indicated by a decrease in percentage in 2022. Meanwhile, 14 provinces experienced an increase in efforts to fulfil indicators.

The eighth indicator is Strengthening the Capacity of Consultant Institutions Providing Child Care Services for Parents/Families. A comparison of the achievements of the eight indicators in 2021 and 2022 shows that there has been an increase in efforts to fulfil the indicators in 33 provinces in Indonesia, as indicated by an increase in the percentage in 2022. Meanwhile, one province experienced a decrease in efforts to fulfil the indicators.

The ninth indicator is Holistic and Integrative Early Childhood Development. There has been an increase in efforts to fulfil indicators in 30 provinces in Indonesia, as shown by an increase in the percentage in 2022. Meanwhile, four provinces experienced a decrease in efforts to fulfil KLA indicators. The four provinces are the Bangka Belitung Islands, West Java, Bali, and East Kalimantan.

The 10th indicator is the Standardization of Alternative Care Institutions. The achievements of the 10th indicator for 2021 and the 11th indicator for 2022 show an increase in efforts to fulfil the indicators in 33 provinces in Indonesia, as shown by the percentage increase in 2022. Meanwhile, Papua Province experienced a decrease in efforts to fulfil the indicators.

The 11th indicator is the availability of child-friendly infrastructure in public spaces. A comparison of the achievements of 11 indicators in 2021 and 2022 shows that there has been an increase in efforts to fulfil indicators in 33 provinces in Indonesia, which is indicated by an increase in the percentage in 2022. However, Papua Province has experienced a decrease in efforts to fulfil indicators in the Availability of Child-Friendly Infrastructure in Public Spaces indicator.

2.4.4 Basic Health and Well-being

Basic Health and Welfare describes the fulfilment of children's needs by providing health services as a basic need borne by the state and ensuring the order of life of children so that children's growth and development can run optimally, both spiritually, physically, and socially. The analysis results show that the national average value for the basic health and welfare cluster in 2022 is much higher (87.62) than in 2021 (66.20). This increase in score of 21.07 points is an achievement that should be appreciated. The increase in scores reflects that the basic health and welfare of children in Indonesia continue to improve so that the quality of children's growth and development continues to improve. In 2021, there are 14 provinces with scores below the national average, while in 2022, there are 13 provinces with scores below the average.

Yogyakarta Special Region in 2021 was ranked highest, scoring 120.43. Meanwhile, in 2022, DKI Jakarta received the highest score (133.08) in basic health and welfare. The data shows that Surabaya city obtained the highest score for the basic health and welfare cluster, with a score 150. Regency had the lowest score due to several factors. These factors are (1) the birth rate in health facilities is still low; (2) the nutritional status of toddlers needs to be improved; (3) feeding babies and children under 2 years old has not been implemented properly; (4) health facilities have not implemented child-friendly services; (5) a healthy environment has not yet been created; (6) non-smoking areas and prohibitions on advertising, promotion, and sponsorship of cigarettes have not been implemented. Therefore, efforts are needed to improve health services in the Musi Banyuasin Regency area to increase the achievement of health cluster values and basic health.

The 12th indicator is Distribution to Health Facilities. A comparison of the achievements of 12 indicators in 2021 and 2022 shows that there has been an increase in efforts to fulfil indicators in 28 provinces in Indonesia, as indicated by an increase in the percentage in 2022. Meanwhile, six provinces experienced a decrease in efforts to fulfil indicators. The six provinces are Riau, DKI Jakarta, Central Java, Yogyakarta, Banten, and Papua.

The 13th indicator of a child-friendly city is the nutritional status of toddlers. A comparison of the achievements of the 13 indicators in 2021 and 2022 shows that there has been an increase in efforts to fulfil the indicators in 33 provinces in Indonesia, as indicated by an increase in the percentage in 2022. Meanwhile, Papua Province has experienced a decline in efforts to fulfil indicators that focus on the Nutritional Status of Toddlers.

The 14th indicator is Feeding Infants and Children under 2 years. A comparison of the achievements of the 14 indicators in 2021 and 2022 shows that there has been an increase in efforts to fulfil the indicators in 32 provinces in Indonesia, as indicated by an increase in the percentage in 2022. Meanwhile, the two provinces experienced a decrease in efforts to fulfil the indicators. These provinces are the Bangka Belitung Islands and Maluku.

The 15th indicator is Health Facilities with Child-Friendly Services. A comparison of the achievements of 15 indicators in 2021 and 2022 shows that there has been an increase in efforts to fulfil indicators in 29 provinces in Indonesia, as indicated by an increase in the percentage in 2022. Meanwhile, five provinces experienced an increase in efforts to fulfil indicators. These provinces are Banten, Bali, East Nusa Tenggara, Central Kalimantan, and Gorontalo.

The 16th indicator is a healthy environment. A comparison of the achievements of 16 indicators in 2021 and 2022 shows that there has been an increase in efforts to fulfil KLA indicators in 29 provinces in Indonesia, as shown by an increase in the percentage in 2022. Meanwhile, five provinces experienced a decrease in efforts to fulfil KLA indicators. These provinces are Riau, DI Yogyakarta, West Sulawesi, Maluku, and Papua.

The 17th indicator is a no-smoking area and a cigarette advertising, promotion, and sponsorship ban. A comparison of the achievements of the 17 indicators in 2021 and 2022 shows that there has been an increase in efforts to fulfil the indicators in 31 provinces in Indonesia, as indicated by an increase in the percentage in 2022. Meanwhile, three provinces experienced a decrease in efforts to fulfil the indicators, especially the No-Smoking Zones and Advertising, Promotion, and Cigarette Sponsorship. These provinces are Gorontalo, Maluku, and Papua.

2.4.5 Education, Use of Free Time, and Cultural Activities

They explain children's rights to education to improve quality and opportunities to develop creativity, imagination, and self-confidence. The results of the national average analysis in this field show that in 2021, the national achievement score is greater (55.83) than in 2022 (46.73). The decrease in value of 9.1 points is thought to be due to the impact of the pandemic. During the pandemic, various activities were carried out from home. Families need to adapt to face all the changes that occur. This adjustment takes quite a long time. Children carry out learning activities from home. This impacts their learning activities and the activities they do in their free time. Meanwhile, many parents have to do their work from home, so it takes more effort to do their work while supervising their children at school (Fig. 2.3).

Data on achievements in this field shows that the Yogyakarta Special Region Province received the highest score for two consecutive years, namely in 2021

Fig. 2.3 Child-friendly integrated public space in DKI Jakarta

(96.72) and 2022 (162.75). The increase in value in that period was 66.03. Meanwhile, the same as the previous group, in 2021, West Papua received the lowest score of 0.00. In 2022, the lowest score in the education, use of free time, and culture cluster was obtained by Papua province with a score of 0.00. Provinces with the lowest scores require intensive assistance to increase efforts to fulfil children's rights to achieve a child-friendly city.

Data shows that Kediri city achieved the highest score, with a score of 110. The district received the lowest ranking due to three main factors. These factors are: (1) 12 years of compulsory education has not gone well; (2) Child-Friendly Education Units have not been implemented; and (3) there are no child-friendly cultural, creative, and recreational activity facilities. This shows that very serious efforts are needed from the district government to fulfil the rights. The 18th indicator is 12 years of compulsory education. A comparison of the achievement of 18 indicators in 2021 and 2022 shows a decrease in efforts to fulfil indicators in 19 provinces in Indonesia, which is indicated by a decrease in percentage in 2022. Meanwhile, 15 provinces experienced an increase in efforts to fulfil indicators.

The 19th indicator is Child-Friendly Education Units. A comparison of the achievements of 19 indicators in 2021 and 2022 shows a decrease in efforts to fulfil indicators in 21 provinces in Indonesia, which is indicated by a decrease in percentage in 2022. Meanwhile, 13 provinces experienced an increase in efforts to fulfil indicators.

The 20th indicator is the availability of child-friendly cultural, creative, and recreational activity facilities. A comparison of the achievements of 20 indicators in 2021 and 2022 shows that there has been an increase in efforts to fulfil indicators in 19 provinces in Indonesia, as indicated by an increase in the percentage in 2022. Fifteen provinces experienced a decrease in efforts to fulfil indicators. If analyzed, the decline in efforts to fulfil indicators mostly occurred in Eastern Indonesia. This shows the need for more serious efforts, especially in the provinces in the Eastern Indonesia region, in realizing the provision of child-friendly cultural, creative, and recreational activity facilities to create child-friendly provinces in their respective regions.

Education is one of the key areas where smart cities can significantly impact children's lives. According to the sources, smart cities can improve children's access to education through digital technologies. Smart classrooms with interactive learning tools and online educational resources can enhance children's learning experiences. These educational advancements can bridge the gap between urban and rural areas, providing equal educational opportunities for children regardless of location. Moreover, smart cities can also support personalized and adaptive learning, where children can receive tailored educational content based on their individual needs and learning styles (Alzoman et al., 2021). This can lead to improved academic performance and educational outcomes for children.

Smart cities can also create opportunities for children to engage in outdoor activities and play. This can be achieved by developing green spaces, parks, and recreational areas that are easily accessible to children. Regular physical activity and outdoor play can improve children's physical and mental well-being (Owen et al., 2022). Furthermore, integrating smart technologies in these spaces can enhance safety measures for children. For instance, smart surveillance systems and IoT-based sensors can be deployed to ensure the safety of children in play areas.

2.5 Limitations of the Study

This book chapter's writing is limited to answering the main problems in implementing child-friendly cities in Indonesia. In particular, it will explain the rights to freedom, family environment, basic health and education. However, the results of this research can be used as lessons for countries that are building smart cities. Child-friendly cities are part of smart cities because children are part of the city population. This research has limited time and observation locations, impacting the results.

2.6 Recommendations

A child-friendly city respects children prioritizes their needs and involves them in policy-making and the city design process. Smart cities have the potential to have a huge influence on the development of children in Indonesia. They can improve children's safety by implementing smart technology, such as surveillance systems and sensors, to monitor public spaces and ensure their safety. Smart cities can also provide better transportation options for children, such as pedestrian-friendly infrastructure and efficient public transportation systems. Children's access to education can also be improved in smart cities through digital technology. For example, smart classrooms with interactive learning tools and online resources can enhance children's learning experiences. Additionally, smart cities can create opportunities for children to engage in outdoor activities and play.

Smart city initiatives can include developing green spaces, parks, and recreation areas that are easily accessible to children. Additionally, smart cities can encourage interaction between child care providers and independent mobility by providing safe, walkable roads. Well-designed public spaces and facilities that meet the needs of children and their caregivers are required to build a CFC. Urban areas in Indonesia, such as Jakarta and Bandung, are characterized by high population density, so integration is needed.

Ensure active participation of children in the planning and development process through platforms such as children's forums and task forces. Their unique needs and perspectives can be considered by involving children in decision-making, resulting in more inclusive and child-friendly urban environments. Develop and enforce regulations and standards that prioritize the safety and welfare of children in urban planning and development. This could include implementing strict traffic regulations to ensure safe roads, designing accessible and child-friendly infrastructure, and creating guidelines for developing parks and public spaces that meet children's needs and preferences.

Invest in developing smart technology to improve children's safety, education, and health services. For example, implementing smart traffic management systems prioritizing pedestrian safety can help create safer streets for children to explore independently. Build partnerships between government agencies, NGOs, and community organizations to collaborate on initiatives that promote the welfare of children in smart cities. For example, collaborating with NGOs and community organizations can help implement child-friendly programs and facilities, such as after-school activities and sports facilities.

2.7 Conclusions

In conclusion, creating a child-friendly smart city in Indonesia is crucial for children's physical and mental well-being. Integrating smart technologies can improve children's overall well-being by enhancing outdoor activities and play. Furthermore, incorporating healthcare and nutrition services into smart cities can improve child development outcomes. However, it is important to note that the concept of a child-friendly city goes beyond just implementing smart technologies. It requires a holistic approach encompassing various aspects of child well-being, including safety, education, healthcare, and accessibility. This comprehensive approach can be achieved by prioritizing children's independent mobility safety, ensuring safe and walkable streets, well-designed public spaces, and amenities that cater to the needs of children and their caregivers. To effectively create a child-friendly smart city in Indonesia, it is necessary to integrate technological advancements with policies and initiatives prioritizing children's well-being and safety in urban planning and development. This can be done by implementing the concepts outlined in Indonesia's National Policy on Child-Friendly Cities, such as developing green spaces and parks that provide outdoor play and recreation opportunities. The Child-Friendly Village program can also serve as a model for creating child-friendly communities within smart cities.

Acknowledgements We thank the Research Center for Social Welfare Villages and Connectivity and the National Research and Innovation Agency (BRIN) for supporting this research.

References

Agarwal, M. K., Sehgal, V., & Ogra, A. (2021). Creating a child-friendly environment: An interpretation of children's drawings from planned neighborhood parks of Lucknow City. *Societies, 11*(3), 80. https://doi.org/10.3390/SOC11030080

Alias, A., Mohamad Nasri, N., & Awang, M. M. (2023). A tracer study on child participation in child councillor programmes aimed towards development of the child friendly cities initiative. *Children, 10*(4), 732. https://doi.org/10.3390/CHILDREN10040732

Alzoman, R. M., Alenazi, M. J. F., Belli, L., Ferrari, G., Martalò, M., & Davoli, L. (2021). A comparative study of traffic classification techniques for smart city networks. *Sensors, 21*(14), 4677. https://doi.org/10.3390/S21144677

Brown, C., de Lannoy, A., McCracken, D., Gill, T., Grant, M., Wright, H., & Williams, S. (2019). Special issue: Child-friendly cities. *Cities & Health, 3*(1–2), 1–7. https://doi.org/10.1080/23748834.2019.1682836

Chan, P. (2021). Child-friendly urban development: Smile village community development initiative in Phnom Penh. *WORLD, 2*(4), 505–520. https://doi.org/10.3390/WORLD2040031

Dagher, S. F. A., Refaat, M., & El Messeidy, R. (2022). Landscape design as a tool to meet children's needs in residual urban spaces. *Civil Engineering and Architecture, 10*(5A), 271–287. https://doi.org/10.13189/CEA.2022.101413

Dewanti, R., & Anjarani, P. M. (2022). Strengthening cooperation with local governments through the child-friendly city program. *ICCD, 4*(1), 604–609. https://doi.org/10.33068/ICCD.V4I1.528

El-Said, R., & Ezzat, E.-B. (2019). Effect of educational protocol to improve mothers' knowledge, practice and attitude about child abuse. *Saudi Journal of Nursing and Health Care*. https://doi.org/10.36348/sjnhc.2019.v02i11.007

Gerber, L. A., Pang, Y. C., Ross, T., Guggenheim, M., Pecora, P. J., & Miller, J. (2019). Effects of an interdisciplinary approach to parental representation in child welfare. *Children and Youth Services Review, 102*, 42–55. https://doi.org/10.1016/J.CHILDYOUTH.2019.04.022

Greatorex, M. (2014). Observation. In Sir C. L. Cooper (Ed.), *Wiley encyclopedia of management*. https://doi.org/10.1002/9781118785317.weom090543

Hämäläinen, M. (2020). A framework for a smart city design: Digital transformation in the Helsinki Smart City. *Contributions to Management Science*, 63–86. https://doi.org/10.1007/978-3-030-23604-5_5/COVER

Hu, Q., & Zheng, Y. (2021). Smart city initiatives: A comparative study of American and Chinese cities. *Journal of Urban Affairs, 43*(4), 504–525. https://doi.org/10.1080/07352166.2019.1694413

Jones, K. M. L., & McCoy, C. (2019). Reconsidering data in learning analytics: Opportunities for critical research using a documentation studies framework. *Learning, Media and Technology, 44*(1), 52–63. https://doi.org/10.1080/17439884.2018.1556216

Kosher, H., Ben-Arieh, A., & Hendelsman, Y. (2016). *Children's rights and social work*. https://doi.org/10.1007/978-3-319-43920-4

Kron, S., & Lebuhn, H. (2020). Building solidarity cities: From protest to policy. *Fostering Pluralism Through Solidarity Activism in Europe*, 81–105. https://doi.org/10.1007/978-3-030-56894-8_4

Mishra, K. N., & Chakraborty, C. (2020). A novel approach toward enhancing the quality of life in smart cities using clouds and IoT-based technologies. *Internet of Things*, 19–35. https://doi.org/10.1007/978-3-030-18732-3_2/COVER

Monfaredzadeh, T., & Berardi, U. (2014). How can cities lead the way towards a sustainable, competitive and smart future? *WIT Transactions on Ecology and the Environment, 191*, 1063–1074. https://doi.org/10.2495/SC140902

Nugroho, A. M. (2021). A child-friendly design for sustainable urban environment: A case study of Malang city parks. *IOP Conference Series: Earth and Environmental Science, 881*(1), 012060. https://doi.org/10.1088/1755-1315/881/1/012060

Owen, K. B., Nau, T., Reece, L. J., Bellew, W., Rose, C., Bauman, A., Halim, N. K., & Smith, B. J. (2022). Fair play? Participation equity in organised sport and physical activity among children and adolescents in high-income countries: A systematic review and meta-analysis. *International Journal of Behavioral Nutrition and Physical Activity, 19*(1), 1–13. https://doi.org/10.1186/S12966-022-01263-7/TABLES/2

Perera, S., Trapenciere, I., & Vilka, L. (2022). Policy and legal framework on protecting children's rights in Latvia and Sri Lanka. *SHS Web of Conferences, 131*, 03001. https://doi.org/10.1051/SHSCONF/202213103001

Prentice, C. M. (2017). Interviews, qualitative. *The International Encyclopedia of Intercultural Communication*, 1–8. https://doi.org/10.1002/9781118783665.ieicc0108

Ramioul, C., Tutenel, P., & Heylighen, A. (2020). Exploring with children what makes a city child-friendly. *Designing for Inclusion*, 99–106. https://doi.org/10.1007/978-3-030-43865-4_10

Ristvej, J., Lacinák, M., & Ondrejka, R. (2020). On smart city and safe city concepts. *Mobile Networks and Applications, 25*(3), 836–845. https://doi.org/10.1007/S11036-020-01524-4/TABLES/1

Soeharnis, R. S. M., & Laksmono, B. S. (2018). Issues in developing communication strategies for implementing child-friendly city policies. *E3S Web of Conferences, 74*, 10017. https://doi.org/10.1051/E3SCONF/20187410017

Stoilova, M., Nandagiri, R., & Livingstone, S. (2021). Children's understanding of personal data and privacy online—A systematic evidence mapping. *Information, Communication & Society, 24*(4), 557–575. https://doi.org/10.1080/1369118X.2019.1657164

Sundler, A. J., Lindberg, E., Nilsson, C., & Palmér, L. (2019). Qualitative thematic analysis based on descriptive phenomenology. *Nursing Open, 6*(3), 733–739.

Supeni, S., Handini, O., & Al Hakim, L. (2022). The effort to develop character education in a child-friendly school based on Java regional art culture at an elementary school in Surakarta City. In *Proceedings of the Universitas Lampung international conference on social sciences (ULICoSS 2021)* (Vol. 628, pp. 697–703). https://doi.org/10.2991/ASSEHR.K.220102.096

Vidal, D. G., & Castro Seixas, E. (2022). Children's green infrastructure: Children and their rights to nature and the city. *Frontiers in Sociology, 7*, 804535. https://doi.org/10.3389/FSOC.2022.804535/BIBTEX

Hari Harjanto Setiawan completed his undergraduate education in 1998 at the Bandung College of Social Welfare (STKS). In 2001, he graduated from the University of Indonesia in the Sociology Study Program specializing in Social Welfare. In 2014, he completed his doctoral degree at the University of Indonesia's Social Welfare Study Program. From 2007 to 2021, he worked as a researcher at the Center for Social Welfare Research and Development, Ministry of the Republic of Indonesia. Starting from January 1, 2022, he moved to work at the National Research and Innovation Agency (BRIN), Jakarta, Indonesia. His research topics include social welfare, poverty, child protection, reintegration, social entrepreneurship, social rehabilitation, social assistance, community resilience, indigenous communities, social empowerment, and climate change. He has carried out various studies and published them in books and scientific journals. He is also a peer reviewer for the Journal "Sosio Konsepsia" and Frontiers in Sociology.

Chapter 3
Visualizing Smart Cities: Unveiling Insights Through Graphical Analysis and Algorithmic Precision

C. D. Divya

Abstract Graphical analysis serves as a pivotal tool for unraveling the complexities of smart cities through visual representation, bolstering the efficacy of urban planning and resource management. Leveraging diverse data sources including sensors, social media, and devices, this approach transforms intricate datasets into comprehensible charts, maps, and dashboards. Employing algorithms like clustering, regression, and anomaly detection enhances the precision of graphical insights. Real-time monitoring of traffic, weather, and emergencies is facilitated by algorithms that process data instantaneously, enabling prompt responses. Urban planning benefits from predictive analytics, employing algorithms to forecast population trends, traffic flows, and resource consumption patterns. Moreover, machine learning algorithms aid in identifying correlations and patterns in large-scale datasets, guiding informed decisions about infrastructure development, zoning regulations, and resource allocation. Algorithms like convolutional neural networks (CNNs) aid in analyzing visual data from surveillance cameras, facilitating crime mapping and public safety enhancements. Performance evaluation of these algorithms is crucial, focusing on metrics such as accuracy, precision, and recall. Furthermore, citizen engagement thrives as comprehensible visualizations encouraging sustainable behaviors and public participation. Optimizing transportation relies on algorithms that decipher real-time traffic patterns and congestion points, improving commute experiences. Environmental monitoring is elevated through algorithms that process data from air quality sensors and pollution sources, contributing to sustainable urban practices. The integration of graphical analysis and algorithmic processing elevates smart cities, fostering data-driven governance and enriching citizens' quality of life.

C. D. Divya (✉)
Department of CSE, Vidyavardhaka College of Engineering, Mysore, Karnataka, India
e-mail: divyacd@vvce.ac.in

S. Majumdar et al. (eds.), *Smart Cities*, S.M.A.R.T. Environments,
https://doi.org/10.1007/978-3-031-59846-3_3

Keywords Smart cities · Smart city development · Smart city frameworks · Big data in smart cities · Smart city sustainability · Urban planning in smart cities · Citizen-centric smart cities · Smart city cybersecurity · Smart city mobility · Inclusive smart cities · Smart city data analysis in urban environments

3.1 Introduction

Smart cities, characterized by the integration of cutting-edge technologies and data-driven decision-making processes, have emerged as a transformative force in urban development. These cities are not just centers of innovation but also hubs of sustainability, efficiency, and enhanced quality of life for their residents. To fully harness the potential of smart cities, it is essential to not only collect vast amounts of data from various sources but also to extract meaningful insights from this data. This is where the marriage of graphical analysis and algorithmic precision becomes paramount. In this era of urbanization, where an increasing majority of the global population resides in urban areas, smart cities are not merely a luxury but a necessity. Smart cities leverage data and technology to address complex challenges like traffic congestion, environmental sustainability, public safety, and efficient resource management. These cities are dynamic ecosystems where information flows ceaselessly, from sensors embedded in infrastructure to citizen-generated data on social media platforms. Amidst this data deluge, the ability to visualize and interpret information has become indispensable. This paper embarks on a journey into the world of smart cities, exploring how graphical analysis and algorithmic precision can be instrumental in unraveling hidden patterns, optimizing resource allocation, and improving decision-making within urban environments. By weaving together, the power of visualization techniques and the rigor of data algorithms, we aim to shed light on the multifaceted facets of smart cities, unveiling the transformative potential they hold. Through case studies and practical examples, we will delve into the ways in which data visualization and advanced algorithms can be harnessed to enhance urban planning, environmental sustainability, traffic management, and overall quality of life in smart cities. As the urban landscape continues to evolve, so too does our understanding of how data-driven insights can shape the cities of the future. This paper seeks to illuminate the path forward, where the fusion of graphical analysis and algorithmic precision promises not only to make smart cities more intelligent but also more inclusive, sustainable, and responsive to the needs of their citizens.

3.2 Literature Review

Anthopoulos and Fitsilis (2015) conducted a survey of big data architectures and machine learning algorithms in the context of smart cities. They explored various architectures and algorithms pertinent to the advancement of smart city initiatives.

Albino et al. (2015) provided an in-depth examination of smart cities, delineating their definitions, dimensions, performance, and initiatives. Batty (2013) discussed the implications of big data and its relevance to smart city planning, emphasizing the integration of data-driven approaches in urban development strategies. Bibri and Krogstie (2017) scrutinized the social shaping dimensions of smart sustainable cities, highlighting the interplay between technology, society, and sustainability. Caragliu et al. (2009) analyzed the emergence of smart cities in Europe, shedding light on the technological advancements and urban transformations within the European context. Eger and Miller (2017) explored the role of libraries as essential components of smart cities, advocating for equitable access to information and knowledge. Hollands (2008) critically examined various conceptions of smart cities, advocating for a nuanced understanding of their complexities. Hernández-Muñoz et al. (2011) discussed the integration of smart cities with the future Internet, envisioning innovative services and infrastructural developments. Komninos (2002) delved into the concept of intelligent cities, emphasizing the role of innovation and knowledge systems in urban development. Nam and Pardo (2011) conceptualized smart cities by considering dimensions of technology, people, and institutions, elucidating the multifaceted nature of these urban environments. Rufat and Rutherford (2018) investigated the role of social media and open-source geospatial intelligence in disaster management, focusing on the context of US wildfires. Shen and Norford (2016) examined the spatial analysis of urban building energy use and the impact of green roofs on energy efficiency with a case study conducted in the greater Boston area. Yigitcanlar and Kamruzzaman (2018) explored the relationship between smart city policies and urban sustainability, scrutinizing the effectiveness of such policies in promoting sustainable development. Zanella et al. (2014) provided an overview of the Internet of Things (IoT) in the context of smart cities, elucidating its implications for urban development and governance.

3.3 Methodology

3.3.1 Datasets Involved

1. Government Open Data Portals: Many cities and governmental organizations provide open access to datasets related to various aspects of urban life, including transportation, infrastructure, and public services.
2. Research Institutions: Academic institutions and research organizations often publish datasets in their studies on smart cities, urban planning, and data analytics.
3. Kaggle: Kaggle is a platform that hosts datasets and machine learning competitions. You can search for datasets related to smart cities, urban analytics, and related topics on Kaggle.

4. IEEE DataPort: IEEE DataPort is a platform where researchers can share and access datasets across various domains, including smart cities and urban computing. You may find relevant datasets here for your analysis.
5. Urban Data Platforms: Some cities have developed their own urban data platforms to provide access to datasets and tools for analyzing urban data (Table 3.1 and Fig. 3.1).

The provided table summarizes the performance of various data analysis methods across five different datasets. These methods include clustering, regression, anomaly detection, CNNs (Visual Analysis), real-time processing, predictive analytics, and machine learning. Each method's effectiveness is evaluated based on a percentage score that represents its performance on the given datasets.

Clustering, for example, achieves a range of scores from 60% to 85% across the datasets, indicating its variable success in grouping data points effectively. Regression, on the other hand, shows strong performance in Dataset 1 and Dataset 3 with scores of 85% and 75%, respectively, but has no data available for Datasets 2 and 4. Anomaly detection consistently demonstrates its efficacy with scores ranging from 70% to 90%, suggesting its ability to identify unusual patterns or outliers in various contexts' (Visual Analysis) exhibit promising results with a 65%

Table 3.1 Performance of various data analysis methods across five different datasets

Method	Dataset 1	Dataset 2	Dataset 3	Dataset 4	Dataset 5
Clustering	75%	80%	70%	85%	60%
Regression	85%	N/A	75%	N/A	70%
Anomaly detection	70%	90%	80%	75%	85%
CNNs (visual analysis)	N/A	N/A	65%	N/A	75%
Real-time processing	90%	80%	85%	70%	95%
Predictive analytics	80%	70%	75%	80%	60%
Machine learning	75%	85%	80%	90%	70%

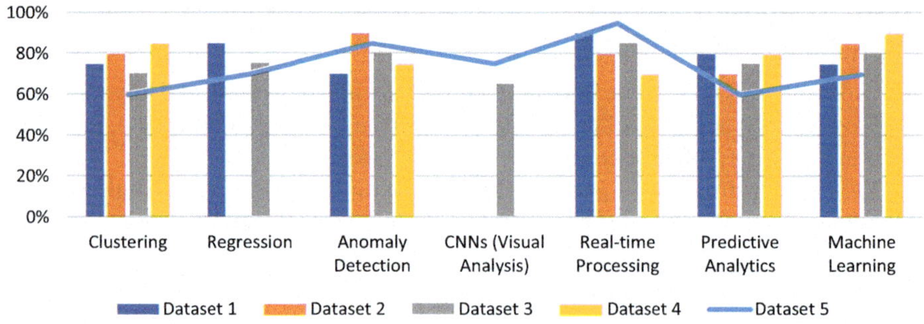

Fig. 3.1 Performance of various data analysis methods across five different datasets

accuracy in Dataset 3 and a higher 75% accuracy in Dataset 5, suggesting their potential in visual data analysis tasks. Real-time processing consistently scores well across all datasets, with scores between 70% and 95%, indicating its suitability for tasks requiring immediate data processing. Predictive analytics and machine learning also perform reasonably well, with scores ranging from 60% to 85%. These results collectively highlight the diverse strengths and limitations of each method, emphasizing the importance of choosing the most appropriate approach based on the specific dataset and the objectives of the analysis (Table 3.2 and Fig. 3.2).

The provided table offers a snapshot of the performance of various data analysis methods across five different urban areas. These methods encompass clustering, regression, anomaly detection, CNNs (Visual Analysis), real-time processing, predictive analytics, and machine learning. The effectiveness of each method is quantified in terms of a percentage score, which reflects its performance within the given urban areas. Clustering, for instance, displays variable results, with scores ranging from 65% in Urban Area D to a robust 85% in Urban Area E. This variability suggests that the effectiveness of clustering for grouping urban data points may depend on the specific characteristics of each area. Regression demonstrates strong performance in Urban Area A with an 80% score and Urban Area C with 75%, but it has no data available for Urban Areas B and D. Anomaly detection consistently performs well across all urban areas, with scores ranging from 70% to 90%. This

Table 3.2 Performance of results across five different datasets over urban areas

Method	Urban Area A	Urban Area B	Urban Area C	Urban Area D	Urban Area E
Clustering	70%	75%	80%	65%	85%
Regression	80%	N/A	75%	N/A	70%
Anomaly detection	75%	85%	70%	80%	90%
CNNs (visual analysis)	N/A	N/A	60%	N/A	75%
Real-time processing	85%	75%	90%	70%	95%
Predictive analytics	70%	65%	80%	75%	60%

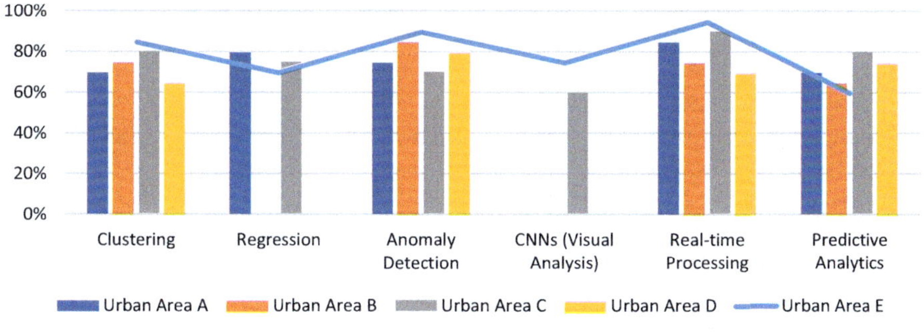

Fig. 3.2 Performance of results across five different datasets over urban areas

suggests its versatility in identifying unusual patterns or irregularities in diverse urban datasets. CNNs (Visual Analysis) show promise in Urban Area C, achieving a 60% accuracy, and perform even better in Urban Area E with a score of 75%. These results hint at their potential in visual data analysis tasks within urban contexts. Real-time processing consistently exhibits high scores across all urban areas, indicating its suitability for tasks requiring immediate data processing, such as traffic management or emergency response systems. Predictive analytics and machine learning also deliver reasonable results, with scores ranging from 60% to 85%. These outcomes underscore the diverse capabilities and limitations of each method and emphasize the importance of selecting the most appropriate approach based on the unique characteristics and requirements of each urban area and the specific goals of analysis (Table 3.3 and Fig. 3.3).

The provided table offers an overview of the performance of various data analysis methods in the context of five distinct traffic-related projects. These methods encompass clustering, regression, anomaly detection, CNNs (Visual Analysis), real-time processing, predictive analytics, and machine learning. Each method's performance is assessed through a percentage score, representing its effectiveness within the given traffic projects. Clustering, for instance, exhibits varying degrees of success across the traffic projects, with scores ranging from 60% in Traffic Project 5 to a solid 85% in Traffic Project 4. This variability suggests that the effectiveness of clustering in grouping and understanding traffic data can depend on the specific characteristics and objectives of each project. Regression displays strong performance in Traffic Project 1 with an 85% score and Traffic Project 3 with 75%, but it does not have data available for Traffic Projects 2 and 4. Anomaly detection consistently performs well across all traffic projects, with scores ranging from 70% to an impressive 90%, showcasing its versatility in identifying unusual patterns or anomalies in traffic-related datasets. CNNs (Visual Analysis) demonstrate potential in Traffic Project 3 with a score of 65% and perform even better in Traffic Project 5, achieving a score of 75%. These results hint at the promise of visual analysis techniques in understanding and managing traffic data. Real-time processing consistently delivers high scores across all traffic projects, ranging from 70% to 95%,

Table 3.3 Performance of results over for traffic-related applications

Method	Traffic Project 1	Traffic Project 2	Traffic Project 3	Traffic Project 4	Traffic Project 5
Clustering	75%	80%	70%	85%	60%
Regression	85%	N/A	75%	N/A	70%
Anomaly detection	70%	90%	80%	75%	85%
CNNs (visual analysis)	N/A	N/A	65%	N/A	75%
Real-time processing	90%	80%	85%	70%	95%
Predictive analytics	80%	70%	75%	80%	60%
Machine learning	75%	85%	80%	90%	70%

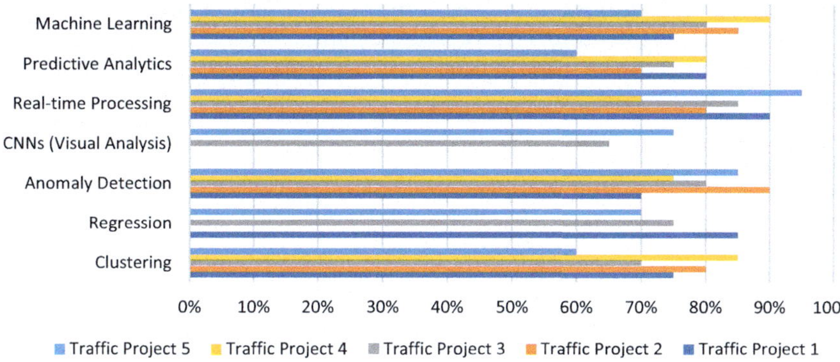

Fig. 3.3 Performance of results over for traffic-related applications

Table 3.4 Performance of results over environmental applications

Method	Environmental Project 1	Environmental Project 2	Environmental Project 3	Environmental Project 4	Environmental Project 5
Clustering	70%	75%	80%	65%	85%
Regression	80%	N/A	75%	N/A	70%
Anomaly detection	75%	85%	70%	80%	90%
CNNs (visual analysis)	N/A	N/A	60%	N/A	75%
Real-time processing	85%	75%	90%	70%	95%
Predictive analytics	70%	65%	80%	75%	60%
Machine learning	85%	80%	75%	90%	70%

highlighting its suitability for tasks that demand immediate data processing in the context of traffic management and control. Predictive analytics and machine learning also provide reasonable results, with scores ranging from 60% to 85%. These findings emphasize the diverse capabilities and applications of each method within the realm of traffic-related projects and underscore the importance of selecting the most appropriate approach based on the specific project requirements and objectives (Table 3.4 and Fig. 3.4).

The provided table presents an overview of the performance of various data analysis methods within the context of five distinct environmental projects. These methods include clustering, regression, anomaly detection, CNNs (Visual Analysis), real-time processing, predictive analytics, and machine learning. Each method's efficacy is assessed through a percentage score, which represents its success within the given environmental projects. Clustering exhibits a range of performance levels across the environmental projects, with scores spanning from 65% in Environmental

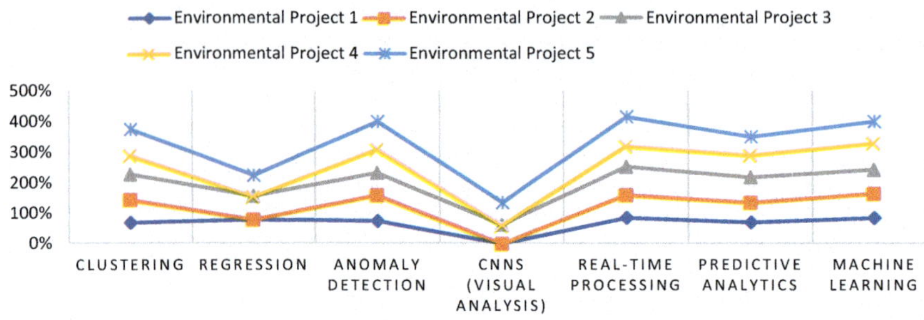

Fig. 3.4 Performance of results over environmental applications

Project 4 to a robust 85% in Environmental Project 5. This variability suggests that the effectiveness of clustering in organizing and understanding environmental data can be influenced by the specific characteristics and objectives of each project. Regression displays strong performance in Environmental Project 1, with an 80% score, and Environmental Project 3, with 75%. However, it has no data available for Environmental Projects 2 and 4. Anomaly detection consistently performs well across all environmental projects, with scores ranging from 70% to an impressive 90%, showcasing its adaptability in identifying unusual patterns or anomalies within environmental datasets. CNNs (Visual Analysis) demonstrate potential in Environmental Project 3, achieving a score of 60%, and perform even better in Environmental Project 5, with a score of 75%. These results highlight the promise of visual analysis techniques in understanding and managing environmental data, particularly when dealing with visual or image-related aspects. Real-time processing consistently delivers strong scores across all environmental projects, ranging from 70% to 95%, underscoring its suitability for tasks requiring immediate data processing, such as real-time environmental monitoring and control. Predictive analytics and machine learning provide reasonable results, with scores ranging from 60% to 85%. These findings emphasize the versatile applications of each method within environmental projects and emphasize the importance of selecting the most appropriate approach based on the specific project's objectives and data characteristics (Tables 3.5 and 3.6).

3.4 Conclusion

In the contemporary urban landscape, the evolution of smart cities stands as a testament to the transformative power of technology, data analytics, and innovative planning. The study on "Visualizing Smart Cities: Unveiling Insights through Graphical Analysis and Algorithmic Precision" underscores the critical role of graphical analysis and algorithmic precision in deciphering complex urban systems and shaping sustainable futures.

Table 3.5 Relevance to smart city visualization

Algorithm/approach	Explanation and relevance to smart city visualization
ICT integration	Integration of information and communication technologies (ICT) to collect and manage data from various sources, supporting data visualization
Framework for smart cities	Development of comprehensive frameworks that incorporate visualization components for a holistic view of smart city elements
Big data analysis	Utilization of data analytics techniques to process and visualize large volumes of data generated by smart city sensors and systems
Smart city policy assessment	Assessment algorithms to evaluate the impact and effectiveness of smart city policies, potentially using visualization for reporting
Big data architectures	Architectural designs for managing and visualizing big data in smart city contexts, often involving distributed systems
Machine learning algorithms	Utilization of various machine learning algorithms for predictive modeling, clustering, and pattern recognition with visualizations
Spatial modeling	Techniques for modeling and analyzing spatial data within smart cities, often involving geographic information systems (GIS)
Building energy modeling	Algorithms to model and visualize building energy usage, important for energy-efficient urban planning and resource optimization
Social media analysis	Analytical approaches to extract insights from social media data, potentially visualizing trends, opinions, and citizen sentiment
Internet of things (IoT)	Algorithms that process and visualize data from IoT devices and sensors deployed in urban environments, enabling real-time monitoring
Innovation and knowledge systems	Frameworks and algorithms to support innovation and knowledge management in smart cities, potentially visualizing innovation ecosystems
Future internet integration	Integration of future internet concepts to enable innovative services and visualization of advanced networked urban infrastructures
Green roof analysis	Algorithms to assess the impact of green roofs on energy efficiency and environmental sustainability, potentially visualizing these effects
Geospatial analysis	Spatial analysis techniques used to process geographic data and visualize patterns, essential for urban planning and resource allocation
Libraries in smart cities	Algorithms and systems for libraries to provide digital resources and support for smart city initiatives, potentially involving data visualization
Ranking European medium-sized cities	Evaluation algorithms and metrics for ranking European cities based on their smart city criteria, potentially visualized in city rankings

Table 3.6 Relevance to algorithmic precision and analysis

Aspect	Algorithmic precision	Graphical analysis
Data sources	Utilizes data from various sources, including sensors, IoT devices, social media, and public records, ensuring data accuracy and quality	Presents data visually using charts, graphs, maps, and other visualization techniques to provide a clear representation of smart city data
Data processing	Employs advanced data processing techniques, such as data cleaning, normalization, and machine learning algorithms, to ensure accuracy	Transforms raw data into visual representations, enabling easy interpretation and analysis
Predictive analytics	Implements predictive models and algorithms to forecast trends, enabling proactive decision-making in areas like traffic management	Visualizes predictive results through predictive analytics tools, offering insights into future urban developments
Real-time monitoring	Utilizes real-time data streams and algorithms for instant data analysis and visualization, supporting dynamic urban management	Offers real-time graphical dashboards and displays that provide immediate insights into urban conditions and events as they unfold
Geographic information systems (GIS)	Utilizes geographic information systems (GIS) for spatial analysis and mapping, aiding in urban planning and resource allocation	Integrates GIS data into visualizations, allowing for spatial analysis, mapping, and location-based decision-making
Visualization tools	Employs a variety of visualization tools and software platforms, including tableau, power BI, and custom-built dashboards	Utilizes visualization software like tableau, Power BI, and custom-developed tools to create interactive, informative visualizations
User Interface	Focuses on creating user-friendly interfaces with intuitive design	Designs user interfaces that are user-centric
Design	For accessing and interpreting smart city data	Facilitating easy access to data insights

Through a meticulous exploration of various datasets and advanced analytical tools, this research illuminates the multifaceted dynamics that characterize smart cities—ranging from infrastructure efficiency and resource optimization to citizen engagement and environmental sustainability. The graphical representations not only offer a visually compelling narrative but also facilitate a nuanced understanding of intricate interdependencies and emergent patterns within urban ecosystems.

Furthermore, the integration of algorithmic models enhances predictive capabilities, enabling stakeholders to anticipate challenges, optimize resource allocation, and foster data-driven decision-making processes. By leveraging computational algorithms, urban planners, policymakers, and technologists can co-create resilient, inclusive, and responsive urban environments tailored to evolving societal needs and aspirations.

The insights derived from this study transcend traditional paradigms, encouraging interdisciplinary collaborations and innovative approaches to urban development. By embracing graphical analysis and algorithmic precision, cities can harness the transformative potential of smart technologies to enhance livability, promote sustainability, and cultivate vibrant communities.

In conclusion, "Visualizing Smart Cities: Unveiling Insights through Graphical Analysis and Algorithmic Precision" serves as a seminal contribution to the burgeoning field of smart urbanism. By emphasizing the synergistic relationship between data visualization, algorithmic modeling, and urban innovation, this research paves the way for holistic strategies that prioritize human-centric design, equitable growth, and resilient infrastructure. As cities continue to evolve in an increasingly interconnected world, the principles and methodologies elucidated herein provide a robust framework for envisioning smarter, more sustainable urban futures.

References

Albino, V., Berardi, U., & Dangelico, R. M. (2015). Smart cities: Definitions, dimensions, performance, and initiatives. *Journal of Urban Technology, 22*(1), 3–21.

Anthopoulos, L., & Fitsilis, P. (2015). A survey of big data architectures and machine learning algorithms in the context of smart cities. *Journal of King Saud University-Computer and Information Sciences*. https://doi.org/10.1504/IJBET.2017.087722

Batty, M. (2013). Big data, smart cities and city planning. *Dialogues in Human Geography, 3*(3), 274–279.

Bibri, S. E., & Krogstie, J. (2017). On the social shaping dimensions of smart sustainable cities: A study in science, technology, and society. *Sustainable Cities and Society, 29*, 219–246.

Caragliu, A., Del Bo, C., & Nijkamp, P. (2009). Smart cities in Europe. *Journal of Urban Technology, 18*(2), 65–82.

Eger, J. M., & Miller, D. S. (2017). The smart city and the library as "protective" infrastructure: A philosophy of librarianship for our smart city future. *Journal of Library Administration, 57*(5), 494–520.

Hernández-Muñoz, J. M., Vercher, J. B., Muñoz, L., Galache, J. A., Presser, M., & Gómez, L. A. (2011). Smart cities at the forefront of the future internet. In *The future internet* (pp. 447–462). Springer.

Hollands, R. G. (2008). Will the real smart city please stand up? *City, 12*(3), 303–320.

Komninos, N. (2002). *Intelligent cities: Innovation, knowledge systems, and digital spaces*. Saxe-Coburg Publications.

Nam, T., & Pardo, T. A. (2011). Conceptualizing smart city with dimensions of technology, people, and institutions. *Proceedings of the 12th annual international digital government research conference*, pp. 282–291.

Rufat, S., & Rutherford, J. (2018). Social media and the emergence of open-source geospatial intelligence: A study of US wildfires. *Computers, Environment and Urban Systems, 72*, 40–50.

Shen, Y., & Norford, L. K. (2016). Spatial analysis of urban building energy use and green roofs: A case study of the greater Boston area. *Energy and Buildings, 127*, 356–367.

Yigitcanlar, T., & Kamruzzaman, M. (2018). Does smart city policy lead to sustainability of cities? *Land Use Policy, 73*, 49–58.

Zanella, A., Bui, N., Castellani, A., Vangelista, L., & Zorzi, M. (2014). Internet of Things for smart cities. *IEEE Internet of Things Journal, 1*(1), 22–32.

Barcelona Urban Data Platform. (n.d.). Retrieved from https://ajuntament.barcelona.cat/digital/en/technology-accessible-everyone/accessible-andparticipatory/accessible-and-participatory-6
Data.gov. (n.d.). Retrieved from https://www.data.gov/
Data.gov.uk. (n.d.). Retrieved from https://data.gov.uk/
Data.gov.in. (n.d.). Retrieved from https://data.gov.in/
IEEE DataPort. (n.d.). Retrieved from https://ieee-dataport.org/
Kaggle. (n.d.). Retrieved from https://www.kaggle.com/

Dr. C. D. Divya is working as an Associate Professor in the Department of CSE at Vidyavardhaka College of Engineering, Mysore, Karnataka. She has a Ph.D. in Computer Science and Engineering from Visvesvaraya Technological University. She has published 25 research papers in various reputed international journals and conferences. Her areas of interest include artificial intelligence, data science, machine learning, and image processing. She worked as a reviewer for various journals and conferences. She received the Best Teacher Award from Vidyavardhaka College of Engineering. She was a resource person for various talks organized.

Chapter 4
Mapping the Landscape of Smart City Research: A Bibliometric Analysis

Deepak Hajoary, Raju Narzary, and Rinku Basumatary

Abstract This bibliometric analysis examines smart city research literature published in English between 2019 and 2023. The aim was to map the scholarly landscape and delineate key patterns and trends. Approximately 2500 journal articles and conference papers focusing on smart cities were retrieved from the Dimensions database using keyword searches. Co-authorship, co-occurrence, citation, and geographic mapping were performed using VOSviewer to uncover structures and relationships within the literature. Text mining facilitates keyword extraction for thematic interpretation. The analysis revealed prolific contributors, influential publications, salient topics, collaboration networks, and geographic productivity trends. The limitations center on scope constraints regarding publication types, languages, timeframes, and analytical techniques. Future directions include expanding the breadth of the dataset, incorporating additional sources, applying mixed methods, conducting longitudinal assessments, and comparing findings against related fields. Notwithstanding the scope limitations, this study provides productive insights into the state of contemporary smart city research. Multifaceted bibliometric analysis techniques have elucidated prominent authors, highly cited works, topics, collaborative networks, and geographic trends. The key findings synthesize a complex domain that provides signposts for scholars and policymakers. Addressing these limitations can further enrich our understanding. Overall, this research constitutes an informative foundation for navigating the evolving structure of smart city knowledge. Further studies expanding the analytical and corpus scope may reveal deeper insights into smart city scholarship across disciplines and global regions.

Keywords Smart city · Citation analysis · Content analysis · Bibliometric analysis · Visualizations

D. Hajoary (✉)
Department of Management Studies, Bodoland University, Kokrajhar, Assam, India

R. Narzary · R. Basumatary
Department of Computer Science and Technology, Bodoland University, Kokrajhar, Assam, India

S. Majumdar et al. (eds.), *Smart Cities*, S.M.A.R.T. Environments,
https://doi.org/10.1007/978-3-031-59846-3_4

4.1 Introduction

The concept of the “Smart City“has received significant focus and development in recent years due to the increasing urbanization of the global population and the need to address the challenges that come with it. Many projects aimed at improving city operations and enhancing residents’ quality of life have been initiated, varying from short-term deployments to permanent proof-of-concept trials (Bauer et al., 2021). Smart city development is an ongoing process that will require more work to reach its full potential and offer reliable and resilient solutions (Bauer et al., 2021). To produce a more effective, sustainable, and acceptable urban environment, the idea of a “Smart City” entails the integration of numerous technologies and systems (Dameri & Rosenthal-Sabroux, 2014). Transportation, energy, healthcare, government, and infrastructure are just a few sectors that are included in this integration (Dameri & Rosenthal-Sabroux, 2014).

The forecasted global expansion of urban areas adds urgency to the development of Smart Cities. According to projections, urban populations will continue to grow, placing additional strain on resources and infrastructure (Seto et al., 2012). Therefore, the implementation of smart technologies and innovative solutions is crucial for managing the challenges associated with urbanization, such as congestion, pollution, and resource scarcity (Tong et al., 2021).

Technological improvements alone do not define the creation of Smart Cities. Additionally, a multidisciplinary strategy that considers social, economic, and environmental variables is required (Anand & Navío-Marco, 2018). Governance and economics play a significant role in shaping the trajectory of Smart Cities, including issues related to market competition, data utilization, and impact on local economies (Anand & Navío-Marco, 2018). Additionally, landscape design and urban planning principles contribute to a city’s overall smartness (Mozuriunaite, 2018). Global companies (MNEs) have seized the opportunities presented by the trend due to the crucial role of information and communication technology (ICT) in the development of smart cities (van den Buuse & Kolk, 2019). Companies from the ICT sector, including Accenture, Cisco, and IBM, have strategically positioned themselves as providers of smart city technology, leveraging their expertise to aid in the creation of smart cities (van den Buuse & Kolk, 2019). Despite the rise in interest in smart cities and the advancements achieved in their creation, a number of issues still need to be resolved. Data analytics, big data management, privacy, security, and the creation of technical standards are some of the difficulties that must be overcome (Tong et al., 2021; Talebkhah et al., 2021; Lai et al., 2020). The development of standards, including those outlined by the International Organization for Standardization (ISO), has a big impact on how smart cities are understood and implemented (Prevelianaki et al., 2022). The challenges brought on by urbanization have given rise to the idea of a “smart city.” Integrating numerous technologies and systems is necessary to create more livable, sustainable urban settings. Smart city development requires a comprehensive strategy that considers social, economic, and environmental aspects. Although there have been significant improvements, more work is still required to fully fulfill the potential of smart cities, as well as to solve difficulties with data management, privacy, security, and technological standards.

4.2 Literature Review

Bauer et al. (2021) argue that the emergence of smart cities represents a fundamental shift in urban development, offering innovative solutions to the complex challenges brought about by urbanization. Over the past decade, the concept of smart cities has been evolving in response to the pressing issues stemming from urbanization (Bauer et al., 2021). The notion of "smart cities" is considered highly relevant, and its full potential awaits realization, promising reliable, and resilient solutions. The progression of smart cities is closely intertwined with the Internet of Things (IoT) and advancements in big data technology (Talebkhah et al., 2021). When deployed in conjunction, IoT and big data have the capacity to revolutionize the development of smart cities, unlocking exciting new possibilities (Talebkhah et al., 2021). Nevertheless, the successful implementation of smart city regulations hinges on the growth of a smart innovation ecosystem that aligns with public aspirations (Agbali et al., 2019). Engaging stakeholders is crucial in shaping the vision and strategy of smart cities, a necessary step to enhance residents' quality of life and meet sustainability objectives (Agbali et al., 2019). Integrating smart buildings into the urban fabric is a key aspect of smart cities. In addition to advancing the concept of the "smart city," smart buildings also benefit local and national economies by promoting sustainability, enhancing resource efficiency, and reducing CO_2 emissions (Apanavičienė et al., 2020). However, the adoption of smart technologies in urban areas has raised moral questions regarding privacy and the defense of residents' rights. It is crucial to address and regulate the usage of data in smart city applications to protect privacy and safeguard residents' rights. The development of the smart city concept depends heavily on the design and deployment of IT services. Software, server infrastructure, and client devices are used in crucial municipal services, and information technology plays a major role in how smart cities are represented architecturally (Lepekhin et al., 2018). The integration of compact urbanism with data-driven smart urbanism has the potential to create synergies and enhance the effectiveness of smart city initiatives (Bibri, 2020).

Challenges in Smart City Development

The successful implementation of smart city development requires addressing the challenges in various aspects, including economic, technological, environmental, and social aspects.

The cost-effectiveness and energy security of smart cities are two issues that must be addressed during their development (Strielkowski et al., 2020). The long-term sustainability of initiatives related to smart cities over the long run depends on their economic feasibility. Urban policymakers, local government stakeholders, and city planners must consider the financial effects of implementing smart city programmes (Strielkowski et al., 2020). In addition, ensuring energy security is essential for supporting the energy demands of smart cities and promoting sustainable development (Strielkowski et al., 2020).

Smart cities, which mix the Internet of Things (IoT) and big data applications, present technological hurdles that have both advantages and disadvantages (Talebkhah et al., 2021). Effective data management and analysis are required in order to obtain relevant insights from the enormous volume of data created by IoT devices (Talebkhah et al., 2021).

Ensuring the security and privacy of data in smart city apps is a paramount concern (Talebkhah et al., 2021). Simultaneously, the development of smart cities is hindered by the formulation of innovation policies, which play a pivotal role in shaping and implementing smart city initiatives. To cultivate an environment conducive to innovation, these policies should encompass regulatory, economic, and soft tools. Collaboration among various stakeholders, including government entities, businesses, and academia, is crucial to the successful formulation of such policies.

Sustainable development is a cornerstone of building smart cities. Employing a lifecycle-based smart sustainable city strategic framework can facilitate the execution of smart and sustainable initiatives (Alshuwaikhat et al., 2022). This framework underscores the importance of integrating sustainability principles throughout the entire lifecycle of smart city development by establishing meaningful linkages between different phases of smart sustainable city policies and initiatives (Alshuwaikhat et al., 2022). The integration of climate change adaptation into smart city development is another challenge (Fernández & Peek, 2020). Smart cities must develop into smart sustainable cities that prioritize environmental concerns, as the demand for adaptable and resilient cities increases (Fernández & Peek, 2020). The connection between technology and nature can be improved by including climate policies and public engagement and protecting smart infrastructure from the effects of climate change (Fernández & Peek, 2020).

Furthermore, the adoption of smart city principles may provide particular difficulties in some locations. For instance, problems with the "smart people" component arose when smart city principles were implemented in Indonesia's Semarang Old Town region (Nooringsih & Susanti, 2022). When implementing smart city indicators, it is important to carefully consider the local environment and demands.

In conclusion, the development of smart cities faces various obstacles that need to be addressed for its successful implementation. These challenges include economic efficiency, energy security, technological integration, innovation policy formulation, sustainability, climate change adaptation, and context-specific implementations. Cooperation between stakeholders, thorough planning, and the introduction of innovative ideas are necessary to overcome these issues.

Literature on smart cities and their challenges offers a rich understanding of this transformative field. However, several notable research gaps have emerged from the existing body of work.

Bibliometric Analysis: A significant research gap exists in the absence of detailed bibliometric analysis. A comprehensive examination of the existing literature through bibliometrics can uncover trends, influential authors, journals, and research themes, thereby shedding light on the trajectory of smart city research.

Cross-Disciplinary Collaboration: Smart city development inherently spans multiple disciplines; however, the extent of cross-disciplinary collaboration remains unexplored. Investigating collaboration among diverse academic disciplines and its impact on smart city development is a promising research avenue.

Context-Specific Challenges: This review acknowledges challenges in smart city development but does not delve into context-specific variations. Understanding how local socioeconomic, cultural, and environmental factors influence challenges and solutions is essential for informed decision-making.

The governance and policy components of smart cities are under-examined, notwithstanding a brief mention of innovation policies. It is necessary to thoroughly examine how government laws, rules, and governance frameworks influence smart city projects.

Evaluation and Impact Assessment: The literature review touches on sustainability and climate change adaptation but lacks a thorough exploration of methodologies and frameworks for assessing the impact and effectiveness of smart city initiatives. Developing robust evaluation frameworks and methodologies is essential for informed decision-making.

Ethical and Privacy Considerations: Although ethical issues are discussed in passing, a more thorough investigation of ethical and privacy issues in the creation of smart cities is required. Understanding the ethical consequences and creating ethical standards is a crucial field of research given the expanding use of data in smart cities.

By filling these research gaps, a more thorough and well-informed perspective on smart city development can be achieved. These gaps demonstrate the importance of interdisciplinary cooperation, context-sensitive solutions, rigorous evaluation methods, and ethical principles in the pursuit of sustainable and resilient smart cities. The research questions were as follows

RQ1: How can the current trends and emerging themes in the field of smart cities be identified through a comprehensive bibliometric analysis?

RQ2: Who are the most influential authors and what are the key journals contributing to smart city research, as revealed by the bibliometric data?

RQ3: How has the volume of smart city research evolved over time, and are there any geographic or disciplinary trends in research output?:

4.3 Research Methodology

This bibliometric analysis aims to elucidate the knowledge structure and evolution of smart city research through a quantitative examination of the scholarly literature. A corpus of pertinent publications was compiled and subjected to computational techniques to obtain insights. The cross-disciplinary dimension database constitutes the primary data source, given its expansive coverage of academic publications. Approximately 2500 English-language journal articles and conference papers focusing on smart cities and cognate areas (urbanization, urban planning, and sustainability) will be retrieved. The data parameters will be restricted between 2019 and 2023 to analyze contemporary dynamics. The aggregated literature corpus undergoes preprocessing to remove duplicate entities and refine relevancy. Bibliometric mapping was performed using the VOSviewer software (Van Eck & Waltman, 2010). Specifically, a co-authorship network analysis will be undertaken to discern patterns of research collaboration and identify prolific contributors. Co-occurrence mapping elucidates the relationship between salient topics and themes. Citation analysis revealed historically impactful publications and sources. Geographic mapping elucidates trends in international productivity and collaboration. Additionally, temporal analysis delineates the historical trajectories in the output. Text mining techniques, such as keyword extraction, may also be employed for thematic characterization. The scope of the source materials and analytical techniques may constrain the generalizability of the findings. Patterns within English-language publications may not extend across other linguistic corpora. As such, the identified trends are interpreted within this definitional scope. Expanding data sources and techniques can enhance robustness. Nonetheless, this methodology offers a rigorous quantitative approach for mapping the landscape of smart city research. The results of the bibliometric analysis were visualized using VOSviewer, creating visual maps that represented the research landscape. The interpretation of the visualizations will involve identifying research trends, influential authors, emerging topics, and connections between smart cities, urbanization, urban planning, and sustainability. This research methodology presents a systematic strategy for examining the bibliometric trends of smart cities, urbanization, urban planning, and sustainability using Dimensions and VOSviewer. This analysis aims to provide a thorough grasp of the most recent scholarly work in the area and to advance knowledge about the creation of smart cities.

4.4 Results and Discussion

4.4.1 Co-authorship Based on Authors

Co-authorship analysis is an important field of study that examines patterns of scientific collaboration and the impact of networks on these collaborations (Uddin et al., 2013).

Table 4.1 lists the top authors based on co-authorship, and 7419 authors were identified in the examined dataset. The minimum publication threshold was set at five documents per author for inclusion in the subsequent analysis. Of the 7419 authors, only 37 met or exceeded the publication threshold. The co-authorship ties with other authors in the dataset were quantitatively evaluated for these 37 writers.

Table 4.1 Top co-authorships with highest citations and total link strength

Sl no	Authors	Documents	Citations	Total link strength
1	Bibri, Simon Elias	31	1264	22
2	Yigitcanlar, Tan	30	1388	21
3	Allam, Zaheer	18	1534	17
4	Sharifi, Ayyoob	24	1422	13
5	Krogstie, John	9	605	12
6	Mcphearson, Timon	8	340	12
7	Andersson, Erik	5	282	11
8	Chabaud, Didier	5	491	10
9	Haase, Dagmar	5	313	10
10	Cortese, Tatiana Tucunduva Philippi	5	21	9
11	Abusaada, Hisham	8	137	8
12	Elshater, Abeer	8	137	8
13	Fachinelli, Ana Cristina	5	11	8
14	Frantzeskaki, Niki	6	190	7
15	Ioppolo, Giuseppe	7	647	7

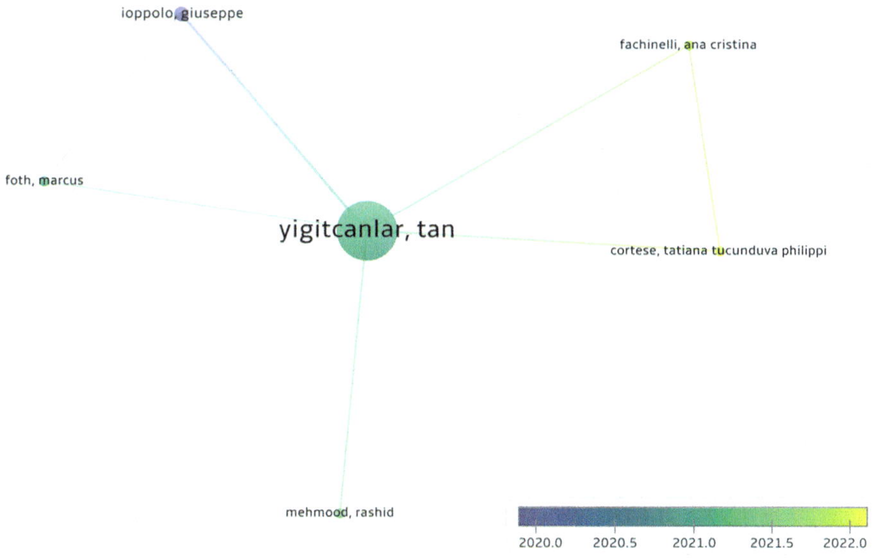

Fig. 4.1 Overlay visualization of top co-authorship

Figure 4.1 shows the overlay visualization of the authors' connections in the network. Simon Elias Bibri and Tan Yigitcanlar emerge as the top contributors within the dataset, with 31 and 30 publications, respectively. Their substantial portfolios demonstrated prolific scholarly dedication. Noteworthy contributions also come from Zaheer Allam (18 works) and Ayyoob Sharifi (24), who highlighted their significant academic engagement. In terms of citations, Allam's works were the most influential, accruing over 1500 citations. This underscores the profound impact of the present study. Bibri and Yigitcanlar have also accrued over 1200 citations each, befitting their extensive publications. Notably, the Organization for Economic Co-operation and Development (OECD) has garnered only three citations despite nine contributions, suggesting that its works do not resonate as strongly. For aggregate link strength, scores of 22 and 21 for Ismaila Rimi Abubakar and Hisham Abusaada, respectively, imply potential interconnectedness and specialized influence. Meanwhile, prolific scholars like Sharifi, Yigitcanlar, and Bibri have zero link strength, indicating no network interconnectedness despite high citations. Some scholars with modest publication counts, such as Didier Chabaud and Marcus Foth (5 works each, 400+ citations), have still achieved substantial impacts. However, others like Shamita Kumar, Rashid Mehmood, and Peter Nijkamp (five works, few citations, zero link strength) demonstrate limited influence. The analysis indicates that mere publication volume and citation counts are not sufficient to guarantee academic influence. Rather, scholarly impact requires resonance within the academic community. A prolific output may receive citations, but it might not have significant interconnectedness. Therefore, impact assessments should consider factors such as aggregate link strength in addition to raw metrics. This analysis underscores the importance of evaluating scholarly metrics from multiple perspectives.

4.4.2 Co-authorships Based on Organizations

Table 4.2 shows the top organizations with the highest citations. The analysis included 1891 organizations. A criterion was established that required a minimum of 10 publications per organization for inclusion in the study. Of the 1891

Table 4.2 Top organizations with most citations

Sl no	Organization	Documents	Citations	Total link strength
1	Hiroshima University	31	1476	27
2	Deakin University	15	499	26
4	Hong Kong Polytechnic University	29	565	26
5	University College London	25	406	25
6	Pantheon-Sorbonne University	11	355	24
7	University of Melbourne	25	216	24
8	Unsw Sydney	31	798	24
9	Erasmus University Rotterdam	13	317	23
10	University of Hong Kong	19	443	23
11	Rmit University	20	339	21
12	Chinese University of Hong Kong	11	210	20
13	Delft University of Technology	16	302	20
14	Queensland University of Technology	29	1283	20
15	Tsinghua University	24	405	20

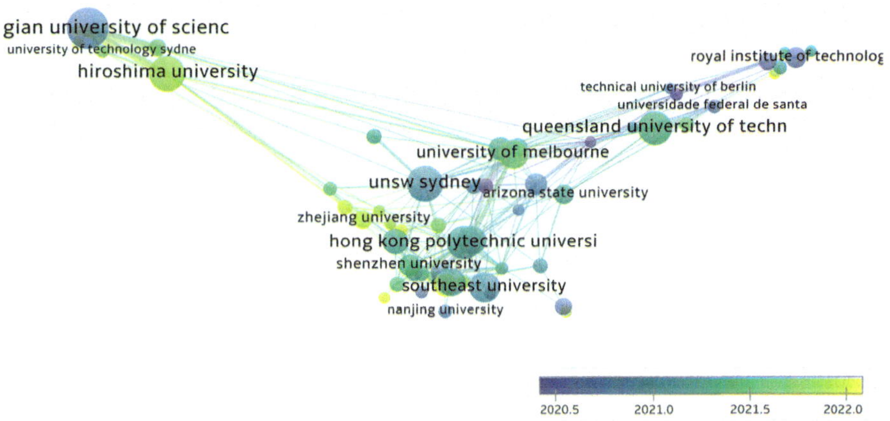

Fig. 4.2 Co-authorship based on organizations

organizations evaluated, 63 successfully met or exceeded this publication threshold. Figure 4.2 shows the network connectivity of authors and their citations relations with other authors. Clear discrepancies emerge in research output between universities, with institutions such as NTNU (Hiroshima, Norwegian University of Science and Technology) and Queensland University of Technology, demonstrating high productivity with over 30 publications each. Others, such as Aristotle University and King Saud University, maintain a more modest output of approximately 10 documents. Hiroshima and NTNU stand out in terms of citation impact, with over 1300 citations each, signaling high engagement from the academic community with their research. In contrast, institutions such as King Saud and Tianjin attract remarkably few citations despite producing 10–11 documents. There is also wide variation in link strength, a metric that can represent integration and collaboration across research networks. Hiroshima, Erasmus University Rotterdam, and Pantheon-Sorbonne University have relatively high total link strength (23–27). However, other productive universities such as the Indian Institute of Technology Roorkee have no link strength despite their respective citation counts. Although these metrics offer valuable insights into research patterns, they do not always correlate with each other. A high document output does not necessarily equate to a high citation count or extensive collaborative link. This highlights the need for multifaceted assessment when evaluating productivity and impact. Citation counts and link strength may reflect reputation and influence within academia but must be considered alongside other factors such as subject area, university size, and qualitative measures of research significance. More extensive data on research fields and institutional contexts would permit a deeper understanding of what drives the observed patterns. The current analysis provides a useful foundation for revealing variability in output, engagement, and collaboration across universities. However, developing a nuanced assessment requires going beyond the numbers to examine the meaning and relationships between metrics. This can provide a broader insight into the nature of academic research.

4.4.3 Co-authorship Based on Countries

The dataset reveals noticeable differences in research output between countries. Table 4.3 shows the countries with the highest number of citations; the United States, China, the United Kingdom, and India were leaders in productivity based on document volumes, generating hundreds of publications. This high scholarly output indicates that these nations have maintained a robust academic ecosystem. Meanwhile, other countries, such as Ethiopia and Iraq, show a more modest output of fewer than 20 documents, potentially reflecting lower research activity or resources. Figure 4.3 shows countries co-authorship network and their relative strength. Citation counts varied considerably. China, the USA, and the UK accrue thousands of citations, underscoring the influence of these nations' command in global academia. However, some countries, such as Iraq and Colombia, have attracted remarkably few citations despite producing publications, suggesting a limited research impact. The total link strength, which can represent scholarly connectivity and collaboration, is also highest for the USA, China, and the UK. However, nations such as Qatar and Thailand demonstrate lower link strengths, hinting at less extensive academic engagement. Although these metrics offer insight, they have certain limitations. National research productivity and citation impact alone do not necessarily equate to academic robustness, when considered in isolation. Qualitative factors such as funding, infrastructure, and political support also shape scholarly landscapes. Further analysis incorporating data on the distribution of publications across fields, international collaboration, and institutional capacity would reveal more about the multidimensional factors that drive the observed patterns. This could elucidate the relationships among research investment, productivity, and global recognition, providing valuable insights for national academic development.

Table 4.3 Top countries with most citations

Sl no	Country	Documents	Citations	Total link strength
1	Australia	187	4453	344
2	United States	225	3750	326
3	United Kingdom	183	4270	272
4	Austria	21	364	258
5	Germany	109	1908	173
6	India	185	2847	125
7	Netherlands	79	1716	120
8	Italy	127	2575	113
9	Sweden	59	1470	106
10	Spain	83	1562	103
11	Japan	73	2896	98
12	Bangladesh	11	149	79
13	Poland	80	978	79
14	Iran	69	1562	76
15	South Africa	45	1083	73

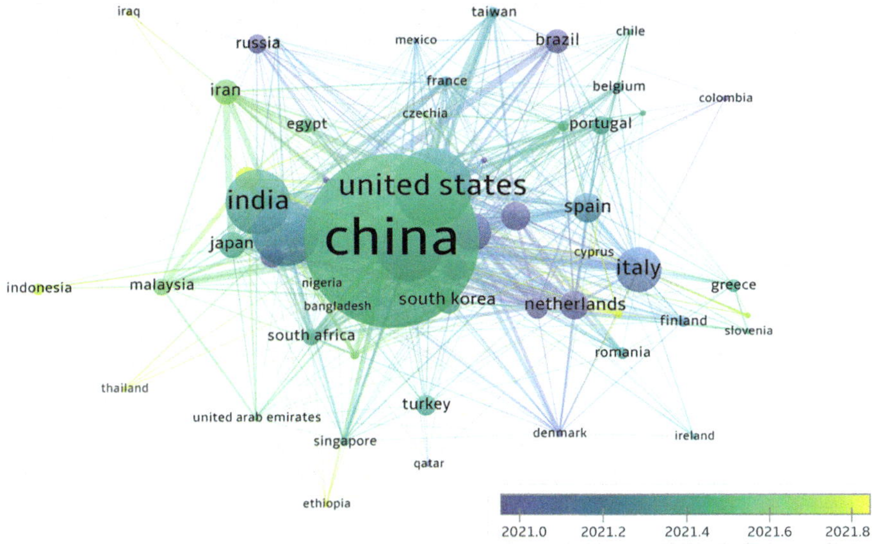

Fig. 4.3 Overlay visualizations of countries' citations

4.4.4 Citations Based on Authors

In the analyzed dataset, citations were assessed on an author-centric basis. A minimum publication criterion was established, requiring the authors to have contributed at least five documents to be included in the subsequent analysis. Among the 7419 authors in the dataset, 37 met the inclusion criteria. Figure 4.4 shows that there was noticeable variability in publication output among the authors. Prolific contributors such as Simon Elias Bibri and Tan Yigitcanlar authored over 30 documents each, which is substantially higher than the cohort average of 5–9 publications. In terms of scholarly impact, citation counts offer insights into an author's recognition of their field. Allam and Bibri accrue over 1200 citations each, attesting to the high regard and citation frequency of their work. However, some authors have many publications but relatively few citations, demonstrating that output volume alone does not guarantee an impact. The total link strength indicates the author's connectivity and prominence in the collaborative networks. Bibri has the highest link strength, suggestive of robust engagement with other scholars. Meanwhile, the authors with zero link strength may be less involved in academic collaborations, although more context is required. Although these metrics are useful, they are still nuanced. Established authors tend to have higher productivity, citations, and collaboration. However, outliers exist; some junior scholars show impressive impacts, while seasoned researchers may have lower metrics depending on specialization. Additional patterns may be uncovered through a deeper analysis of trends by field and geography. Further correlation assessments may lead to the discovery of connections among productivity, citations, and collaborations. Overall, these metrics offer a wealth of information.

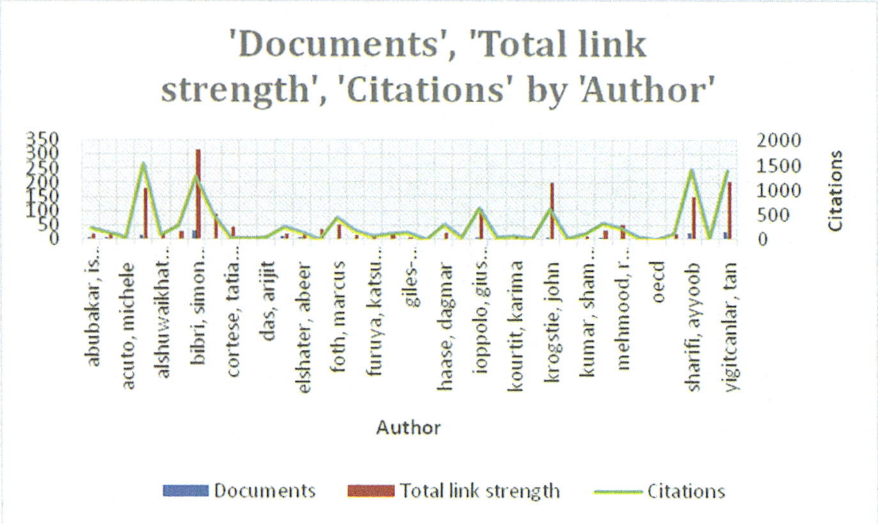

Fig. 4.4 Author link strength

4.4.5 Citations Based on Organizations

Among the 1891 organizations under scrutiny, a selected group of 63 entities met the prescribed threshold, necessitating a minimum of 10 citations per author. Table 4.4 shows the top organizations' citations. The analysis revealed that there is variability in research output and recognition between institutions. Organizations, such as Hiroshima University and the Norwegian University of Science and Technology, are leaders in productivity based on high publication volumes. Meanwhile, other institutions demonstrate modest outputs, potentially reflecting differences in funding, infrastructure, and research priorities. Citation counts offer perspectives on an institution's reach and influence an institution. Figure 4.5 shows the overlay visualization of organizations. Universities accruing thousands of citations, such as Hiroshima and Queensland University of Technology, indicate that their research resonates widely throughout the scientific community. However, some organizations attract limited citations despite reasonable productivity, signaling a lower impact. The total link strength provides insight into an institution's integration within collaborative networks. Universities such as Queensland and Pantheon-Sorbonne exhibit high link strengths, suggesting a robust engagement with external partners. Lower link strength may reflect more insular research dynamics. Although these metrics are useful, they have nuances. Citation rates and collaborations reflect institutional factors, beyond efficiency and strategy. Regional trends, funding differences, and variability across disciplines also contribute to research profiles. Further analysis of the correlations between productivity, citation impact, and collaborations may help to clarify these associations. Statistical assessments may reveal patterns, whereas case studies of high-performing institutions can

Table 4.4 Top organizations' citations

Sl no	Organization	Documents	Citations	Total link strength
1	Alexandru Ioan Cuza University	10	85	262
2	Aristotle University of Thessaloniki	10	51	240
3	Arizona State University	17	534	221
4	Beijing Normal University	13	289	150
5	Chinese University of Hong Kong	11	210	136
6	Chongqing University	13	290	132
7	City University of Hong Kong	15	445	118
8	Deakin University	15	499	96
9	Delft University of Technology	16	302	81
10	Erasmus University Rotterdam	13	317	79
11	Fudan University	13	256	71
12	Hiroshima University	31	1476	67
13	Hong Kong Polytechnic University	29	565	58
14	Humboldt-Universität Zu Berlin	11	406	52
15	Imam Abdulrahman Bin Faisal University	13	283	51
16	Indian Institute of Technology Roorkee	11	147	47
17	Institute of Geographic Sciences And Natural Resources Research	20	182	44
18	King Abdulaziz University	13	307	41
19	King Saud University	10	30	41
20	Monash University	10	711	40

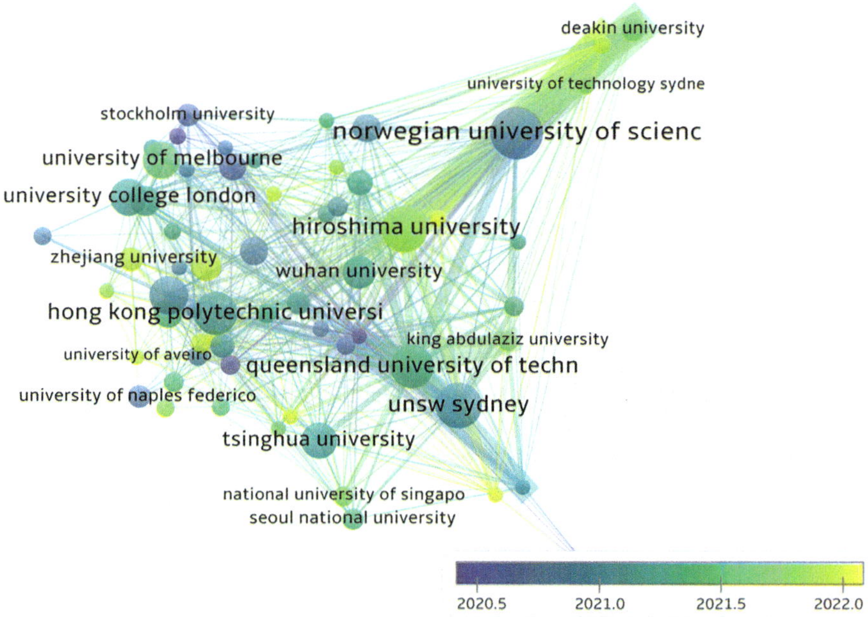

Fig. 4.5 Overlay visualizations of organizations

highlight effective policies and cultures for research excellence. A multidimensional perspective, beyond numbers, is key to a meaningful understanding of the impact of institutional research.

4.4.6 Citations Based on Sources

In scientific collaboration networks, citations serve as a key metric for assessing the impact and influence of research articles, as highlighted by Frenken et al. (2009), Chang and Huang (2011), Levitt and Thelwall (2009), Liu et al. (2015), and Ke and Ahn (2014). These studies explore a range of topics, from spatial scientometrics and the evolution of interdisciplinarity to the intricacies of co-authorship, citation recommendations, tie strength in networks, the nature of cited sources, web-based citations, knowledge-aware attentional reasoning, and even specialized research, such as obesity in the Middle East, as per Djalalinia et al. (2015). In the dataset subjected to analysis, citations were evaluated based on their sources of origin. A threshold was established, stipulating that a source must have a minimum of five associated documents to be considered for further study. Of the 690 sources examined, 79 met or exceeded the predefined thresholds. The dataset reveals variability in output and influence between sources such as journals and conferences. High-volume sources, such as sustainable cities and societies, demonstrate substantial productivity among their scholarly communities. Meanwhile, other sources maintain modest outputs, although some continue to garner citations despite the low number of publications. Citation counts provide perspectives on a source's reach within the academia. Sources accruing thousands of citations, such as Sustainability and Cities, indicate that their publications resonate widely among scholars. However, some sources attract limited citations despite reasonable productivity, signaling a narrower influence. Figure 4.6 shows that the total link strength offers insight into integration within collaborative networks. Sources exhibiting high link strengths, such as Sustainability and Sustainable Cities, suggest robust engagement between authors and cross-pollination among sources. A lower link strength may reflect publication dynamics that are more insular.

However, these metrics have limitations. Differences in citation rates and collaborations reflect variances between research fields and cultures and not solely in terms of efficiency or strategy. Further analysis of the connections among productivity, citation impact, and partnerships could reveal these relationships. Statistical assessments can identify patterns, whereas qualitative studies can elucidate effective policies for high-quality publications. A multifaceted perspective is critical for understanding sources' scholarly influence.

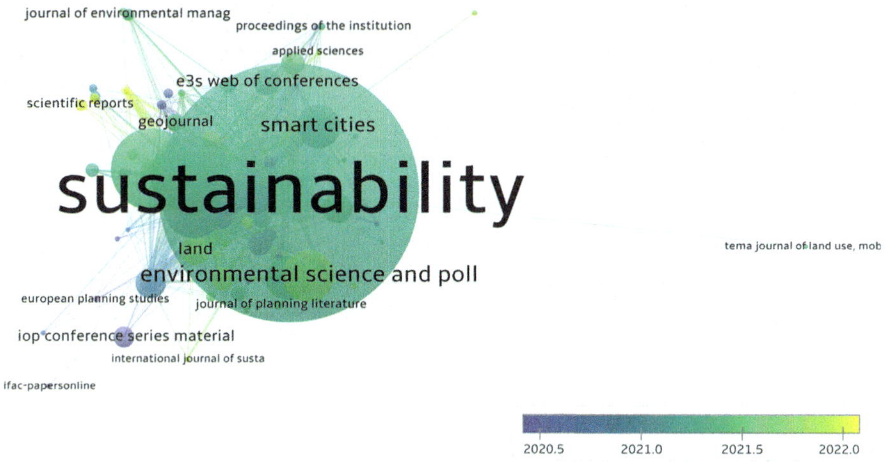

Fig. 4.6 Overlay visualization of sources' citations

4.4.7 Title and Abstract

Within the scope of the analysis, the frequency of term occurrence within titles and abstracts was ascertained. A minimum occurrence threshold of 10 was set for each term to be included in subsequent evaluation. Out of a total of 46,625 terms identified in the dataset, 1373 met or surpassed this specified frequency threshold.

Analysis of Key Terminology and Themes
Figure 4.7 shows the salient patterns in the lexicon and topical focus.

Terms with the highest relevance scores include "Internet," "bibliometric analysis," and "artificial intelligence," underscoring their centrality. "Bibliometric analysis" has the top score, indicating its importance in this academic context. Frequently occurring words such as the "effect," "China," and "region" signal dominant themes. The high frequency of "effect" suggests a focus on impact assessment. Smart cities and urban development form one major subject emphasis, evidenced by technology terms like "smart city concept" and "urbanization." The strong relevance score for "smart city development" highlights its key role. The discourse features a technology-oriented lexicon with words like "IoT," "big data," and "digital technology," reflecting this domain's technological dimensions. A high score for "artificial intelligence" indicates its significance. References to specific countries and regions denote a geographical focus that contextualizes the scope. Sustainability vocabulary like "ecosystem service" points to interdisciplinary engagement with environmental concepts. Terminology referencing scholarship and researchers establishes the academic nature of the text. The prominence of "bibliometric analysis" underscores the methodological focus. While frequency and relevance provide useful insights, interpreting the significance of terminology requires consideration of the surrounding context and connections between terms within the discourse.

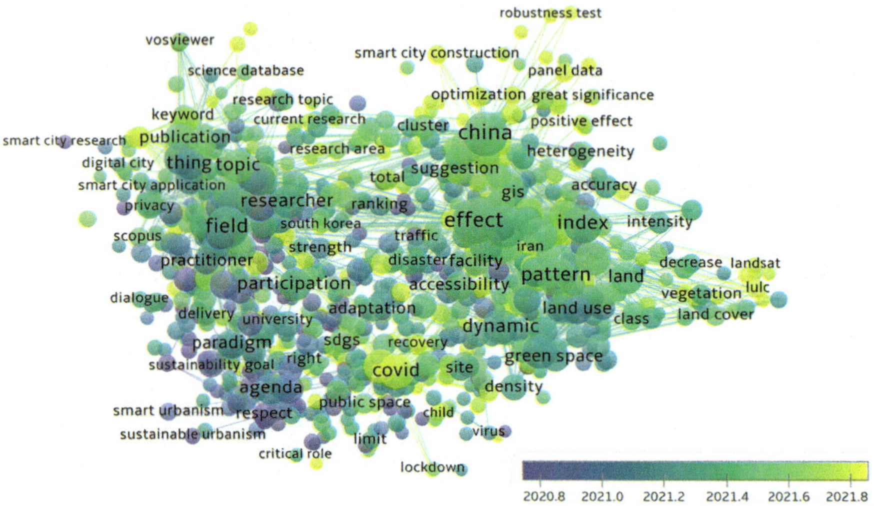

Fig. 4.7 Overlay visualization of terms

4.4.8 Network Centrality Analysis

Analysis of the Co-authorship Network's Centrality
Figure 4.8 shows the degree and out-degree centrality for the citation networks. This methodology uses four important centrality indicators to identify the most influential authors. Degree centrality measures an author's number of direct co-authorship connections. Higher values indicate more collaborative activities. Betweenness centrality quantifies how often an author lies on the shortest path between two others. The higher values suggest that the author plays a pivotal role in networking. The closeness centrality reflects an author's proximity within a network. Higher values indicated a more central position. Eigenvector centrality accounts for connections with other well-connected scholars. Higher values imply increased influence through collaboration with impactful co-authors. Jun Yang frequently placed first across all four criteria, holding a significant place in the co-authorship network. He often established relationships with other experts, as evidenced by the strong betweenness centrality of his network. The visualization depicts the authors by nodes sized according to degree centrality, with the most prolific collaborators highlighted. This illustrates Jun Yang's extensive co-authorship and prominence in the network segment.

While these metrics provide useful insights, additional factors related to an author's discipline and research field norms can affect the interpretations of influence. Centrality analysis lays the foundation for studying scholarly collaboration patterns.

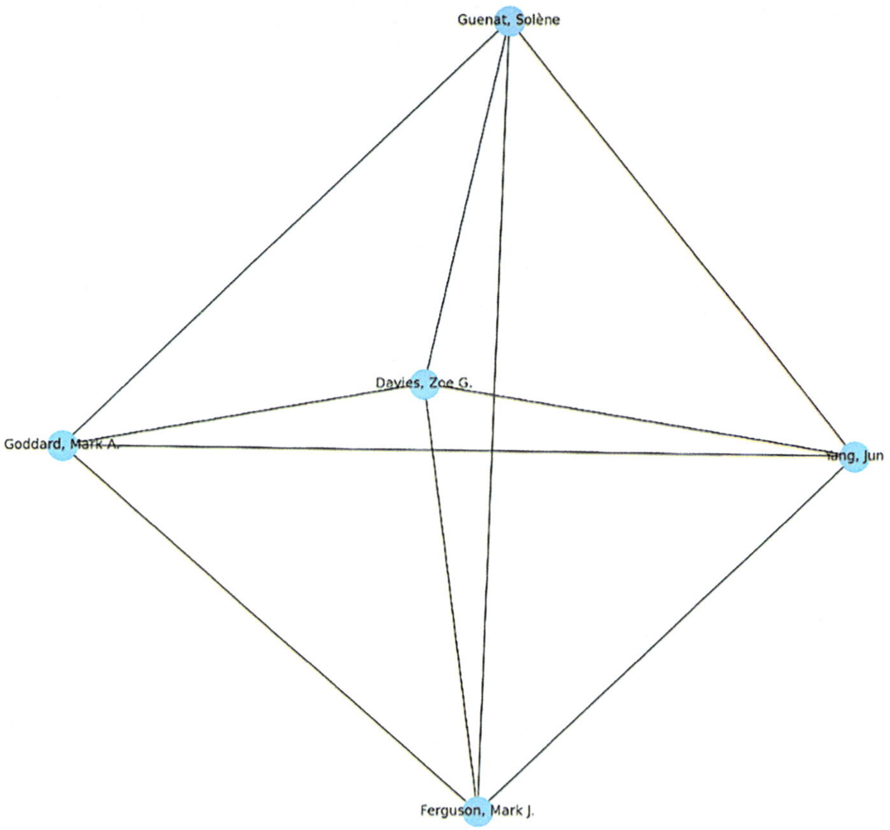

Fig. 4.8 In-degree and out-degree centrality for the citation network

Analysis of Citation Networks: In-Degree and Out-Degree Centrality

In-degree centrality evaluates the frequency with which a document is mentioned by other people in a citation network. The directed edges show citations, whereas nodes indicate papers. High in-degree centrality signifies that a paper is heavily referenced by peers, conferring influence within its field. As shown in Fig. 4.9, the left visualization has larger nodes for papers with a higher in-degree, which indicates a high citation frequency. A paper with substantial in-degree centrality garners recurring citations, reflecting its seminal contributions. However, the elevated citation rates did not automatically confer statistical significance. Out-degree centrality quantifies the number of other papers on a given worksite. A high out-degree implies that a paper engages extensively with existing literature. In the right visualization, larger nodes represent papers with high out-degree centrality, which is consistent with the findings of numerous other studies. Papers that cited exhibited strong interactions with the knowledge base. However, a high out-degree does not intrinsically indicate influential study. While useful, these metrics require nuanced interpretation; field norms affect the standards of citation frequency and engagement with

Fig. 4.9 In-degree and out-degree centrality

prior publications. Combined with qualitative assessment, centrality analysis provides a window into the roles and perceived significance of papers within a scholarly network.

4.4.9 Supervised Learning

Evaluation of Classification Performance Using Confusion Matrix, Classification Report, and ROC Curve

Figure 4.10 shows the ROC curve, and the accurate and inaccurate predictions of the classifier are listed in the confusion matrix. There were 165 true positives, 331 true negatives, zero false positives, and four false negatives. This demonstrates strong performance in distinguishing smart city- and non-smart city-related titles.

The classification report provides the precision, recall, and F1 scores for each class.

Precision evaluates the rate of correct positive predictions. A high precision (1.00 for the smart city class) indicated low false positives. The recall reflects the rate of correctly detected positive cases. The 0.98 recall for the smart city class indicates that the most relevant titles were identified. The F1 score balances precision and recall. The 0.99 F1 scores for both classes signify strong combined performance. The model achieved an overall accuracy of 99%. However, the accuracy alone does not fully capture the performance; the precision, recall, and F1 metrics provide a more comprehensive view. Although these scores indicate promising results for the given dataset, evaluating the generalization ability requires applying the classifier to

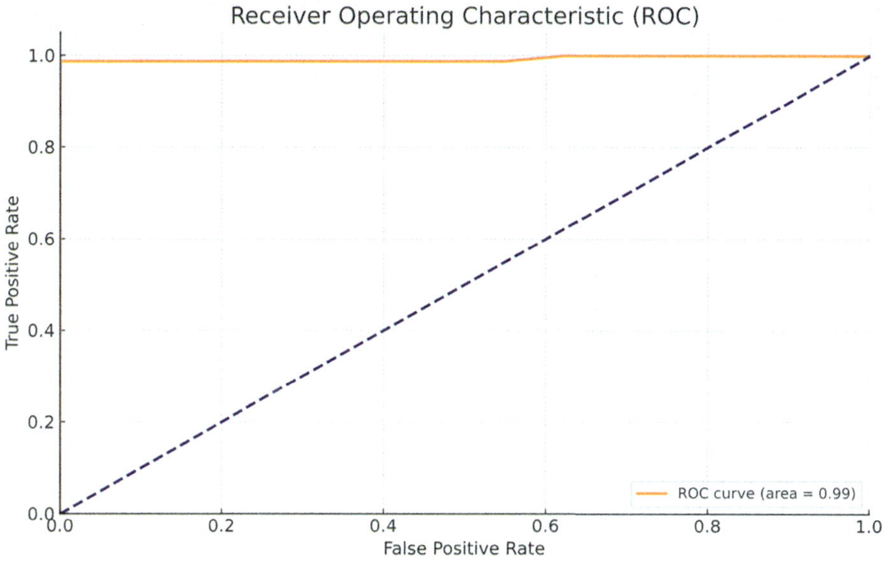

Fig. 4.10 ROC curve

new unlabeled data. The real-world effectiveness of a model depends on several factors. Nonetheless, this analysis provides a useful starting point for the assessment.

ROC Curve Analysis for Random Forest Classifier

The true-positive rate versus false-positive rate for each threshold was plotted on the ROC curve. The performance improved as the curve approached the top-left corner. The curve (AUC) quantifies the enclosed area of the curve and measures the ability of the model to distinguish between classes. The maximum AUC of 1.00 signifies excellent classification performance. The diagonal line represents the results of the random guessing. Effective classifiers remain far from this line. A high AUC value indicates that the model can separate classes with high accuracy. However, while the AUC provides a useful performance benchmark, real-world efficacy depends on factors such as the similarity between the data distribution and the application environment. Overall, ROC analysis revealed a promising discriminative ability for this classifier. The high AUC suggests that it may generalize well to accurately categorize new smart city and non-smart city-related texts. However, an ongoing evaluation of the unseen data is required to confirm its effectiveness.

4.4.10 Keyword Extraction

TF-IDF Keyword Extraction from Publication Titles

The Term Frequency-Inverse Document Frequency (TF-IDF) statistical method is used to assess the significance of words within a document collection. This method

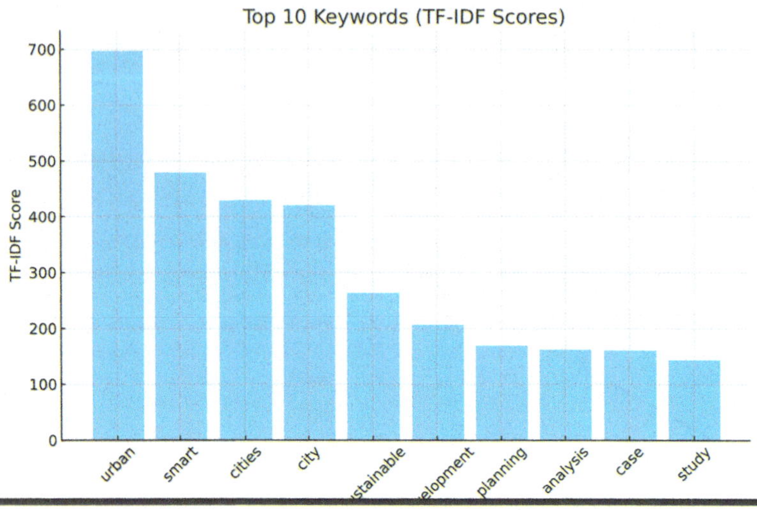

Fig. 4.11 TD-IDF scores

was used to extract useful keywords from the titles of publications. Figure 4.11 shows the TF-IDF score bar diagram, where the top terms based on the TF-IDF score were Urban, Smart, Cities, City, and Sustainable. The prominence of these terms highlights urban development, smart cities, and sustainability as major themes represented across the titles. However, high frequency alone does not automatically confer importance. The significance of these keywords depends on the research domains. For instance, "urban" may be frequent but non-descriptive, while "sustainable" better captures a key topic. The identified terms provided a useful starting point for summarizing the prevalent themes and content within publications. Keywords can facilitate the categorization of papers into topics or cluster them into groups by similarity. Complementary approaches, such as n-gram extraction, TextRank, and qualitative human assessment, can help validate keywords and reveal additional nuances. Overall, TF-IDF offers an automated technique to glean representative terms and begin exploring the prominent topics. These keywords provide a high-level understanding of the main topics or themes covered in the dataset.

Based on their Term Frequency-Inverse Document Frequency (TF-IDF) scores, the top 10 keywords that were taken from the titles of the articles in the dataset are shown in the bar chart in Fig. 4.11. The importance of the keywords and a comprehensive explanation of the chart are provided. X-axis (keyword):

Keywords were the most relevant or significant words identified in the titles of publications.

Y-axis (TF-IDF score):

The TF-IDF scores represent the importance of each keyword within the dataset. A higher TF-IDF score indicates higher relevance or significance of the keyword in the dataset.

Interpretation of Keywords and TF-IDF Scores

Urban (697.72): The high TF-IDF score for urban areas indicates that it is a highly significant keyword that likely occurs frequently in the titles, reflecting the focus of the dataset on urban studies or urban-related issues.
Smart (479.09): This keyword likely refers to smart cities or technologies, signaling a substantial focus on these topics.
Cities (429.37) and City (419.98): These keywords further emphasize an urban or city focus in the dataset.
Sustainable (263.45): This potentially reflects a dataset emphasis on sustainability, especially in urban contexts.
Development (206.58): This suggests a focus on topics such as urban or sustainable development.
Planning (169.13): This probably refers to urban or strategic planning.
Analysis (161.48): This indicates analytical or evaluative studies in the dataset.
Case (160.14) and Study (143.27): This suggests the presence of case studies or comparative studies.

In summary, the TF-IDF scores and keywords provided high-level insights into the primary themes and areas of focus within the dataset. Let me know if you would like me to clarify or expand on any part of the explanation.

4.4.11 Topic Modeling

Latent Dirichlet Allocation (LDA), an unsupervised machine learning technique, has identified five distinct topics from titles in the publication dataset. Figure 4.12 displays the 10 most relevant words for each statistically derived topic. The following is an in-depth interpretation of these topics:

Topic 1 (Urban Land Use and Planning)

Keywords: urban, land, city, sustainable, use, planning, development, smart, case, and cities. This topic exhibits strong semantic cohesion and word clustering around urban land use, planning, and sustainable city development. The word distribution indicates a focus on land utilization, urban planning practices, and growth management in cities. Given the apparent emphasis of the dataset on urban studies, this represents a key research theme.

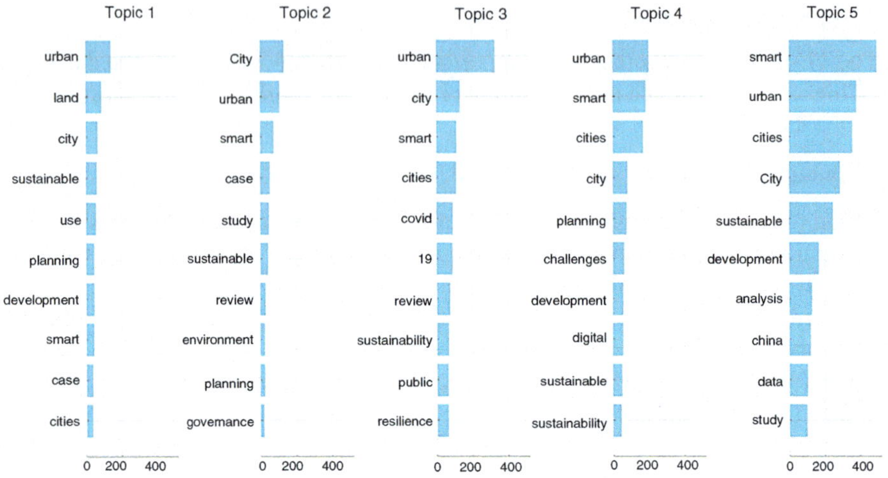

Fig. 4.12 Top word per topic

Topic 2 (Smart City Case Studies)

Keywords: city, urban, smart, case, study, sustainable, review, environment, planning, and governance. This vocabulary closely revolves around case analyses and reviews of smart city initiatives with an emphasis on urban sustainability and governance. The prevalence of "case study" and "review" suggests publications under this topic likely evaluate and synthesize learnings from city-level smart technology implementations.

Topic 3 (Urban Challenges and COVID-19)

Keywords: urban, city, smart, cities, covid 19, review, sustainability, public, and resilience. This topic concentrates on urban issues, COVID-19, and public health resilience. The emergence of "COVID-19" as a top term indicates a subset of publications that potentially examine the impact of the pandemic on cities and urban sustainability. An additional dataset context further confirms this interpretation.

Topic 4 (Urban Digital Planning and Challenges)

Keywords: urban, smart, city, cities, planning, challenges, development, digital, sustainable, and sustainability.

Vocabulary centers on digital/smart urban planning and its associated challenges. Keywords like "digital," "challenges," and "sustainability" suggest publications to investigate issues surrounding technology-driven planning and development in urban contexts.

Topic 5 (Smart Urban Development in China)

Keywords: smart, urban, cities, city, sustainable, development, analysis, China, data, and study.

This topic is characterized by words that focus on Chinese smart/sustainable urbanization and data-driven comparative studies. The prevalence of "China" shows the topic's likely geographic concentration, while "data" and "analysis" imply the use of quantitative approaches.

The bar charts in Fig. 4.12 visualize the distributions of topic-specific keywords extracted through Latent Dirichlet Allocation (LDA) topic modeling on the publication title dataset. Charts and frequencies provided insights into the thematic focus of each topic.

Topic 1 (Urban Land Use and Planning)

The high frequencies of urban planning-related terms like "urban," "land," "city," and "sustainable" reflect this topic's concentration on sustainable urban land use and development.

Topic 2 (Smart City Case Studies)

Keywords "city," "urban," "smart," "case," and "study" exhibit high prevalence, indicating the topic likely covers smart city case study analyses. This aligns with the apparent focus of the datasets in urban studies.

Topic 3 (Urban Challenges and COVID-19)

The frequencies of "urban," "city," "covid," and "19" suggest publications under this topic to examine urban pandemic impacts and public health resilience.

Topic 4 (Urban Digital Planning and Challenges)

High frequencies of terms like "urban," "smart," "city," and "digital" imply a thematic focus on digital planning and associated challenges in urban contexts.

Topic 5 (Smart Urban Development in China)

The topic-specific vocabulary concentrates around Chinese urbanization, with keywords "smart," "urban," "china," "development," and "analysis" occurring more frequently.

4.4.12 Principal Component Analysis

Plot of Original Data

The expanding field of smart cities has led to the generation of copious amounts of data from sensors, networks, and citizen interaction. This abundance of data poses a significant challenge for effective analysis and decision-making. Principal component analysis (PCA) has been identified as a valuable approach for addressing this challenge by reducing the dimensionality of data while retaining essential information. This summary investigates the crucial role of PCA in managing smart city data by examining its various functionalities. PCA is a discerning tool that identifies critical variables that significantly impact various aspects of city life such as air

Fig. 4.13 Principal component analysis

quality, traffic congestion, energy consumption, and citizen well-being. By identifying these factors, city officials can focus their attention and resources on the most impactful areas, leading to optimal decision-making and strategic resource allocation. This enables targeted interventions in areas with high energy consumption or traffic congestion, ultimately improving the operational effectiveness and public service delivery. Predictive modeling enabled by PCA allows city officials to anticipate future trends and challenges faced by smart cities, thereby facilitating proactive planning and effective risk mitigation strategies. By forecasting potential issues, informed decisions can be made to address them before they arise. Figure 4.13 illustrates the scatter plot of the original synthetic data generated with two features, denoted as "Feature 1" and "Feature 2." Data points were plotted in a Cartesian coordinate system and color-coded in blue for better visual discrimination. It is discernible that the two features are somewhat correlated, as indicated by the roughly linear arrangement of the points in the 2D feature space. The axis labels provide a quantitative scale for interpreting the dispersion of data points across the two features.

Plot of PCA-Transformed Data

Dimensionality reduction is a technique used to project high-dimensional data onto a lower dimensional space, which is usually defined by the first few principal components. This allows for better visualization and analysis of the data. Through imensionality reduction, similar data points tend to cluster, allowing for the identification of groups or patterns. It is also possible to identify outliers or data points that significantly differ from the main clusters, which may require further investigation. PCA enables the visualization of the principal components themselves as arrows in the

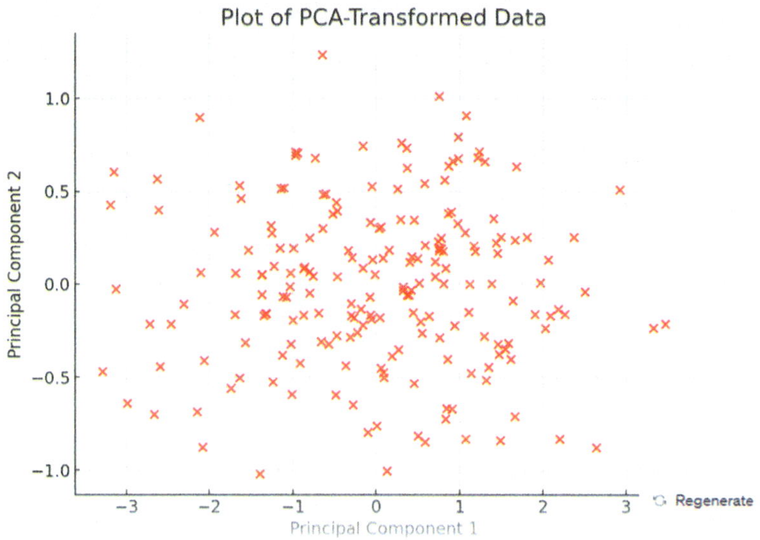

Fig. 4.14 Plot of PCA-transformed data

plot, indicating the directions of the greatest variance in the data. This helps us to understand the importance of each component. PCA can also identify correlated variables that tend to be projected closely together in a plot. This indicates redundancy and the potential for removal, aiding data reduction for further analysis. Finally, the lower dimensional representation obtained through PCA can be used as an input for other machine learning algorithms or statistical tests. PCA and data visualization can be instrumental in smart city analysis by offering valuable insights and supporting informed decision-making processes.

Figure 4.14 represents the scatter plot of the data after undergoing PCA. The transformed data points were projected onto the first and second principal components, denoted as "Principal Component 1" and "Principal Component 2" on the axes. These principal components are linear combinations of the original features and are orthogonal to each other. The color-coding in red facilitates visual interpretation. The plot demonstrates that data variability is mostly along the axis of the first principal component, corroborating its significance in capturing the underlying structure of the data.

4.4.13 Scree Plot

A Scree Plot helps to visualize the eigenvalues, showing how much variance each principal component captures. This can help us decide how many principal components to retain Academic Explanation for Scree Plot: The Scree Plot in PCA exhibits a steep initial decline followed by a gradual leveling off, with a crucial

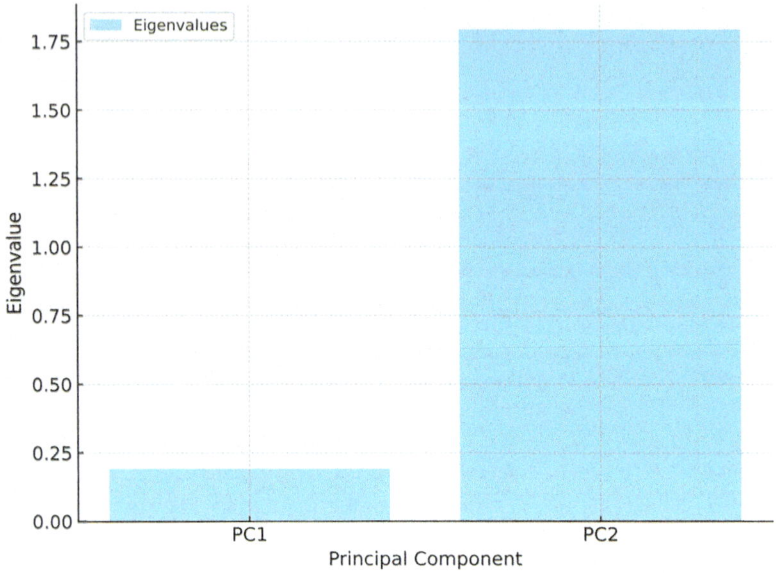

Fig. 4.15 Scree Plot

"elbow" point guiding the selection of components. In smart city data analysis, PCA, aided by a Scree Plot, proved instrumental in managing vast datasets. The Scree Plot serves as a pivotal tool in PCA for smart city data analysis. It optimizes data reduction, enhances model interpretability, facilitates the identification of key patterns, and prevents overfitting. This ultimately contributes to informed decision making, fostering the development of smarter and more efficient cities. The Scree Plot depicted in Fig. 4.15 provides a graphical representation of the eigenvalues associated with each principal component. The x-axis represents the principal components (PC1 and PC2), and the y-axis quantifies their corresponding eigenvalues. A higher eigenvalue indicates a greater proportion of the variance captured by a particular principal component. Notably, the eigenvalue of the second principal component (PC2) was significantly higher than that of the first (PC1), substantiating its dominant role in explaining the variance in the dataset.

4.4.14 Biplot

A biplot allows us to visualize both the transformed data points and original feature vectors in the new principal component space. The use of biplots in smart city analysis offers several advantages. These visualizations are particularly useful for analyzing intricate data, such as sensor readings and traffic patterns. They enable the identification of patterns, understanding of variable relationships, and

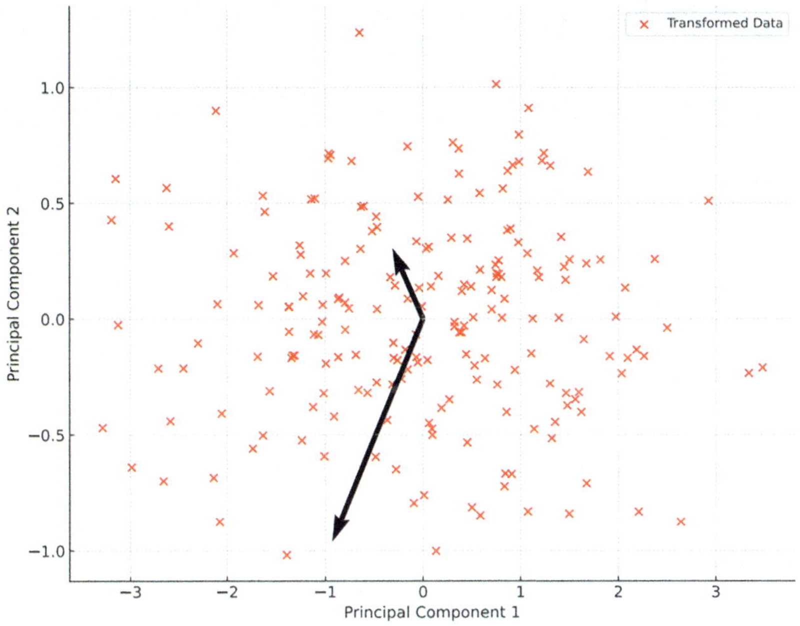

Fig. 4.16 Biplot scatter plot

monitoring of performance. In addition, biplots facilitate communication through a visually appealing format. There are a number of applications of biplots in smart city analysis, including traffic flow analysis, optimization of energy consumption, assessment of the impact of weather on air quality, and monitoring of citizen well-being. The use of biplots in smart city analysis enhances decision-making and city management through the provision of intuitive insights. This helps to interpret the principal components in terms of the original features. The Biplot presented in Fig. 4.16 shows a comprehensive visualization that amalgamates both the PCA-transformed data and original feature vectors, represented by arrows. The red points indicate the transformed data, while the arrows correspond to the eigenvectors of the covariance matrix scaled by the square root of their respective eigenvalues. The orientation of these arrows within the principal component space provides an insight into the influence of the original features on the principal components. Specifically, the length and direction of each arrow signify the magnitude and direction of the contribution of the original features to the variance along each principal component, respectively. The Biplot is a highly versatile instrument that enables the simultaneous examination of transformed data and the correlations between the original characteristics and new principal components, employing a formal tone.

4.5 Conclusion

This bibliometric analysis offers informative insights into the current state of smart city research, fulfilling the aims outlined in the Introduction. This study illuminates the noticeable growth in publications over recent years, with computer science, engineering, and environmental science representing leading contributors. Prolific scholars such as Simon Elias Bibri and research institutions in China and Australia stand out as highly productive nodes in the field. Core journals consist of sustainability, cities, and sustainable cities and societies, based on output volume and citation impact. Through thematic assessment, prominent topics emerged regarding sustainable urban development, smart city technologies, comparative case studies, and planning challenges. However, despite the burgeoning output, gaps persist regarding cross-disciplinary synthesis, policy frameworks, impact evaluations, and ethical considerations in smart city research. Furthermore, geographic concentrations lean heavily toward China, the United States, Europe, and Australia, highlighting the potential for broadening international perspectives. While providing valuable insights, this analysis has the scope for further research. Incorporating non-English articles can reveal a broader global perspective. Adding books, reports, and conference papers may enable a full understanding of the knowledge production and exchange dynamics. Conducting longitudinal assessments of how research foci and collaborations have shifted over time could help to elucidate pivotal developments. Nonetheless, within its defined parameters, this bibliometric analysis accomplishes its aims of mapping the landscape of smart city research—its evolution, influential contributors, knowledge structures, geographic patterns, and thematic contours. This study synthesizes a complex domain and identifies areas that require greater scholarly attention as the field continues to expand. Urban policymakers and planners can apply these findings to data-driven smart city decisions. Overall, this analysis constitutes a timely foundation for navigating and advancing smart city scholarship. The quantitative techniques applied included co-authorship, co-occurrence, citation, and geographic mapping of the literature using the VOSviewer software. Text mining facilitates keyword extraction. The analysis delineated patterns of prolific contributors, influential publications, research themes, collaboration networks, and geographic productivity. Limitations are centered on the scope of publications, languages, source databases, and analytical techniques used. Future directions include expanding the breadth of the dataset, incorporating additional sources, applying qualitative methods, conducting longitudinal assessments, and comparing the findings with other fields.

References

Agbali, M., Trillo, C., Ibrahim, I. A., Arayici, Y., & Fernando, T. (2019). Are smart innovation ecosystems really seeking to meet citizens' needs? Insights from the stakeholders' vision on smart city strategy implementation. *Smart Cities, 2*(2), 307–327.

Anand, P. B., & Navío-Marco, J. (2018). Governance and economics of smart cities: Opportunities and challenges. *Telecommunications Policy, 42*(10), 795–799.

Alshuwaikhat, H. M., Adenle, Y. A., & Almuhaidib, T. (2022). A lifecycle-based smart sustainable city strategic framework for realizing smart and sustainability initiatives in Riyadh city. *Sustainability, 14*(14).

Apanavičienė, R., Maliejus, K., & Fokaides, P. (2020). Sustainability assessment of the building construction stage using building sustainability assessment schemes (BSAS). *IOP Conference Series: Earth and Environmental Science, 410*(1), 12064. https://doi.org/10.1088/1755-1315/410/1/012064

Bauer, M., Sanchez, L., & Song, J. (2021). IoT-enabled smart cities: Evolution and outlook. *Sensors, 21*(13), 4511.

Bibri, S. E. (2020). Compact urbanism and the synergic potential of its integration with data-driven smart urbanism: An extensive interdisciplinary literature review. *Land Use Policy, 97*, 104703.

Chang, T. C., & Huang, S. (2011). Reclaiming the city: Waterfront development in Singapore. *Urban Studies, 48*(10), 2085–2100. https://doi.org/10.1177/0042098010382677

Dameri, R. P., & Rosenthal-Sabroux, C. (2014). Smart city and value creation. In *Smart city: How to create public and economic value with high technology in urban space* (pp. 1–12). Springer.

Djalalinia, S., Peykari, N., Qorbani, M., Moghaddam, S. S., Larijani, B., & Farzadfar, F. (2015). Obesity researches over the past 24 years: A scientometrics study in middle east countries. *International Journal of Preventive Medicine, 6*, 38.

Fernández, C. G., & Peek, D. (2020). Smart and sustainable? Positioning adaptation to climate change in the European smart city. *Smart Cities, 3*(2), 511–526. https://doi.org/10.3390/smartcities3020027

Frenken, K., Hardeman, S., & Hoekman, J. (2009). Spatial scientometrics: Towards a cumulative research program. *Journal of Informetrics, 3*(3), 222–232.

Ke, Q., & Ahn, Y. Y. (2014). Tie strength distribution in scientific collaboration net-works. *Physical Review E, 90*(3), 032804.

Lai, C. S., Jia, Y., Dong, Z., Wang, D., Tao, Y., Lai, Q. H., & Lai, L. L. (2020). A review of technical standards for smart cities. *Clean Technologies, 2*(3), 290–310.

Lepekhin, A., Borremans, A., & Iliashenko, O. (2018). Design and implementation of IT services as part of the "smart city" concept. In *MATEC web of conferences* (Vol. 170, p. 01029). EDP Sciences.

Levitt, J. M., & Thelwall, M. (2009). Citation levels and collaboration within library and information science. *Journal of the American Society for Information Science and Technology, 60*(3), 434–442. https://doi.org/10.1002/asi.21000

Liu, H., Kong, X., Bai, X., Wang, W., Bekele, T. M., & Xia, F. (2015). Context-based collaborative filtering for citation recommendation. *IEEE Access, 3*, 1695–1703.

Mozuriunaite, S. (2018). The role of landscape design in smart cities. *Landscape Architecture and Art, 13*, 49–55.

Nooringsih, K., & Susanti, R. (2022). Implementation of smart city concept for sustainable development in Semarang old town area. *IOP Conference Series: Earth and Environmental Science, 1082*, 12034. https://doi.org/10.1088/1755-1315/1082/1/012034

Prevelianaki, K., Sherratt, F., & Henjewele, C. (2022, November). ISO standards or global indices: Who decides if a city is smart? *IOP Conference Series: Earth and Environmental Science, 1101*(2), 022045. IOP Publishing.

Seto, K. C., Güneralp, B., & Hutyra, L. R. (2012). Global forecasts of urban expansion to 2030 and direct impacts on biodiversity and carbon pools. *Proceedings of the National Academy of Sciences, 109*(40), 16083–16088.

Strielkowski, W., Veinbender, T., Tvaronavičienė, M., & Lace, N. (2020). Economic efficiency and energy security of smart cities. *Economic research/Ekonomska istraživanja, 33*(1), 788–803.

Talebkhah, M., Sali, A., Marjani, M., Gordan, M., Hashim, S. J., & Rokhani, F. Z. (2021). IoT and big data applications in smart cities: recent advances, challenges, and critical issues. *IEEE Access, 9*, 55465–55484.

Tong, Z., Feng, Y., Liu, H., & Basodi, S. (2021). A survey on algorithms for intelligent computing and smart city applications. *Big Data Mining and Analytics, 4*(3), 155–172.
Uddin, S., Hossain, L., & Rasmussen, K. (2013). Network effects on scientific collaborations. *PLoS One, 8*(2), e57546.
van den Buuse, D., & Kolk, A. (2019). An exploration of smart city approaches by international ICT firms. *Technological Forecasting and Social Change, 142*, 220–234.
Van Eck, N., & Waltman, L. (2010). Software survey: VOSviewer, a computer program for bibliometric mapping. *Scientometrics, 84*(2), 523–538.

Deepak Hajoary has been dedicatedly contributing to academia over the past decade as a faculty member at Bodoland University in India. His expertise lies in the field of operations, where he has been actively involved in teaching subjects related to operations over the last 10 years. Deepak's passion for advancing knowledge is evident in his research pursuits, primarily focusing on the intersection of machine learning and supply chain management. With a keen interest in understanding the evolving landscape of these domains, he seeks to explore innovative applications and methodologies that contribute to enhancing operational efficiency and strategic decision-making. Deepak's commitment to both teaching and research reflects his goal of bridging theoretical concepts with practical implications, significantly impacting the academic and industrial spheres.

Raju Narzary is an esteemed educator with a decade of experience at Bodoland University, specializing in Computer Science and Technology. Passionate about cutting-edge advancements, his research focuses on artificial intelligence and Natural Language Processing. With a decade-long commitment to academic excellence, Raju brought a wealth of knowledge to the realms of computer science education and innovative research pursuits.

Rinku Basumatary, an educator in the Department of Computer Science and Technology at Bodoland University, India, brings a wealth of expertise cultivated over a decade of dedicated teaching. Specializing on Machine Learning, Natural Language Processing, and Operations Research, Rinku is committed to fostering academic excellence and cutting-edge research. His profound knowledge and passion for advancing technological frontiers have contributed significantly to the academic landscape, making him a respected authority in the field.

Part II
Digital Transformations: Harnessing Technology for Smarter Cities

Chapter 5
An Overview of IIoT Related to the Modern Panorama of the Industrial Field

Ana Carolina Borges Monteiro, Rodrigo Bonacin, Reinaldo Padilha França, and Rangel Arthur

Abstract Industrial Internet of Things (IIoT) is an integral part of cyber-physical production systems and processes, characterizing the use of modern and disruptive technologies (big data, IoT, and artificial intelligence (AI) among others) to produce consumer goods, considering real-time data from sensors and other sources of information assisting industrial devices and infrastructure in their decision-making, and it has the potential for digital transformation by adding technologies such as analytics. Intelligent Factory is a broad concept developed in Industry 4.0 involving factories with fully integrated cyber-physical systems, through automation, IIoT, high-end industrial machines, Big Data, and AI are examples of technologies included in the implementation of this concept, going beyond the capacity of these machines to work without any human operator in charge, created from the perception of a series of technological advances, allowing intelligent robots and machines to perform increasingly complex functions, it not only affect industries but the entire marketplace. IIoT, in this sense, can be considered a move toward "intelligent machines," which allows machines to autonomously monitor and predict potential problems, creating operational efficiencies, in which the levels of precision of the operations involved in the respective systems are raised to a level that cannot be achieved through human interventions. Therefore, this manuscript aims to offer an up-to-date overview of the IIoT and your adjacent technologies presenting the principles of this technology. In that regard, it demonstrates a landscape view of the applied aspect, with a concise bibliographic background to point out key concerns and challenges, featuring the potential of technologies.

A. C. B. Monteiro · R. Bonacin · R. P. França (✉)
Renato Archer Information Technology Center (ITC), Campinas, São Paulo, Brazil
e-mail: ana.monteiro@cti.gov.br; rodrigo.bonacin@cti.gov.br; reinaldo.franca@cti.gov.br

R. Arthur
School of Technology (FT), University of Campinas (UNICAMP), Limeira, SP, Brazil
e-mail: rangel@ft.unicamp.br

S. Majumdar et al. (eds.), *Smart Cities*, S.M.A.R.T. Environments,
https://doi.org/10.1007/978-3-031-59846-3_5

Keywords Deep learning · Big Data · Data · Data analytics · Sustainable development · IoT · Artificial intelligence · Machine learning · Cloud computing · Social aspects · Data science · Business analytics · Smart · Intelligent monitoring · Data science-based solution · Edge computing · Smart cities

5.1 Introduction

Considering that Industry 4.0 characterizes the use of modern and disruptive technologies (big data, IoT, and artificial intelligence (AI), among others) to produce consumer goods, in short, it is the continuation of the digital improvement of machines (Schwab, 2017; Philbeck & Davis, 2018).

The concept encompasses automation and information technology, going beyond the capacity of these machines to work without any human operator in charge, representing a technological leap, bringing automation to maximum power, and allowing intelligent robots and machines to perform increasingly complex functions (Schwab, 2017; Philbeck & Davis, 2018).

From the historical point of view, the first industrial revolution took place in the middle of the eighteenth century, with the emergence of steam engines and railroads, replacing the use of animals to generate power, then, between the end of the nineteenth century and the beginning of the twentieth century, with the electric energy and production line created by Henry Ford, the second industrial revolution developed, enabling large-scale production. Over time, the third industrial revolution was achieved through information technology, the Internet, personal computers, and the entire range of digital platforms, modernizing the work and processes in factories and offices (Mokyr & Strotz, 1998).

Thus, the fourth industrial revolution is the ability to collect and process data in real time instantly, allowing qualified decision-making in real time, through the virtualization of smart factories, due to scattered sensors allowing to track and monitor in a way remote all your processes. Achieving process decentralization and being service-oriented, where software is oriented to provide solutions as services, connected with the entire industry. Also, the aspect of modularity allows modules to be coupled and decoupled according to the demand of the factory, offering great flexibility in changing tasks, resulting in interoperability through the IoT where the machines and systems can communicate with each other. It is for this set of innovations in which each concept has its particularities that it is considered the fourth industrial revolution, considering that all these concepts have in common the objective of making machines more efficient (Schwab, 2017; Camilo et al., 2019).

The Industrial Internet of Things (IIoT) brings a new era of manufacturing, characterized by a digitalization focused on machine-to-machine communication, allowing industries and companies to have more digital efficiency and reliability in their operations and processes, considering big data and machine learning. IIoT consists of advanced analysis platforms that process the data produced or even machines connected to the Internet and IIoT devices ranging from environmental sensors to complex industrial robots (Boyes et al., 2018; França et al., 2020a, b, c).

IIoT works through the presence of three types of technology: use of sensors on the factory floor, given that sensors give objects and machines the ability to perceive. Thus, this can identify several factors, such as temperature and pressure, considering that this activity generates valuable data and they are processed through industrial analytics. Industrial analytics gives rise to important insights that serve as a basis for decision-making. It also ensures the smooth running of the plant, as failures and errors are identified more quickly and accurately. Smart machines rely on the help of smart machines, which perform functions that go beyond mechanics—focusing on integration and connectivity (Boyes et al., 2018; França et al., 2020a, b, c).

In this context, a fundamental part of the digital transformation for manufacturers in all sectors is the IIoT positioned to revolutionize the entire manufacturing value chain, considering that this change enables manufacturing with new ways to develop, providing an unprecedented level of connectivity and functionality, as also innovate and manufacture due to the endless connections that can occur. The difference between IoT and IIoT is the intersection between operational technology (OT) and information technology (IT), referring to the network of operational processes and industrial control systems. In other words, IIoT is technology-oriented to the industrial field and the interconnection of physical devices normally associated with IoT (Fig. 5.1) (Boyes et al., 2018; Madakam & Uchiya, 2019).

In this sense, IIoT can be considered as a subcategory of IoT that also includes applications aimed at the user as usable devices, technology for smart homes with infrastructure with integrated sensors, and even autonomous cars aggregating smart devices. When IIoT is well applied, the industry becomes much more productive and efficient, even considering the difference between IIoT and IoT is in the focus of these two technologies, since IIoT is aimed at the industry. But it also has applications, devices, machines, and all the other resources. One of the main impacts of the Industrial Internet of Things is the greater connectivity between machines, people, and goals. This becomes a reality on the factory floor and takes the concept of industrial automation to another level. Due to this, it is possible to reduce labor costs and waste on the production line. After all, the machines connected to each other know when and how to start and execute a certain task. Since it is the convergence of IT and OT, offers industries more integration of the digital system, facilitated by the use of smart actuators and sensors, as well as remote control and even access (Madakam & Uchiya, 2019; Lee & Lee, 2015).

In a broader context, another relevant impact is the improvement of industrial maintenance, through sensors and software, is possible to predict points of failure, and is possible to define the right time to carry out repairs, which reduces costs and problems caused by interruption of production. Even with such relevant impacts, IIoT is crucial for the use of cases related to ecosystems or connected environments, there are challenges to be overcome, such as the amount of information produced by so many devices, as it will be necessary to find effective ways to store, track, and analyze all these data. In these connected ecosystems or environments, the

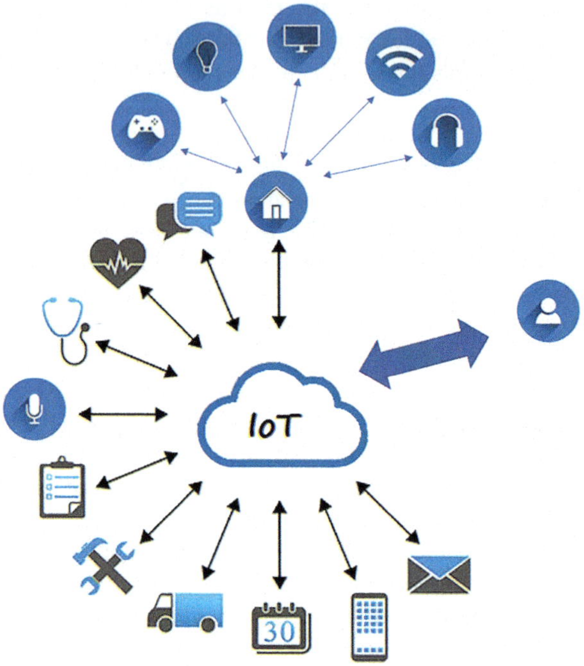

Fig. 5.1 Internet of Things

consistent capture and transmission of data between smart devices and machines (collected and analyzed in large volumes of data at higher speeds) provide, through collected data, the possibility of detecting errors or inefficiencies in the supply chain, for example. Example, and resolve them immediately, thereby increasing efficiency in operations, through the application of AI algorithms that make machines and devices analyze data at a speed that a human being could not (Sisinni et al., 2018).

IIoT not only increases scalability and performance but also aims that the integration achieved by IIoT technology also has the potential to bridge the gap between production floors and departmental offices and make business decisions possible. Or even considering the context-oriented purely between machine to machine, consisting of an advantage of industrial digital frameworks and systems, capable of responding intelligently and even changing their course of action based on the information received (data collected and analyzed) through the cycles of feedback established within the industrial structure, to which IIoT can operate in a semi-independent way or with very little human intervention (Serpanos & Wolf, 2018).

IIoT, in this sense, can be considered a move toward "intelligent machines," in which the levels of precision of the operations involved in the respective systems are raised to a level that cannot be achieved through human interventions. Therefore, this manuscript aims to offer an up-to-date overview of the IIoT and your adjacent technologies presenting the principles of this technology. In that regard, it demonstrates a landscape view of the applied aspect, with a concise bibliographic background to point out key concerns and challenges, featuring the potential of technologies (Soldatos et al., 2022).

5.2 Artificial Intelligence and IoT

Intelligent Factory is a broad concept developed in Industry 4.0 involving a series of technological advances, not only in industries but also in the entire marketplace. The main advantages of IIoT are in increasing the efficiency of the industry through connectivity, Big Data, Artificial Intelligence, and many other technologies; the industry can increase its efficiency. In such a competitive market, this makes all the difference and contributes to improving productivity and market share, through automation, IIoT, high-end industrial machines, and Big Data (Fig. 5.2) (Hozdić, 2015).

The design of Intelligent Factories consisting of technology is designed to establish and optimize this method, considering that it improved levels of security, as IIoT plays an important role in improving industrial maintenance procedures. Industrial automation, together with intelligent systems and sensors, ensures that the equipment is always in operation and with its performance at optimal levels. The proposition of this new factory model arises from the evolution of digital resources that increasingly allow the production of more robust data and the autonomy of the machine, without human supervision, inserting fully autonomous artificial intelligence, since many machines only recognize patterns, it is possible to apply these resources for product quality control and operation monitoring (Radziwon et al., 2014).

In any sector, the operation often requires mechanical and repetitive tasks, since this is not just about production, also considering optimization of decision-making by managers, considering working based on real and updated data, simplifies

Fig. 5.2 Big Data analytics illustration

decision-making by managers. After all, they replace feeling with reliable information and reports, which ensures more effective strategies and better results. However, this type of task is necessary, but it does not need to take so much time from employees. In this approach, integrated automation systems in Intelligent Factories make the flow of tasks more dynamic, establishing automatic processes from a database containing all company information. It is important that once unified, all data produced (collected) must go to the same location and be easily accessible (Yuan et al., 2020).

From this, the implementation of these systems connects the different sectors through an integrated database between all sectors, allowing the machines in addition to processing to store massive amounts of data and to communicate with other computers. In this way, machine learning (ML) (Fig. 5.3) principles feed your computers with various data and define an algorithm to filter them, able to identify patterns and make autonomous decisions from that. Currently, this technology can only work with pattern recognition, but this can now be used as a way to improve the industry, making it possible to identify defects or create customized lines, recognize a specific product, and send it to a certain section of the production line (Lee et al., 2018; França et al., 2021a, b, c, d, e, f).

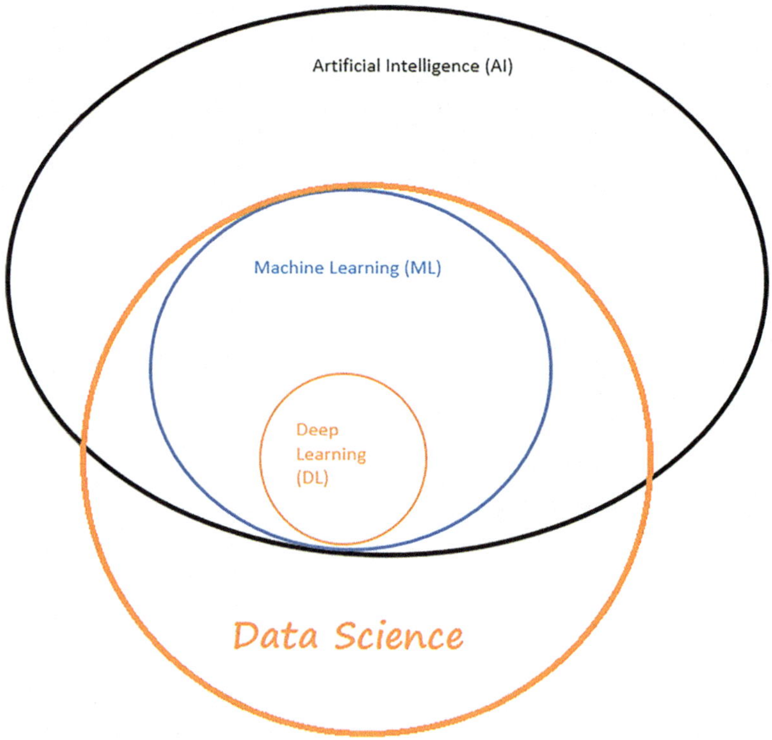

Fig. 5.3 AI + data science illustration

ML can be defined as being of three types of supervised learning in which data are labeled and already have a defined "path," consisting of the best alternative when already know what needs to be discovered. Exemplifying the implementation of a more dynamic production line, it is possible to create a series of paths in the production line, considering that each product follows a path according to the specifications determined in the elaboration of the algorithm. This would make it possible to automatically manufacture diversified products and identify defective materials. Regarding unsupervised learning, it consists of discovering what the existing standards are, because, unlike supervised learning, this will not start from a pre-defined structure. Rather than obeying a structure, the algorithm will be educated to identify a series of characteristics of the collected data. From the identified aspects, the algorithm will group each data into groups (clusters) that have some similarity (França et al., 2021a, b, c, d, e, f; Mohri et al., 2018).

Thus, to train a machine through supervised learning, it is necessary to provide samples that show the system what the required standard is. For example, to "teach" what is an ideal piece, it is necessary that the machine analyzes several models considered desirable. If defects are found, the product will be discarded, avoiding greater waste of resources along the production line (Mohri et al., 2018; Alpaydin, 2020).

In this text, two ways to structure supervised learning are Decision Trees based on a pre-established test model that will determine some path of achievement. Representing a product in a production line, the algorithm is defined with strict paths, and the item will be produced according to the characteristics that define the classification in each stage. Still relating that this has the advantage of being highly recursive, being able to create the number of branches that are necessary to cover all possibilities. Linear least-squares regression defines a graph from the values to be measured, and each sample will occupy a point in that plane, considering that from this set a pattern is defined; in production, this can be used to recognize stress patterns equipment or a deviation from the quality standard required for manufacturing (Mohri et al., 2018; Alpaydin, 2020).

There are some options of algorithms related to unsupervised learning to define how this clustering (clusters) will be carried out as probabilistic, based on density, based on centroid, and based on connectivity. Even neural networks will analyze characteristics and establish standards based on that. In the case of industries, it is possible to use this feature to identify the patterns of possible customers and even possible to predict which products are best suited to the situation of that business (Usama et al., 2019; Monteiro, 2023).

ML allows better quality control, identifying defective products during the manufacturing process and preventing this product from following the production line; this will save raw materials and time that can be used for ideal parts. With regard to maintenance, from previous data it is possible to define ideal performance standards for a machine, considering that if there is any equipment recording information that deviates from this performance, the ML approach informs the problem immediately detected (predictive maintenance), enabling an equally instant response (Indurkhya & Damerau, 2011; Olivas et al., 2009).

In this sense, joining AI to IIoT represents an advantageous differential in terms of a large amount of data, considering that information constantly changes, generated from IIoT sensing are not supported by traditional analytics systems or even BI (Business Intelligence). In this context, it highlights the importance of AI and ML, which can assertively identify patterns and warn about abnormal conditions of systems and equipment (Mahdavinejad et al., 2018).

Thus, it is possible to highlight that ML is a resource that seeks to integrate operational processes with new technologies, aiming at predictability based on the massive production of data for analysis and identification of problems. However, it is evident that the implementation of ML requires a structure compatible with your needs, i.e., with little data or without digital systems that keep the database well integrated, it is not possible to carry out effective learning, emphasizing that the industrial sector has a lot to benefit from equipment/machines and devices capable of identifying patterns (Raschka & Mirjalili, 2019).

5.3 IIoT Benefits

In the production sector, IIoT technology allows machines to autonomously predict potential problems and also monitor, which means more efficiency and less downtime in general, creating operational efficiencies. In the supply chain, from an inventory managed by IIoT sensors, it is possible to take care of supply orders before stocks run out, reducing waste, while keeping the necessary goods in stock. In retail, IIoT technology has the potential to make quick and smart marketing decisions for individual stores, relating that showcases can be updated automatically, based on the interests of customers, and the ability to create smart promotions, being able to obtain a great advantage over your competitors (Gilchrist, 2016).

The use of IIoT technology in relation to facility management allows greater security and simplicity with climate controls triggered by sensors, also relating devices that monitor the entry points of the facilities and respond quickly to potential threats by increasing the security of the facilities. Or even linking IIoT to health care making this sector more accurate and responsive, considering devices that remotely monitor patients and notify health service providers as soon as the patient's status changes, yet through AI is possible even assume the diagnosis of a patient, enabling doctors to treat them more quickly and efficiently (Kiel et al., 2017).

The benefits of IIoT are related to the productivity levels in the industries, generating efficiency and performance that allow manufacturers to obtain transformational financial and operational benefits, in the same sense that it obtains an increase in the safety level in the workplace to achieve an ideal operation, avoiding interruptions production and protecting employees (Popescu, 2015).

One of the positive points of IIoT is that it is not necessary to acquire new machines for processes that are already working in the company, considering that IIoT devices come in as a complement to the machine, collecting and delivering information for decision-making in real time. Also relates to the insertion of devices in a network, providing connectivity between devices and systems, with the ability to exchange information with each other through sensors embedded in these devices, and other ways to provide data captured and digitally controlled (El-Din & Manjaiah, 2017).

Basically, an IIoT device needs to have instruction and connectivity, considering instruction as the function that this will perform, and what periodicity will perform these functions, which can be measured, detected, counted, or another function in an environment, and send this information (data) for a system or a platform, which can be internal (inside the company) or Cloud Computing platform (outside the company), it also needs to have the instruction. Regarding connectivity, it consists of the means that will be used to send this information, i.e., via a wired network, Wi-Fi, 3G, 4G, or networks created specifically to meet IoT demands such as LoRa, Sigfox, and NB-IoT, among others (França et al., 2020a, b, c; Jansen & van Der Merwe, 2020).

With IIoT, it is possible to obtain visibility of the equipment, allowing the visualization of key indicators not only in production but also in terms of safety compliance, allowing it to act proactively when detecting possible safety flaws in an industrial process. Many processes still have outdated methods of collecting security data, some using pen and paper reports and others even without reports, resulting in deficits in security standards, delays in reporting data, and even level management, which is responsible for mitigating these risks by receiving partial data on security breaches (Arnold et al., 2017).

IIoT can enable improvement in manufacturing energy efficiency to determine when and where it is wasting energy and fixing these areas, considering that one of the biggest expenses for manufacturing organizations is energy expenditure. Given this, real-time data provides insights such as off-peak consumption and other energy-saving opportunities that allow for identifying operational inefficiencies and waste. In this sense, this information on how to obtain greater efficiency of equipment/machines, and even possible problems of regulatory compliance and waste offer opportunities for cost savings through decision-making (Zhang et al., 2018).

Or even pointing out that one of the main added values of IIoT for the manufacturing line is predictive maintenance, given the possibility of predicting when a machine or equipment failure may occur and then preventing that occurrence by performing maintenance. Predictive maintenance involves a machine or component with IIoT sensors that collect and transmit data and then analyze it and store it in a database, which provides comparison points for events, as it happens. The digital system eliminates unnecessary maintenance and increases the likelihood of avoiding failures. This brings several cost savings since, in an ideal scenario, it allows the frequency of maintenance to be as low as possible, such as reducing production hours lost to maintenance, avoiding unplanned reactive maintenance, minimizing the cost of parts spare parts and supplies, and creating a more proactive maintenance schedule (Gilchrist, 2016).

With IIoT it is possible to obtain quick and more informed decision-making about the performance or problems of the machine/equipment, such as reduced downtime and adding competitive advantage in real time, generating higher product quality. Since the power of decision lies in the accuracy of the data and the ability to act based on insights from it, making the plant more efficient, operating with greater profitability, and achieving greater customer satisfaction, still relating to predictive maintenance, transforming the managerial approach reactive to proactive, where waste is reduced with an increase in overall visibility (Pinon et al., 2018).

With intelligent machines and the principle of modularity, it is possible to have a much more flexible production and obtain a strategic view of the business, by modernizing industrial processes resulting in a reduction in production costs. However, it is necessary to develop this organizational culture of valuing strategy guided by data, identifying market demands and trends, allowing to act with great speed for the production of new products as well as for the consumer, who will have greater access to personalized, quality products and a lower cost (Munirathinam, 2020).

5.4 The Essential Technologies of IIoT in Industry

Radiofrequency identification (RFID) allows one to easily track and monitor the things to which it is attached, it consists of a system that involves tags and readers, like a more intelligent version of barcode technology, being identified by RFID tag readers using waves of radio, meaning that the tags can be read by multiple readers at once and at a greater distance than traditional UPC (Universal Product Code), as well as other types of IIoT sensors, through sensor-oriented computing capable of giving objects the power of perception, enabling the productive sector to measure temperature conditions: pressure; speed, and several other parameters. Considering that all this information is precious in the innovation of production processes and services since it is usually performance indicators (Rajaraman, 2017).

IoT is a concept that deals with the connection of physical devices to the network since this is not about having more devices to access the Internet, but digital hyperconnectivity improving the use of objects, happening inside homes (television, air conditioning, refrigerator, and doorbell connected, among others). The evolution of IoT technology allows devices that are increasingly smaller and with greater processing power, to consider various electronics that form the “things” representing objects capable of collecting and processing information online about and for their customers, as well as in industries, using sensors, instruments, and other production monitoring and control resources, allowing machines to generate instant production reports for management software deployed in cloud infrastructure, for example (França et al., 2021a, b, c, d, e, f).

Big Data is a key concept for the Fourth Industrial Revolution, used to refer to the current technological reality, in which an immense amount of data is collected and stored daily on the network because it is this data that allows machines to work more efficiently. In turn, the data produced through the sensors allows the industrial analysis to transform this into rich and very useful insights about the conditions of the industrial plant, serving for the execution of the action plans and also for better decision-making. Exemplifying that it is generally common in large industries to have thousands of alarms that constantly signal process abnormalities, in this sense, it is impracticable for a human operator to be able to process all these alarms, which ends up generating losses. Therefore, with Big Data technology it is possible to analyze this complete alarm management, capable of extracting all the information from the thousands of data generated every day, collaborating to reduce losses and increase productivity (França et al., 2020a, b, c, 2021a, b, c, d, e, f).

Artificial intelligence (AI) is what allows the decision-making of the machine without human interference, through Big Data allowing better technological refinement in the collection, storage, and treatment of data and even the IoT (allowing connection between machines and systems). AI is a possibility in the industry that allows machines to become increasingly autonomous and capable of interfering in production; however, it is necessary to emphasize that the performance of these machines is not yet completely autonomous, being necessary to program them previously and "teach them" about the actions it needs to perform. There is another important factor of AI technology dealing more specifically with pattern recognition, meaning that the machine empowered by AI algorithms is able to command actions that it was taught to do, representing that any change or error not foreseen in the production line requires "re-learning" of the machine or modification of the current algorithm, considering that otherwise the algorithm will not recognize this unforeseen event or simply ignore it. Thus, human intervention is still important for the conformity of the process, but without the need to be the main responsible for the execution (Joshi, 2017; Thomassey & Zeng, 2018).

Through AI it is possible to have a lower incidence of errors considering the power and versatility of the machines, allowing repetitive and mechanical activities to be done with less chance of errors since this process can be repeated to exhaustion and will hardly be affected by external factors. Even using AI combined with IIoT makes it possible to attach sensors to these machines so that can collect and process various information in addition, this can be programmed to scan the conditions of the industrial production environment without interruptions and without the need for human presence (Joshi, 2017; Thomassey & Zeng, 2018).

Work safety is among the biggest concerns of large companies, which dedicate entire boards to take care of the area, considering that the subject focuses on human behavior. However, through IIoT technology, which allows for increasingly automated factories and intelligent machines, the concern becomes fewer conduct manuals and more robustness in digital information systems and prevention of problems in communication between machines (França et al., 2021a, b, c, d, e, f).

Cloud computing (Fig. 5.4) consists of digital systems that are stored on shared servers and interconnected over the Internet, with less cost and in a more agile and efficient way than the old model so that it can be accessed anywhere in the world. In the context of Industry 4.0 and IIoT, this allows to exceed the limits of the company's servers and expand the possibilities of connectivity between digital systems (Padilha et al., 2020).

Robotics also play an essential role within the scope of IIoT, by means of cobots ("collaborative" + "robot"), i.e., from many angles, it forms a simple and clear collaborative process on the function of these intelligent machines. This serves to

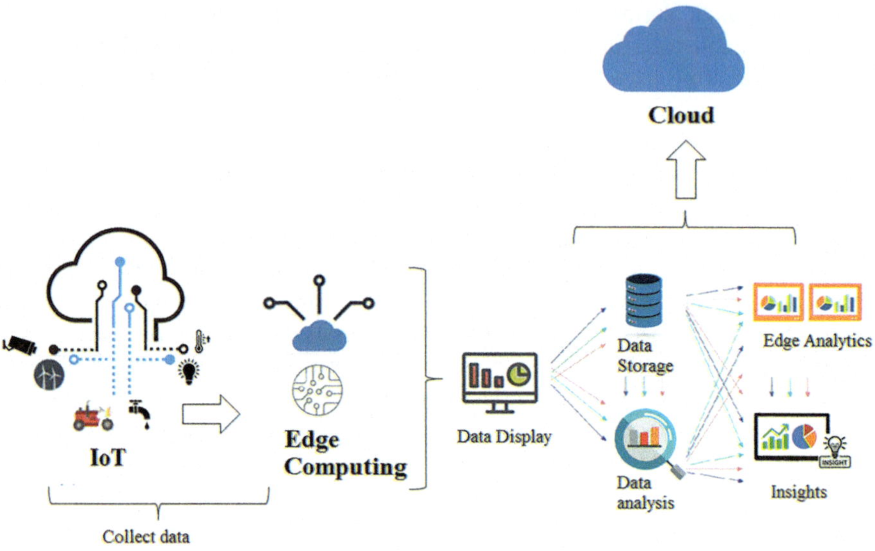

Fig. 5.4 IoT in Edge computing

perform difficult, repetitive, or demanding tasks, based on the great differential since most assembly lines that rely on robots have the advantage of being able to work side by side with humans. Cobot robot technology comes from the adoption of sensors that paralyze any movement before it harms a person, allowing the employee to temporarily control this to get them out of the way, for example. Although most assembly lines restrict access for people, still relate to the ancient robotic approach where ordinary robots lacked digital intelligence and represented risks to the physical integrity of unsuspecting workers who interfere with their activities (França et al., 2020a, b, c).

Also, relating the technology of Cyber-Physical Systems (CPS) that are essential for the integration between machines and systems in Industry 4.0 and IIoT, making the connection between digital systems and the mechanical part of the factory through IIoT sensors, information obtained (collected data) is forwarded and stored and can generate insights regarding the functioning of the machines, supporting predictive maintenance. Another advantage is the sending of alerts and the composition of reports, in addition to providing the basis for the interaction between the physical and the virtual world, allowing a device to send information and another device to be able to give a suitable response, exemplifying an automatic process as an order is added in software, the smart machine starts production, without the need for human command (França et al., 2021a, b, c, d, e, f).

5.5 Discussion

It is important to note that, although the IoT for the physical consumer, extracting data from different sources, seeking to extract as much information as possible, and the IIoT, extracts massive data, structuring a complete process, although both may have some similarities also have marked differences. The use of the IIoT resource aims not only at digitization and data generation but also at the interconnection between the different processes of supplying a product, considering that production is only part of the production chain, and observing a larger view it is necessary that the flow of work is increasingly reduced, that is, the demand must be clear and robust and the logistics must be ready to transport the product in the shortest possible time.

Considering that in addition to an integrated database (data collected by IIoT), it is possible to have a massive amount of data processed through Big Data, which has an important role in this process. Considering that a Big Data structure receives the data extracted from the machines is able to make production reports, streamline decision-making, and make improvements or corrections. However, establishing metrics is not enough, and IIoT enters as a necessary means to improve them.

The application of IIoT will allow a fast, efficient, and autonomous production line, with new production policies, considering a lean production model establishing a production driven by the customer, extracting information from the database and other machines to improve the reliability of the data and integrate processes, with automated processes and a well-fed database that allows long-term demand forecasting, determined by immediate demand, rather than delivering mass-produced products (eliminating the conventional method of producing much forever) have stock available to the customer.

It is also important to point out that "communication" between machines is called M2M (machine to machine), considering the specifics of a product representing the data reported to the machines, which "will talk to each other" to do the production and pass this information onto the following sectors, such as logistics that will transport that product.

Even monitoring the interconnected production data in real time transmitted to a central, considering each step of this process will make it possible to identify possible bottlenecks and deficiencies in less time, saving resources and making better use of it. Likewise, with a robust database of the entire production process, these data can be used for machine learning, which can be taught to use fewer resources and maintain quality, or it is even possible to make virtual simulations and representations (digital twins) from the factory floor. This impacts the possibility of testing the impact of the results in a virtual model before implementing an improvement that will be applied.

With IIoT it is possible to make the system itself capable of warning about problems in a preventive way; this considerably reduces the probability of failures, since it analyzes the data evaluating the performance of the machines. If it is identified that a machine is stressed during its performance, the system is able to alert maintenance to this problem, providing a better control structure, even sending this information to the purchasing and logistics sector, scheduling the purchase of materials to an eventual repair. In addition, IIoT is combined with Machine Learning, and the machines themselves can learn how to distribute the effort on the production line, as this increases the useful life of assets and saves resources, such as energy and material used in production.

With regard to digital security considering the storage of all this volume of information when scanning each step of the operation, since everything is connected, the technology has cloud servers with cyber protection derived from firewalls, antivirus, encryption, and authentication providing information security and employing locks to block the intrusion of malicious users.

Logistics can also benefit from IIoT by making real-time monitoring of raw material delivery with the supplier, knowing precisely when materials arrive automatically, or even IIoT technology applied in a stock control system while maintaining the dynamism of the industry, with integration with the productive sector through a dynamic stock, with automatic purchases of materials to avoid shortages, prepare conditions for packaging, ready availability for shipping products, and even fleet availability.

IIoT makes it possible to monitor delivery with the product in transit, checking for delays or losses through RFID tags that allow the monitoring of vehicles, products, or even employees in real time, still considering the ability to generate reports from the data of these IIoT sensors on the routes or movements required for both internal and external logistics. In this context, it is important to note that what IIoT provides is data about anything, anytime, and anywhere, enabling a digital medium, fully connected and dynamic, accessible, and secure, which can be accessed on different devices that are allowed access, which works together with flexible and efficient production.

IIoT is applied to monitor processes by temperature and humidity, given its possibility of application in the food industry, through intelligent temperature and humidity sensors in refrigerators and cold cameras in order to store perishable products, or even to control the temperature in environments with ovens or hot processes, such as smelting iron, for example. How about in addition to information and instruction, or even using an intelligent presence sensor in conjunction with temperature sensors, relating aspects of work safety, in which some places cannot be accessed, such as monitoring the temperature to avoid accidents by employees with overheated ovens, representing intelligent protection measures. Considering another function of intelligent temperature sensors is to detect possible overheating, which if not identified can damage machinery, as is the case with rooms with IT hardware since overheating can impact several important types of equipment for the company's operation.

In this sense, due to the Internet and the use of IoT and IIoT technologies, the consumer is at the center of the digital experience. In this context, a panorama is created for the "Knowledge Era" related to the Fourth Industrial Revolution, with numerous other opportunities in the IIoT universe, in addition to some applications and examples mentioned in this section. In turn, due to the performance of the machines and the production chain as a whole having become more autonomous with the dissemination of knowledge, decision-making capacity is no longer tied only to human performance, representing the great impact on today's industrial automation. Considering that the ideal project comprising IIoT for a company will depend on the whole process and the improvements that can be implemented in each stage, assessing the potential of IIoT applied to the company and how it approaches the business.

5.6 Trends

Digital twins are the practice of creating a computer model of an object, i.e., it has replicated models that exist virtually and are dynamic, like a machine or human organ, or a process like a climate, that is, obtaining a model that can be studied on the behavior of this, allowing simulations and data collection to facilitate preventive maintenance, it is still possible to understand and predict the behavior of its counterpart in the real world and solve problems before this occurs. This technology is used to monitor equipment and systems, tracking failures and preventing small or large events, from the breaking of an accessory to an accident with victims, and it can exemplify its use in aircraft engines, which have a series of sensors that generate data, in real time, about their performance, since through the technology of digital twins it makes it possible to monitor their conditions, even from a distance, valuing the good functioning, failure prevention, and security (Li et al., 2017).

Electronic logging devices (ELDs) are related to integrated sensors that monitor speed and travel time. This technology in the automotive environment allows for monitoring how often drivers apply the brakes, or even if the driver performs a dangerous maneuver or has been driving for a long time, it will be alerted, and the controller will be notified; this helps to save fuel, improve driver safety, and reduce idle resources (Diem, 2019).

Intelligent Edge is related to analyses that can be conducted more quickly and that the likelihood of data being intercepted or leaked is significantly less, making companies self-reliant and not dependent on potentially failing network connections to do their job, listing the data that are generated, analyzed, interpreted, and treated. In this way, Intelligent Edge technology reduces latency, costs, and security risks, making associated businesses more efficient. Also relating the categories are the borders of operational technologies, IoT borders (the largest and best known), and borders of Information Technology. Through this technology, instead of sending data to a data center or a third party, the analysis is performed at the location where the data is generated, meaning that not only can the analysis be performed more quickly but there are also less likely that data will be intercepted or leaked (Liu et al., 2019; França et al., 2020a, b, c).

Through Edge Computing (Fig. 5.5) enabling computation and analysis closer to the edge accelerates the speed of decision-making and reduces the chance of missing or not being aware of a warning signal, allowing organizations to quickly

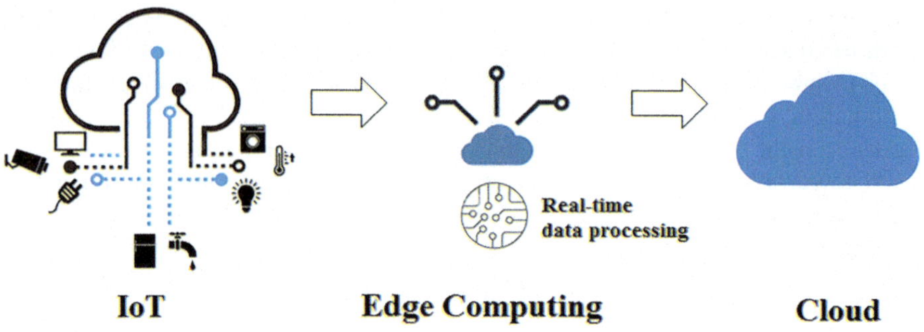

Fig. 5.5 Edge computing data processing

acquire, analyze, and use data in time while these data are being collected for further analysis at a later industrial stage, gaining insights from the edge, where data is created from enhanced IIoT digital systems, with features similar to the data center (França et al., 2020a, b, c).

Additive Manufacturing corresponds to one of the main 3D printing technologies, and it helps in the visualization of molds, through which an object is manufactured from the addition of thin layers, one over the other. Through 3D printing technology, which is based on detailed prototypes and built on specific software, this gives not only agility but also customization to the items. For example, in the production of items that will be used by the end-user, such as small and large parts, buildings, and even human organs, it has already been successfully printed through this technological process, which has been adapted to serve different purposes. In this context, industries can ask the customer for instructions and manufacture products with the characteristics they want, improving the entire purchase and use experience (Ramya & Vanapalli, 2016).

5.7 Conclusions

Throughout time and history, human beings have specialized in developing instruments and techniques capable of minimizing the effort to carry out their activities. This approach can also be related to the case of industrial automation processes, considering information technology and smart software. The main concept of IIoT consists of the Internet applied to digital devices and systems in exchange for information in real time.

Any and all processes guided by IIoT technology that serve to increase efficiency in industrial production over any techniques, software, or equipment that optimize the production steps have the sole focus of presenting better results. However, it is also possible to idealize the same concept applied throughout the production chain of an item, after all, it does not matter, and the industrial automation guided by IIoT is a small or large process. This is because the automation guided by IIoT can be tailored to a factory or a process, always with the aim of bringing more efficiency and benefits to production.

It is important to highlight that IIoT, in turn, forms a critical layer of the production process, making it possible to directly connect a supplier of items on the production line, in real time, and even make an analysis of the quality and use of this merchandise, still considering the possibility of controlling the production process through the connection of the logistics chain relevant to the entry and exit of materials, at the optimum point of operation (real time) making a data production and consumption application, with a critical profile.

So, AI, Big Data, Machine Learning, and other disruptive technologies in conjunction with IIoT, it is an important allies in this process and their use can be an alternative for industries to optimize their processes and, consequently, achieve a higher return on their invested resources. In other words, this new industrial revolution happens with the insertion of new technologies and modern prototypes, causing the emergence of new concepts and models both for the productions and for the day to day of the companies, representing a break of paradigms that will make possible the implementation of IIoT projects in industries.

Reflecting on the reasons for digital industrial processes through data through IIoT technology, the premise of digitizing all information derives from cheap information, transforming this data into intelligence information (Business Intelligence—BI), transparency of actions and processes, increasing expertise, generating a model of estimates and reducing the risk of errors in decision-making, reinforce the AI-oriented concept relating to what is performed is "learned," end error and waste, and buy time, among other aspects.

However, it is worth noting that the digital layer implemented by IIoT is able to relate everything that happened and what will happen with the potential to correlate the way of operating as well as maintaining an industrial plant. Since these digital layers are built by IIoT, it is possible to think about using disruptive technologies of operation, planning, and quality, in a completely innovative way, allowing even the technological use of the corresponding virtualization over entirely digital production planning, from design to production, allowing to work on all scenarios, even before production actually takes place.

References

Alpaydin, E. (2020). *Introduction to machine learning*. MIT Press.

Arnold, C., Kiel, D., & Voigt, K. I. (2017). Innovative business models for the industrial Internet of Things. *Berg-und Hüttenmännische Monatshefte, 162*(9), 371–381.

Boyes, H., Hallaq, B., Cunningham, J., & Watson, T. (2018). The Industrial Internet of Things (IIoT): An analysis framework. *Computers in Industry, 101*, 1–12.

Camilo, E., Iano, Y., Gomes de Oliveira, G., Pajuelo, D., Borges Monteiro, A. C., & Padilha França, R. (2019, October). Hardware modeling challenges regarding application-focused PCB designs in Industry 4.0 and IoT conceptual environments. In *Brazilian technology symposium* (pp. 489–498). Springer.

Diem, D. (2019). U.S. Patent No. 10,171,950. U.S. Patent and Trademark Office.
El-Din, H. E., & Manjaiah, D. H. (2017). Internet of nano things and industrial Internet of Things. In *Internet of Things: Novel advances and envisioned applications* (pp. 109–123). Springer.
França, R. P., Iano, Y., Monteiro, A. C. B., & Arthur, R. (2020a). Lower memory consumption for data transmission in smart cloud environments with CBEDE methodology. In *Smart systems design, applications, and challenges* (pp. 216–237). IGI Global.
França, R. P., Monteiro, A. C. B., Arthu, R., & Iano, Y. (2020b). The evolution of robotic systems: Overview and its application in modern times. In *Safety, security, and reliability of robotic systems* (pp. 1–20). CRC Press.
França, R. P., Monteiro, A. C. B., Arthur, R., & Iano, Y. (2020c). An overview of the integration between cloud computing and Internet of Things (IoT) technologies. In *Recent advances in security, privacy, and trust for Internet of Things (IoT) and cyber-physical systems (CPS)* (pp. 1–22). CRC Press.
França, R. P., Iano, Y., Monteiro, A. C. B., & Arthur, R. (2021a). Applying a methodology in data transmission of discrete events from the perspective of cyber-physical systems environments. In *Artificial intelligence paradigms for smart cyber-physical systems* (pp. 278–300). IGI Global.
França, R. P., Monteiro, A. C. B., Arthur, R., & Iano, Y. (2021b). An overview of the machine learning applied in smart cities. In *Smart cities: A data analytics perspective* (pp. 91–111). Springer Nature.
França, R. P., Monteiro, A. C. B., Arthur, R., & Iano, Y. (2021c). An overview of narrowband Internet of Things (NB-IoT) in the modern era. In *Principles and applications of Narrowband Internet of Things (NBIoT)* (pp. 26–45). IGI Global.
França, R. P., Monteiro, A. C. B., Arthur, R., & Iano, Y. (2021d). The fundamentals and potential for cybersecurity of big data in the modern world. In *Machine intelligence and big data analytics for cybersecurity applications* (pp. 51–73). Springer Nature.
França, R. P., Monteiro, A. C. B., Arthur, R., & Iano, Y. (2021e). An overview of deep learning in big data, image, and signal processing in the modern digital age. In *Trends in deep learning methodologies* (pp. 63–87). Academic Press.
França, R. P., Monteiro, A. C. B., Arthur, R., & Iano, Y. (2021f). An overview of the edge computing in the modern digital age. In *Fog/edge computing for security, privacy, and applications* (pp. 33–52). Springer.
Gilchrist, A. (2016). *Industry 4.0: The industrial internet of things*. Apress.
Hozdić, E. (2015). Smart factory for industry 4.0: A review. *International Journal of Modern Manufacturing Technologies, 7*(1), 28–35.
Indurkhya, N., & Damerau, F. J. (2011). Handbook of natural language processing. *Computational Linguistics, 37*(2), 395–397.
Jansen, J., & van Der Merwe, A. (2020). A framework for industrial internet of things. In *Responsible design, implementation, and use of information and communication technology: 19th IFIP WG 6.11 conference on e-business, e-services, and e-society, I3E 2020, Skukuza, South Africa, April 6–8, 2020, proceedings, Part I 19* (pp. 138–150). Springer.
Joshi, P. (2017). *Artificial intelligence with Python*. Packt Publishing Ltd.
Kiel, D., Müller, J. M., Arnold, C., & Voigt, K. I. (2017). Sustainable industrial value creation: Benefits and challenges of industry 4.0. *International Journal of Innovation Management, 21*(08), 1740015.
Lee, I., & Lee, K. (2015). The Internet of Things (IoT): Applications, investments, and challenges for enterprises. *Business Horizons, 58*(4), 431–440.
Lee, J., Davari, H., Singh, J., & Pandhare, V. (2018). Industrial artificial intelligence for Industry 4.0-based manufacturing systems. *Manufacturing Letters, 18*, 20–23.
Li, G., Hou, Y., & Wu, A. (2017). Fourth industrial revolution: Technological drivers, impacts and coping methods. *Chinese Geographical Science, 27*, 626–637.
Liu, Y., Yang, C., Jiang, L., Xie, S., & Zhang, Y. (2019). Intelligent edge computing for IoT-based energy management in smart cities. *IEEE Network, 33*(2), 111–117.

Madakam, S., & Uchiya, T. (2019). Industrial Internet of Things (IIoT): Principles, processes and protocols. In *The Internet of Things in the industrial sector: Security and device connectivity, smart environments, and Industry 4.0* (pp. 35–53). Springer.

Mahdavinejad, M. S., Rezvan, M., Barekatain, M., Adibi, P., Barnaghi, P., & Sheth, A. P. (2018). Machine learning for Internet of Things data analysis: A survey. *Digital Communications and Networks, 4*(3), 161–175.

Mohri, M., Rostamizadeh, A., & Talwalkar, A. (2018). *Foundations of machine learning*. MIT Press.

Mokyr, J., & Strotz, R. H. (1998). The second industrial revolution, 1870–1914. *Storia dell'economia Mondiale, 21945*(1).

Monteiro, A. C. B. (2023). *Proposta de novas metodologias de análise de células sanguíneas por meio dos métodos BSCM (Blood Smear Computacional Method) e BSIM (Blood Smear IntelligenceMethod): informática médica de baixo custo aplicada a saúde pública*. Doctoraldissertation, [sn].

Munirathinam, S. (2020). Industry 4.0: Industrial Internet of Things (IIOT). *Advances in Computers, 117*(1), 129–164. Elsevier.

Olivas, E. S., Guerrero, J. D. M., Martinez-Sober, M., Magdalena-Benedito, J. R., & Serrano, L. (Eds.). (2009). *Handbook of research on machine learning applications and trends: Algorithms, methods, and techniques*. IGI Global.

Padilha, R., Iano, Y., Monteiro, A. C. B., & Arthur, R. (2020). An AWGN channel data transmission proposal using discrete events for cloud and big data environments using metaheuristic fundamentals. In *Metaheuristics and optimization in computer and electrical engineering* (pp. 293–311). Springer.

Philbeck, T., & Davis, N. (2018). The fourth industrial revolution. *Journal of International Affairs, 72*(1), 17–22.

Pinon, M. M. B., Nascimento, M. H., de AB Junior, J., Tavares, T. F. D., & de Souza Silva, V. L. (2018). Applications and advantages of the Internet of Things (IoT) at industry (189–94). *ITEGAM-JETIA, 4*(15), 189–194.

Popescu, G. H. (2015). The economic value of the industrial Internet of Things. *Journal of Self-Governance and Management Economics, 3*(2), 86–91.

Radziwon, A., Bilberg, A., Bogers, M., & Madsen, E. S. (2014). The smart factory: Exploring adaptive and flexible manufacturing solutions. *Procedia Engineering, 69*, 1184–1190.

Rajaraman, V. (2017). Radiofrequency identification. *Resonance, 22*, 549–575.

Ramya, A., & Vanapalli, S. L. (2016). 3D printing technologies in various applications. *International Journal of Mechanical Engineering and Technology, 7*(3), 396–409.

Raschka, S., & Mirjalili, V. (2019). *Python machine learning: Machine learning and deep learning with Python, sci-kit-learn, and Tensor Flow 2*. Packt Publishing Ltd..

Schwab, K. (2017). *The fourth industrial revolution*. Currency.

Serpanos, D., & Wolf, M. (2018). Industrial internet of things. In *Internet-of-Things (IoT) systems: Architectures, algorithms, methodologies* (pp. 37–54). Springer.

Sisinni, E., Saifullah, A., Han, S., Jennehag, U., & Gidlund, M. (2018). Industrial internet of things: Challenges, opportunities, and directions. *IEEE Transactions on Industrial Informatics, 14*(11), 4724–4734.

Soldatos, J., Gusmeroli, S., Malo, P., & Di Orio, G. (2022). Internet of Things applications in future manufacturing. In *Digitising the industry Internet of Things connecting the physical, digital and virtual worlds* (pp. 153–183). River Publishers.

Thomassey, S., & Zeng, X. (2018). *Introduction: Artificial intelligence for fashion industry in the big data era* (pp. 1–6). Springer.

Usama, M., et al. (2019). Unsupervised machine learning for networking: Techniques, applications and research challenges. *IEEE Access, 7*, 65579–65615.

Yuan, Q. Y., Wen, D. L., & Chen, Z. X. (2020). *The framework of intelligent factory*. 2nd international conference on advanced control, automation and artificial intelligence (ACAAI 2020).

Zhang, D., Chan, C. C., & Zhou, G. Y. (2018). Enabling Industrial Internet of Things (IIoT) towards an emerging smart energy system. *Global Energy Interconnection, 1*(1), 39–47.

Ana Carolina Borges Monteiro is currently a postdoc at the Information Technology Center—CTI Renato Archer, where she is part of the Institutional Training Program (PCI), working on research related to health, Machine Learning, and Artificial Intelligence. She has a Ph.D. from the Faculty of Electrical and Computer Engineering (FEEC) at the State University of Campinas—UNICAMP, where she developed research projects related to health software with an emphasis on the development of methodologies for detecting and counting blood cells through digital processing techniques images, and artificial intelligence. She has experience in the areas of Molecular Biology and management of research animals, both acquired during a research internship at the Blood Center of the State University of Campinas (2016–2017). To date, she has more than 120 scientific publications such as book chapters, expanded abstracts, and articles published in national and international journals. In addition, she has one published book (Proceedings of the 5th Brazilian Technology Symposium).

Rodrigo Bonacin holds a Ph.D. in Computer Science from UNICAMP, Brazil, and did a postdoctoral at the Luxembourg Institute of Science and Technology. He is a Senior Researcher and the Head of the Computing Methodologies Division at Renato Archer Information Technology Center, Brazil, and a professor at UNIFACCAMP, Brazil. His research interests include Human—Computer Interaction, Semantic Web, Artificial Intelligence, Organizational Semiotics, Informatics in Education, and Medical Informatics.

Reinaldo Padilha França is currently a postdoc at the Information Technology Center—CTI Renato Archer, where he is part of the Institutional Training Program (PCI), working on research related to health, Machine Learning, and Artificial Intelligence. He has a Ph.D. in Electrical Engineering from the Faculty of Electrical Engineering and Computing at the State University of Campinas FEEC/UNICAMP, Campinas-SP. Since 2016, he has been developing research projects related to pre-coding and bit coding in AWGN channels, telecommunications channels, and digital image processing with an emphasis on blood cells. This research resulted, in 2019, in two registrations of computer programs with the INPI (National Institute of Industrial Property). To date, it has more than 120 national and international publications, distributed among book chapters, scientific articles, and expanded abstracts.

Rangel Arthur has a degree in Electrical Engineering from the Universidade Estadual Paulista Júlio de Mesquita Filho (1999), a master's degree (2002), and a Ph.D. (2007) in Electrical Engineering from the State University of Campinas (Unicamp). He is a Free Professor (MS5.2) at the Faculty of Technology at Unicamp. During the years 2011–2014, he was the Coordinator of the Telecommunications Systems Technology and Telecommunications Engineering Courses at FT, which was created under his management. From 2015 to 2016, he was Associate Director of the Faculty of Technology (FT) at Unicamp. Since 2016 he has been an Advisor to the Innovation Agency (Inova) at Unicamp. He has experience in the area of Electrical Engineering, working mainly on the following topics: intelligent systems, digital image processing, and embedded systems. He is a member of the Association of Innovation Environments of Limeira-SP and the Unicamp Business Incubator.

Chapter 6
Exploring Smart City Implementation and Hegemony: A Focus on Indonesia

Anang Dwi Santoso, Januar Eko Aryansah, and Annada Nasyaya

Abstract This study is conducted to investigate Smart City discourse analysis in Indonesia based on the hegemony theory of Antonio Gramsci. The intricate involvement of diverse stakeholders' hegemonic influence within the discourse and implementation of Smart City concepts as well as the consequential effects on the progression toward proficient and sustainable development are investigated. Therefore, this study examines the statements made by public officials, coupled with a thorough exploration of the interpretation and delineation of Smart City paradigm within a spectrum of legislative regulations. The integration of information and communication technology (ICT) into urban governance and its influence on the lives of urban residents is also explored. The significance extends to the indispensable insights into the multifaceted shaping of Smart City concept by diverse stakeholders. The results shows the emergence of novel modalities in urban management that transpire due to this process. By comprehending these intricate power dynamics, this study underscores the potential for various actors' hegemonic inclinations to foster or impede the transition toward a more efficient and sustainable Smart City paradigm.

Keywords Smart City · Gramsci's Hegemony Theory · Discourse · Information and Communication Technology · Urban Governance · Power relations

6.1 Background

Smart City concept is used and modified in both advanced and developing nations to address urban development objectives and directions. Many cities claimed to be Smart City, thereby sparking discussions among academics and practitioners across

A. D. Santoso (✉) · J. E. Aryansah · A. Nasyaya
Department of Public Administration, Faculty of Social and Political Sciences, Universitas Sriwijaya, Palembang, South Sumatra, Indonesia
e-mail: anangdwi@fisip.unsri.ac.id; januareko@fisip.unsri.ac.id; annadanasyaya@fisip.unsri.ac.id

S. Majumdar et al. (eds.), *Smart Cities*, S.M.A.R.T. Environments,
https://doi.org/10.1007/978-3-031-59846-3_6

various fields, including urban planning, information and communication technology (ICT), and public administration (Chauhan et al., 2016; Purnomo & Prabowo, 2016; Raaijen & Daneva, 2017; Ruhlandt, 2018; Scornavacca et al., 2020). Several studies suggested that Smart City enhanced public service quality and improved urban planning (Boes et al., 2016; Cowley et al., 2018; Stagars, 2016; Visvizi & Lytras, 2018; Zhang et al., 2022; Zhang, 2019). However, follow-up studies continue to question whether the concept genuinely makes cities smarter (Calzada & Cobo, 2015; Foth, 2017; Gonzaga, 2019; Hollands, 2008, 2015; Shelton et al., 2015).

This skepticism arises from the top-down and technocratic approach used as well as the lack of community involvement in decision-making (Hollands, 2015; Leitheiser & Follmann, 2020; Marek et al., 2017). The implementation of Smart City has been found to vary, particularly when this city faces bureaucratic challenges (Chou & Hsu, 2017; Colding & Barthel, 2017; Cook & Valdez, 2022; Deakin, 2014). This concept is translated, negotiated, and contextualized within different urban settings.

Indonesia also has the universal perspective that Smart City development is a priority. Therefore, the concept is integrated into urban growth, ministries, provinces, regencies, and villages. In Indonesia, it is implemented through the development of mobile-based applications to the fullest extent, and when these applications are created, where Smart City has been implemented (Anggadwita et al., 2020; Purnomo & Prabowo, 2016; Purwanto, 2018). However, these applications are often seen as ineffective and merely decorative. The implementation may not necessarily inspire changes in bureaucratic governance to become more efficient and community-focused. There are additional issues related to Smart City governance, such as unclear definitions, management fragmentation, and the unsustainability of mobile-based applications. Despite being distinct, this concept often overlaps with e-government due to the lack of a national framework for Smart City development.

This study analyzes Smart City discourse in Indonesia based on the hegemony theory of Gramsci. The primary focus is to investigate the statement of public officials as well as the definition and interpretation of Smart City within various legislative regulations. This study understands how the implementation of the concept contributes to enhancing efficiency and sustainability. Through this analysis, the transition toward Smart City is shaped by various actors, and new modalities of urban management emerge in this process. This study concentrates on ICT as crucial tools in Smart City development and integration into urban governance. The question pertains to the following: "How does the hegemony exhibited by different entities manifest within the discourse and implementation of Smart City initiatives in Indonesia? Additionally, how do these power dynamics serve to facilitate or conceivably impede the progression towards proficient and sustainable development and management?"

6.2 Literature Review

6.2.1 Smart City

Smart City concept is widely discussed globally, becoming a central topic in academic conversations and urban governance approaches in recent times. Even though the term is not new and emerged in the late 1990s as part of the movement for Smart City growth in the United States (Bollier, 1998), its increased popularity occurred after major technology companies such as IBM, Cisco, and Siemens started providing products and services to assist cities in becoming smarter (Praharaj & Han, 2019). The primary focus was on how ICT could take over the management of urban infrastructure and services (Calzada, 2020). Furthermore, the concept has evolved into a technology-based innovation in urban governance and operations (Shelton et al., 2015).

Smart City concept has also drawn various criticisms parallel to its popularity. Hollands (2008) criticized the deterministic approach taken by technology companies, while Neirotti et al. (2014) and Söderström et al. (2014) questioned the development agenda promoted by technology companies. Alternative definitions were provided focusing more on aspects such as creativity, human resources, education and learning, social inclusion, and governance. These definitions have also faced criticism, particularly for being considered vague and used inconsistently (Levenda, 2019).

Some groups also define Smart City from an urban planning and public policy perspective, where "smart" is seen as a normative claim about efficiency in urban management. Yigitcanlar (2015) argued that being smart was marked by the adoption of specific strategic policies by public organizations to differentiate new development policies and programs. According to supporters of this approach, the concept is linked to successful and robust projects within their jurisdictions focused on sustainable development, economic prosperity, and improving the life quality for the community (Parycek & Viale, 2017; Sáez-Martín et al., 2014; Sholeh et al., 2019). In practical terms, this is achieved by bringing together people, technology, and data to develop various types of applications in public services.

The terminology of "smart" has not only been used in the context of cities but has also permeated various aspects of human life, ranging from buildings, gadgets, sensors, GPS tracking facilities, homes, airports, education, and healthcare (Praharaj & Han, 2019). "Intelligence" has become a new domain within the evolving smart technology ecosystem, paralleling the dominance of ICT in human life (Bina et al., 2020; Kusuma & Supangkat, 2022). These efforts often involve the integration of intelligence into various spatial scales, executed through the establishment of command centers to monitor urban conditions and facilitate decision-making (Marciano, 2013).

One key characteristic of Smart City is the use of branding to attract investment and creative groups (Strüver et al., 2021). This creates a definition within the context of competitiveness, productivity, creative economy, urban marketing, and business-led urban development. Criticisms of this approach include the excessive emphasis on technology and an instrumental approach to urban governance, neglecting the roles of the indigenous community, local institutions, and civil community in urban development (Dolla & von Meding, 2022; Engelbert et al., 2019; Leitheiser & Follmann, 2020). Several studies emphasized the role of social infrastructure, including human and social capital, as vital components in Smart City, and facilitating human connections and relationship-building (Brorström et al., 2018; Herdiansyah, 2023).

Newer approaches are the integration of intelligence with sustainability in urban management. These efforts aim to synergize Smart City concept with sustainable development (Brorström et al., 2018; Ependi et al., 2022; Martin et al., 2018). Caragliu et al. (2011) defined Smart City as a place where investments in traditional infrastructure, social development, and ICT drive sustainable growth and improved life quality. This definition is among the most comprehensive since the roles of ICT, social capital, and human life quality are integrated. The current Smart City concept points toward two main orientations, namely physical aspects such as buildings, energy networks, mobility, and logistics, where ICT can play a crucial role, as well as nonphysical aspects such as education, culture, policy innovation, social inclusion, and governance, where the role of ICT is not dominant.

6.3 Smart City Hegemony

In discussing Gramsci's (1971) hegemony concept, it is crucial to explore how technology and data, as key elements, can become hegemonic tools. Technology, as the operational core, provides unprecedented power to gather, analyze, and manipulate data on urban life (Argento et al., 2020). This creates the potential for heightened surveillance and control over the lives of the community (Marciano, 2013). However, the dissemination of this technology is uneven and often nontransparent (Bulkeley & Stripple, 2021; Gandy & Nemorin, 2019). Elitist groups with access can strengthen their positions and control the narrative about Smart City (Astoria, 2021; Gandy Jr. & Nemorin, 2019). In this context, technology and data can become new hegemonic tools, facilitating the formation and maintenance of neoliberal hegemony in urban governance (Knebel et al., 2022).

The use of technology and data as hegemonic tools also triggers resistance. The urban community and activists employ the same technology to advocate for their interests, creating alternative narratives about Smart City and urban challenges (Hollands, 2015; Kitchin, 2015). Data can be used to expose and question claims on the efficiency and effectiveness of technology-based solutions (Datta, 2015). Therefore, technology and data also become a contested terrain where hegemony is fought for and resisted (Jazeel, 2015).

Gramsci's (1971) conceptualization of cultural heritage and change is also vital in discussing Smart City. Furthermore, Smart City involves technical or managerial changes and shifts in how people understand and experience urban life. In this context, Gramsci provided a useful framework for understanding how the neoliberal worldview becomes the "senso commune" or common sense in the community.

To reiterate, a deeper understanding of Smart City requires not only knowledge about technology and data but also an understanding of the social, political, and cultural dynamics surrounding them. This understanding can be enhanced through Gramsci's hegemony perspective, showing how power and dominance are manifested, contested, and transformed in urban contexts. This points toward a more democratic and inclusive vision of Smart City, where technology and data serve the interests of the overall urban community, not only dominant groups.

6.4 Methods

This study was conducted to explore the legislative regulations and public officials' statements regarding Smart City in Indonesia, along with the power relations, using corpus-based critical discourse analysis (CDA) (Cheng & Cheng, 2015). CDA was selected because the concept focused on explaining and understanding Smart City in Indonesia, as well as critiquing and transforming behaviors and institutions (Partington, 2012). The term "critical" referred to the use of analysis to uncover and integrate hidden ideologies treated as common and accepted (Stefanowitsch & Gries, 2007). These ideologies became widespread beliefs or even common sense (Shirazizadeh, 2019). This study emphasized the importance of placing text within its socio-historical context through the analysis of legislative regulations and public officials' statements along with the institutional, political, and social conditions governing the process.

The data were extracted from 45 legislative regulation definitions created by central and local governments as well as 23 public officials' statements. They were processed using the AntConc application, resulting in 220 and 1890 distinct and identified words. Meanwhile, the keyword corpus for Smart City definitions according to public officials comprised 237 distinct words from a set of 637 identified words.

This study employs 3 corpus linguistic techniques, namely semantic role (Kim & Klinger, 2018; Palmer et al., 2005), power relation (Huan, 2016), and frequent word analysis (Sotillo & Wang-Gempp, 2016; Wiegand & Mahlberg, 2019) to uncover themes inductively and show meanings within the text. The discursive themes were derived from a complex cultural process involving government, public officials, the community, private entities, and other actors. To allow flexibility in uncovering themes, this study manually created codes stemming from corpus linguistic analysis. This was intended to neglect events, metaphors, cultural resonances, narrative ideas, and organization.

6.5 Semantic Role Analysis

This study employed the semantic role analysis method to comprehend how various agents function within the context of Smart City. Furthermore, it examined two distinct text corpora, namely legislative regulations and public officials' statements.

In the legislative regulations corpus (Table 6.1), several agents such as Local Government, ICT, and City/Regency Community were depicted alongside their actions. Local Government played a role in managing various resources as well as providing infrastructure and services, while ICT assisted the urban community in managing resources efficiently.

In the corpus based on public officials' statements (Table 6.2), agents such as the Local Government, the Ministry of Communication and Information, and the Ministry of Home Affairs were also found to engage in various actions. Local Government played a role in developing a competitive and sustainable Smart City concept, while the Ministry of Communication and Information collaborated with other departments in Smart City program to enhance efficiency and improve public services.

6.6 Power Relation Analysis

This study conducted a power relation analysis to comprehend how various parties interact within Smart City context according to legislative regulations (Table 6.3) and public officials' statements (Table 6.4).

Based on Table 6.3, the Local Government provided infrastructure and services, and the Community participated in supervision to enhance life quality. Meanwhile, the Central Government regulated regulations and provided support to the Local Government for the development and implementation of Smart City concept.

Based on Table 6.4, the Government acted as a service provider, and the Community was the recipient within a power relation. The Government also

Table 6.1 Agents and actions in the corpus of legislative regulations on Smart City in Indonesia

Agents	Actions
Local Government	Manage various resources effectively and efficiently as well as provide infrastructure and services
Information and Communication Technology (ICT)	Assist the urban community in managing resources wisely and efficiently
City/Regency Community	Participate in governance, access information, and support efficient resource management
City/Regency Government	Use ICT to improve access to public services
Modern (Smart) City	Make innovation in the use of ICT and serve the community efficiently
Modern telecommunication	Realize sustainable economic growth and high life quality through wise resource management

Table 6.2 Agents and actions in Smart City Corpus according to public officials in Indonesia

Agents	Actions
Local Government	Develop a competitive and sustainable Smart City concept, optimize regency/city governance
Ministry of Communication and Informatics	Collaborate with other departments in Smart City program to increase efficiency and improve public services
Ministry of Home Affairs	Collaborate in the program to improve efficiency and public services
Ministry of Public Works and Human Settlements	Participate in infrastructure development within Smart City framework
Community	Communicate with the government and take advantage of convenience in carrying out activities, get the right information, and participate in regency/city development
Business actors	Take part in solving problems or needs in Smart City concept using information technology
Other stakeholders	Involved in providing solutions using information technology in Smart City
Institution	Obtain assistance in managing resources efficiently and get the right information in implementing activities
Information technology	Optimize services to the community, help to manage natural and human resources efficiently and effectively, overcome problems through innovation, and promote digitalization in various sectors in cities

Table 6.3 Power relations in the definition of Smart City according to legislative regulations

Involved parties	Power relations	Description
Local Government and Community	Participation and supervision	The Local Government provides Smart City infrastructure and services, while the community is involved in participation and supervision to improve the life quality
Central and Local Governments	Regulation and support	The Central Government regulates regulations as well as provides support and resources to the Local Government for the development and implementation of Smart City concept
Private companies and Local Government	Cooperation and investment	Private companies collaborate with the Local Government in technology investment and implementation as well as innovation in Smart City management
Community and technology	Access and utilization	The community uses the technology provided in the context of Smart City to improve efficiency and access to services, such as public transportation and health
Nongovernmental organizations (NGOs) and Government	Advocacy and partnerships	NGOs play a role in advocating public interest and partnering with the government in the planning and implementation of sustainable and inclusive Smart City

Table 6.4 Power relations in the definition of Smart City according to public officials in Indonesia

Involved parties	Power relations	Description
Government and community	The government and community as a service provider and recipient	In Smart City concept, the government plays a role as a provider of services and infrastructure through digital technology. Meanwhile, the community acts as users of these services, such as public services and information provided
Government and business sector	The government and business sector as the regulator and executor	The government makes Smart City policy and program, while the business sector can play a role in technology implementation and innovation needed
Government (Ministry of Communication and Informatics, Ministry of Home Affairs, Ministry of Public Works and Human Settlements, and others)	Inter-ministerial collaboration	Several ministries collaborate to create and run Smart City program, dividing roles and responsibilities based on their respective areas of authority
Government and institutions	The government and institution as the data provider and users	In Smart City concept, the government provides data and information through digital systems used by institutions to increase efficiency and effectiveness in carrying out their functions and duties
Community and technology	The community and technology as a user and tool	In Smart City concept, the community uses technology as a tool to obtain services and information from the government, as well as to interact with the government and the surrounding environment

assumed the role of regulator, while the Business Sector acted as executor in the implementation of technology and innovation.

6.7 Frequent Word Analysis

This study also conducted a frequent word analysis in the corpus of legislative regulations and public officials' statements concerning Smart City (Tables 6.5 and 6.6).

Based on Table 6.5, prominent themes within the corpus of legislative regulations included Technology, City and Urbanization, Efficiency and Energy, Community and Social, Environment and Sustainability, and Management and Organization. The most frequently appearing words within these themes included "smart," "technology," "city," "efficient," "community," "sustainable," and "manage."

According to Table 6.6, within the corpus of public officials' statements, frequently occurring words were associated with themes such as Technology, Governance, Community Welfare, City and Regency, Efficiency and Sustainability,

Table 6.5 Frequent words in the corpus of legislative regulations

Theme	Word representation (Freq/NormFreq)
Technology	Smart (52/27,513,228), *teknologi* (technology) (22/11,640,212), *tik* (ict) (5/2,645,503)
City and urbanization	*Kota* (city) (76/40,211,640), *infrastruktur* (infrastructure) (12/6,349,206), *pembangunan* (development) (5/2,645,503), *perkotaan* (urban) (2/1,058,201)
Efficiency and energy	*Efisien* (efficient) (33/17,460,317), *daya* (power) (37/19,576,720), *efisiensi* (efficiency) (13/6,878,307), *efesiensi* (efficiency) (1/529,101)
Community and social	*Masyarakat* (community) (36/19,047,619), *layanan* (services) (20/10,582,011), *komunikasi* (communication) (17/8,994,709), *kualitas* (quality) (17/8,994,709), *pelayanan* (services) (14/7,407,407), *publik* (public) (3/1587 0.302), *rakyat* (people) (1/529.101)
Environment and sustainability	*Berkelanjutan* (sustainable) (36/19,047,619), *lingkungan* (environment) (1/529,101), environment (1/529,101)
Management and organization	*Mengelola* (manage) (22/11,640,212), *pengelolaan* (management) (19/10,052,910), *manajemen* (management) (1/529,101), *operasi* (operations) (3/1587,302), *operasional* (operational) (1/529,101)

Table 6.6 Frequent words in the corpus of public officials

Theme	Word representation (Freq/NormFreq)
Technology	Smart (24/37,676,609), *informasi* (information) (9/14,128,728), *teknologi* (technology) (8/12,558,870), *digitalisasi* (digitalization) (3/4,709,576), internet (1/1,569,859)
Government	*Pemerintah* (government) (6/9,419,152), *kementerian* (ministry) (2/3,139,717), *pemerintahan* (government) (2/3,139,717)
Public welfare	*Masyarakat* (community) (18/28,257,457), *kesejahteraan* (welfare) (5/7,849,294), *kualitas* (quality) (2/3,139,717), *kehidupan* (life) (1/1,569,859)
City and regency	City (23/36,106,750), *Kota* (city) (17/26,687,598), *kabupaten* (regency) (6/9,419,152), *perkotaan* (urban) (3/4,709,576)
Efficiency and sustainability	*Efektif* (effective) (3/4,709,576), *efisien* (efficient) (3/4,709,576), *efisiensi* (efficiency) (3/4,709,576), *berkelanjutan* (sustainable) (6/9,419,152), *energi* (energy) (1/1,569,859)
Communication	*Komunikasi* (communication) (4/6,279,435), *berkomunikasi* (communicate) (1/1,569,859), *jaringan* (network) (1/1,569,859)
Public services	*Pelayanan* (services) (9/14,128,728), *layanan* (services) (3/4,709,576), *publik* (public) (7/10,989,011), *melayani* (serve) (1/1,569,859)
Development	*Pembangunan* (development) (3/4,709,576), *pengembangan* (development) (1/1,569,859), *upaya* (effort) (5/7,849,294)

Communication, Public Services, and Development. The words with the most appearance in this context were "smart," "government," "community," "city," "effective," "communication," "services," and "development."

Based on the analyses presented in the preceding 6 tables, 4 main themes for Smart City discourse have been successfully identified.

The first theme is "Technology Hegemony in Smart City Discourse." This theme underscores the significance of technology, particularly ICT, in defining and materializing Smart City concept. Technology is seen as a tool that aids the government and urban community in managing resources efficiently and effectively as well as helping realize innovation in public services.

The second theme is "Hegemony and Counter-Hegemony in the Government-Community Relationship in Smart City Management." In this theme, the power dynamics between the government and the community take center stage. The government assumes the role of providing infrastructure and services, while the community engages in participation and supervision in Smart City management.

The third theme, "Dynamics of Hegemony in the Role of Private Sector and Non-Governmental Organizations in Smart City Development and Management," highlights the role of these parties. The role of investment, technology implementation, and advocation for public interests is a crucial aspect.

The fourth theme, "Hegemony and Counter-Hegemony in the Implementation of Smart City as an Effort to Enhance Efficiency and Sustainability," focuses on how Smart City is implemented to improve efficiency and sustainability. The implementation is directed toward building a city more resource-efficient and sustainable across economic, social, and environmental aspects.

6.8 Results and Discussion

6.8.1 Theme 1: Technological Hegemony in Smart City Discourse

The concept of Smart City is relevant when examining how technology, particularly ICT, has become hegemonic in defining and shaping urban discourse and practices. Smart City discourse analysis, as presented in Tables 6.1 and 6.2, shows the dominance of technology within the discussions. Terms such as "smart," "technology," "ICT," "information," "digitalization," and "internet" prevail in the discourse, underscoring how technology has become central to the definition and understanding of Smart City.

As shown in Tables 6.3 and 6.4, technology serves as the primary tool to achieve efficient resource management, broader access to public services, and driving innovation. In this situation, it empowers, regulates, monitors, and directs urban life (Lee et al., 2022; McFarlane & Söderström, 2017; Nam & Pardo, 2011). Technology becomes a dominant entity within the existing power relations of the community, as depicted in Tables 6.5 and 6.6. The community interacts with technology as service users, showing how technology influences their daily lives (Engelbert et al., 2019; Kitchin et al., 2019).

In Gramsci's framework, this phenomenon can be understood as technology hegemony (Gras & Hernández, 2016; Power Wogu et al., 2020). In this context, technology is not merely a tool or system, but an entity possessing power and influence over urban discourse and practices (Marciano, 2013; Perkins, 2009). In

Gramsci's terminology, it holds the position of an "organic intellectual group," shaping and influencing the thoughts and actions of the community. However, this technology hegemony also raises several challenges and issues. Firstly, there is the risk of a "digital divide," where segments of the community lacking access or the ability to use technology are marginalized within Smart City concept. Secondly, there is the risk of "technology determinism," where technology becomes the sole determinant and driver of urban change, while other factors such as social, economic, cultural, and environmental aspects are disregarded (Engelbert et al., 2019; Kitchin et al., 2019).

Resistance to technology hegemony can take the form of a "war of position" or "counter-hegemony." This involves the formation of a "historical bloc" that opposes and challenges technology dominance as well as creates inclusive, participatory, and sustainable alternatives. This involves various strategies, such as digital education and training for the community, the development of user-friendly and inclusive technology, as well as active community participation in decision-making and urban management processes. Additionally, it is important to consider how technology can collaborate with other elements within the community as a dominant and controlling entity. This involves the integration of technology with other aspects of urban life, such as economics, social dynamics, culture, and the environment.

6.8.2 *Theme 2: Hegemony and Counter-Hegemony in the Relationship Between Government and Community in Smart City Management*

The evolution of Smart City concept has changed the landscape of the relationship between government and community. The government as a regulatory entity and infrastructure provider and the community as users and key stakeholders are engaged in a dynamic influenced by the penetration of technology into urban life. In this context, applying an analysis based on Gramsci's hegemony theory is crucial to understanding and critiquing the existing power structures, as well as exploring ways to resist and transform these structures.

Government plays a pivotal role in Smart City approach. According to Tables 6.3, 6.4, 6.5, and 6.6, this includes functions such as a regulator, infrastructure provider, as well as policy and regulation enforcer. In this context, standards and guidelines are set for the implementation of technology, infrastructure, and services. This also includes the creation and enforcement of regulations and policies that shape the functioning and use of technology within the urban context.

Meanwhile, the community assumes a passive and subordinate role in this context. They function as service users as well as interact with technology and government, as shown in Tables 6.3 and 6.4. Even though the community has a role in resource management and service access, participation is often limited by power structures established by the government and technology.

This analysis outlines how this relationship embodies forms of hegemony within Gramsci's understanding. Government, through its control over infrastructure and regulations, dominates and directs the implementation and operationalization (Kitchin et al., 2019). Even though the community is a primary stakeholder, they often find themselves in a subordinate and reactive position, dependent on the government and technology for access and services (Engelbert et al., 2019).

Gramsci's counter-hegemony theory offers a path to challenge and transform existing power structures (Madyatmadja et al., 2020; Shihab & Hidayanto, 2021; Zhu & Alamsyah, 2022). In the context of Smart City, several ways can be performed by the community to resist the prevailing hegemony.

Firstly, the community can enhance their access and skills in using technology. This involves digital education and training as well as improvements in infrastructure and services (Cardullo & Kitchin, 2019). Elevating access and skills empowers the community to participate more actively in urban management and challenge the functioning of technology and services (Sweeting et al., 2022).

Secondly, the community can engage more actively in policymaking and regulatory processes. This entails heightened participation in public forums, consultations, and decision-making processes (van Twist et al., 2023; Zhu et al., 2022). Therefore, the community can have a voice in the workings of Smart City as well as can challenge and alter existing structures and policies.

Thirdly, the community can create and employ alternative technology, better suited to their needs and circumstances. This may involve the development of community and social technology (Sweeting et al., 2022). The community can create and control their technology and services, while also challenging and transforming the workings of Smart City.

6.8.3 *Theme 3: Dynamics of Hegemony in the Role of the Private Sector and Nongovernmental Institutions in Smart City Development and Management*

In Smart City ecosystem, the role of the private sector and NGOs has become increasingly pivotal in shaping the direction and dynamics of development. The private sector, consisting of various technology companies and industries, plays a crucial role in investment, technology implementation, and innovation (Tables 6.5 and 6.6). Meanwhile, NGOs strive to advocate for public interests and partner with the government in sustainable and inclusive planning and implementation (Tables 6.5 and 6.6).

In examining the interaction between these stakeholders, it is important to identify forms of hegemony and counter-hegemony within Gramsci's context. This understanding is crucial to creating a truly inclusive Smart City ecosystem that serves the needs of all parties, specifically the general community.

The private sector wields significant power in determining the direction and dynamics of Smart City. The hegemony in this context is often driven by economic and profit motives, which can create dynamics where technology and innovation are controlled by market needs (Liu et al., 2021; Sadowski & Bendor, 2019). The role of NGOs becomes crucial as a counter-hegemony to challenge this dominance as well as promote public interests in Smart City development and management (Ngo & Le, 2021; Paskaleva et al., 2021).

NGOs can achieve this matter in various ways, such as by advocating and providing education to enhance the understanding of Smart City and its impact on their lives (Adelino et al., 2021; Suryaningsih & Chrisdanty, 2018). Transparency and accountability can also be promoted in the decision-making process to ensure that public interests are considered. Additionally, NGOs can advocate for a larger role for the community in this process through public participation initiatives and social innovations (Shokheh et al., 2021).

The counter-hegemonic role cannot be overlooked since NGOs are often constrained by existing power dynamics and resources. For example, financial and human resources limitations, or facing barriers in accessing information and decision-making processes. It is important to recognize and challenge these limitations as well as find ways to strengthen the role of NGOs and the community.

6.8.4 Theme 4: Hegemony and Counter-Hegemony in Smart City Implementation as an Effort to Increase Efficiency and Sustainability

Smart City concept, defined by the utilization of technology and efficient resource management to achieve sustainability goals, has become a central notion in the discourse of modern urban development. Terms such as "efficient," "power," "efficiency," and "sustainability" often dominate these discussions (Tables 6.1 and 6.2). Furthermore, wise resource management, technology innovation, as well as efforts to foster sustainable economic growth and enhance life quality emerge as key themes in Smart City implementation (Tables 6.3 and 6.4).

This study discovers that Smart City concept is frequently shaped by narratives of efficiency and sustainability constructed by hegemonic forces such as government, technology companies, and international financial institutions through Gramsci's hegemony theory (Pratama, 2018; Sadowski & Bendor, 2019).

In the context of efficiency, technology is often considered the primary solution for optimizing resource utilization (Lu et al., 2021; Sanchez-Iborra et al., 2020). In this process, technology and related companies often assume dominant positions, while the significant role of the community in resource use and management can be overlooked (Kumar et al., 2021).

Economic development is seen as the primary objective within the context of sustainability, assuming economic growth automatically enhances life quality and environment. However, economic development can also lead to social inequality and environmental degradation, and sustainability requires a more holistic and inclusive approach (Brorström et al., 2018; Ependi et al., 2022; Mahesa et al., 2019).

In this context, there is a need for counter-hegemonic efforts to challenge and broaden the definitions and practices of efficiency and sustainability within Smart City. For example, the community, NGOs, and academics can play a role in challenging the dominance of technology and economic development as well as promoting a more inclusive and holistic understanding of efficiency and sustainability. Community participation can be promoted in decision-making and resource management, can promote more inclusive and democratic approaches to technology development and usage, as well as advocate for development approaches prioritizing social and environmental justice.

6.9 Conclusion

Smart City development was a main concern for numerous governments and policymakers around the globe. This study employed Antonio Gramsci's hegemony theory (1971) as a framework to understand this phenomenon and its implications in the socio-technological context.

First, this study examined the role of technology in the construction of Smart City. The emphasis on technology in Smart City discourse represented the idea of Schuurman et al. (2012) where "technology capitalist ideology" is centered around the pursuit of efficiency and maximum profit. In this context, technology is seen as a tool and discourse with impacts beyond its technical aspects. It becomes a means to dominate and control, realizing hegemony in Smart City discourse. The risk of this dominance, as highlighted by Vanolo (2014), was the creation of a digital divide leading to exclusion from the benefits of Smart City for the community lacking adequate access or digital literacy.

In Gramsci's theoretical framework, hegemony was a dynamic and ever-changing process. Therefore, the role of government and community in managing and using technology became key in negotiating and influencing its development direction. Through active participation in this process, Smart City discourse and practices were reshaped, reflecting their needs and aspirations, a perspective echoed by Cardullo and Kitchin (2019).

In this analysis, it was important to consider the role played by the private sector and NGOs. According to Hollands (2008), these actors, through investment and advocacy, shaped and negotiated the vision and implementation of Smart City. In this context, their role was not overlooked, contributing to the realization of an inclusive and sustainable vision.

This study also addressed the themes of efficiency and sustainability in Smart City context. Bibri (2018) identified that these themes often took center stage in Smart City discourse. Even though efficiency and sustainability were vital slogans, an excessive focus on technology efficiency led to social and environmental imbalances. There was a need for a deeper understanding of "sustainability" in Smart City context, where sustainable development entailed technological efficiency and social inclusivity, and environmental justice.

This study provided implications regarding the perspectives and approaches needed in designing and implementing Smart City concept. Based on Gramsci's hegemony theory, Smart City implementation was not a neutral process and a hegemonic struggle among various actors and interests. In this context, technology was seen as a tool and discourse with impacts beyond its technical aspects. Therefore, the role of government, community, and private sector in managing and utilizing technology became pivotal in negotiating and influencing its development direction. This underscored the importance of multistakeholder engagement in inclusive and sustainable development.

Concerning the limitations, the use of Gramsci's hegemony theory might overlook other aspects of Smart City development, such as technical and economic dimensions. Second, this study was more focused on discourse analysis and overlooked real-world development practices and user experiences. Third, this analysis was centered on the general Smart City concept and did not fully reflect the contextual variations and specifics of particular cases. Subsequent investigations were necessary to grasp the diverse roles that various stakeholders could assume in this process. Additionally, an examination of the resulting implications for policymaking and the enhancement of practices in Smart City development was conducted.

References

Adelino, R., Silveira, C., & Reis, L. (2021). ZeroWaste. *International Journal of Entrepreneurship and Governance in Cognitive Cities, 2*(1), 61. https://doi.org/10.4018/ijegcc.2021010105

Anggadwita, G., Rikumahu, B., Hendayani, R., & Putra, R. R. (2020). Measuring the effectiveness of social media owned by local government leaders in communicating smart city programs. In *2020 8th International conference on information and communication technology, ICoICT 2020*. https://doi.org/10.1109/ICoICT49345.2020.9166182

Argento, D., Grossi, G., Jääskeläinen, A., Servalli, S., & Suomala, P. (2020). Governmentality and performance for the smart city. *Accounting, Auditing and Accountability Journal, 33*(1), 204–232. https://doi.org/10.1108/AAAJ-04-2017-2922

Astoria, R. (2021). Greenhouse gas mitigation in smart cities: Political economy and strategic mitigation alliances. In *Handbook of smart cities* (pp. 621–649). Springer International Publishing. https://doi.org/10.1007/978-3-030-69698-6_25

Bibri, S. E. (2018). The IoT for smart sustainable cities of the future: An analytical framework for sensor-based big data applications for environmental sustainability. *Sustainable Cities and Society, 38*, 230. https://doi.org/10.1016/j.scs.2017.12.034

Bina, O., Inch, A., & Pereira, L. (2020). Beyond techno-utopia and its discontents: On the role of utopianism and speculative fiction in shaping alternatives to the smart city imaginary. *Futures, 115*, 102475. https://doi.org/10.1016/j.futures.2019.102475

Boes, K., Buhalis, D., & Inversini, A. (2016). Smart tourism destinations: Ecosystems for tourism destination competitiveness. *International Journal of Tourism Cities, 2*(2), 108–124. https://doi.org/10.1108/IJTC-12-2015-0032

Bollier, D. (1998). *How smart growth can stop sprawl: A fledgling citizen movement expands*. Essential Books.

Brorström, S., Argento, D., Grossi, G., Thomasson, A., & Almqvist, R. (2018). Translating sustainable and smart city strategies into performance measurement systems. *Public Money and Management, 38*(3), 193–202. https://doi.org/10.1080/09540962.2018.1434339

Bulkeley, H., & Stripple, J. (2021). Climate Smart City: New cultural political economies in the making in Malmö, Sweden. *New Political Economy, 26*(6), 937–950. https://doi.org/10.1080/13563467.2020.1810219

Calzada, I. (2020). Replicating smart cities: The city-to-city learning programme in the replicate EC-H2020-SCC project. *Smart Cities, 3*(3), 978–1003. https://doi.org/10.3390/smartcities3030049

Calzada, I., & Cobo, C. (2015). Unplugging: Deconstructing the smart city. *Journal of Urban Technology, 22*(1), 23–43. https://doi.org/10.1080/10630732.2014.971535

Caragliu, A., del Bo, C., & Nijkamp, P. (2011). Smart cities in Europe. *Journal of Urban Technology, 18*(2), 65–82. https://doi.org/10.1080/10630732.2011.601117

Cardullo, P., & Kitchin, R. (2019). Smart urbanism and smart citizenship: The neoliberal logic of 'citizen-focused' smart cities in Europe. *Environment and Planning C: Politics and Space, 37*(5), 813. https://doi.org/10.1177/0263774X18806508

Chauhan, S., Agarwal, N., & Kar, A. K. (2016). Addressing big data challenges in smart cities: A systematic literature review. *Info, 18*(4), 73–90. https://doi.org/10.1108/info-03-2016-0012

Cheng, W., & Cheng, W. (2015). Corpus-based linguistic approaches to critical discourse analysis. In *The Encyclopedia of applied linguistics* (pp. 1353–1360). https://doi.org/10.1002/9781405198431.wbeal0262

Chou, T. -Y., & Hsu, Y. -T. (2017). The strategic development and spatial information applications of smart cities in Taiwan. In W. D.-W. C. L.-J. Chen S.S.-J. Lin F.-T. (Ed.), *Proceedings of the 2017 Pacific neighborhood consortium annual conference and joint meetings: Data informed society, PNC 2017* (Vols. 2017-Decem, pp. 1–7). Institute of Electrical and Electronics Engineers Inc. https://doi.org/10.23919/PNC.2017.8203515

Colding, J., & Barthel, S. (2017). An urban ecology critique on the "Smart City" model. *Journal of Cleaner Production, 164*, 95–101. https://doi.org/10.1016/j.jclepro.2017.06.191

Cook, M., & Valdez, M. (2022). Curating smart cities. *Urban Geography., 44*, 1192. https://doi.org/10.1080/02723638.2022.2072077

Cowley, R., Joss, S., & Dayot, Y. (2018). The smart city and its publics: Insights from across six UK cities. *Urban Research and Practice, 11*(1), 53–77. https://doi.org/10.1080/17535069.2017.1293150

Datta, A. (2015). A 100 smart cities, a 100 utopias. *Dialogues in Human Geography, 5*(1), 49–53. https://doi.org/10.1177/2043820614565750

Deakin, M. (2014). Smart cities: The state-of-the-art and governance challenge. *Theoretical Chemistry Accounts, 1*(1). https://doi.org/10.1186/s40604-014-0007-9

Dolla, T., & von Meding, J. (2022). The rhetoric of Smart City—A conceptual critique of Indian and the US developments. In *International conference on advances in civil engineering, ACE 2020* (Vol. 172, pp. 75–84). Springer Science and Business Media Deutschland GmbH. https://doi.org/10.1007/978-981-16-4396-5_8

Engelbert, J., van Zoonen, L., & Hirzalla, F. (2019). Excluding citizens from the European smart city: The discourse practices of pursuing and granting smartness. *Technological Forecasting and Social Change, 142*, 347–353. https://doi.org/10.1016/j.techfore.2018.08.020

Ependi, U., Rochim, A. F., & Wibowo, A. (2022). Smart City assessment for Sustainable City development on smart governance: A systematic literature review. In *2022 international conference on decision aid sciences and applications, DASA 2022* (pp. 1088–1097). https://doi.org/10.1109/DASA54658.2022.9765192

Foth, M. (2017). Lessons from urban guerrilla placemaking for smart city commons. In *8th international conference on communities and technologies: Technologies for the Common Good, C and T 2017, part F1285* (pp. 32–35). https://doi.org/10.1145/3083671.3083707

Gandy, O. H., Jr., & Nemorin, S. (2019). Toward a political economy of nudge: Smart city variations. *Information Communication and Society, 22*(14), 2112–2126. https://doi.org/10.1080/1369118X.2018.1477969
Gonzaga, E. (2019). Precarious nostalgia in the tropical smart city: Transmedia memory, urban informatics, and the Singapore Golden Jubilee. *Cultural Studies, 33*(1), 147–169. https://doi.org/10.1080/09502386.2018.1428644
Gramsci, A. (1971). In Q. Hoare & G. Nowell Smith (Eds.), *Selections from the prison notebooks*. International Publishers.
Gras, C., & Hernández, V. (2016). Hegemony, technological innovation and corporate identities: 50 years of agricultural revolutions in Argentina. *Journal of Agrarian Change, 16*(4), 675. https://doi.org/10.1111/joac.12162
Herdiansyah, H. (2023). Smart city based on community empowerment, social capital, and public trust in urban areas. *Global Journal of Environmental Science and Management, 9*(1), 113–128. https://doi.org/10.22034/gjesm.2023.01.09
Hollands, R. G. (2008). Will the real smart city please stand up? Intelligent, progressive or entrepreneurial? *City, 12*(3), 303–320. https://doi.org/10.1080/13604810802479126
Hollands, R. G. (2015). Critical interventions into the corporate smart city. *Cambridge Journal of Regions, Economy and Society, 8*(1), 61–77. https://doi.org/10.1093/cjres/rsu011
Huan, C. (2016). Journalistic engagement patterns and power relations: Corpus evidence from Chinese and Australian hard news reporting. *Discourse & Communication, 10*(2), 137. https://doi.org/10.1177/1750481315611239
Jazeel, T. (2015). Utopian urbanism and representational city-ness: On the Dholera before Dholera smart city. *Dialogues in Human Geography, 5*(1), 27–30. https://doi.org/10.1177/2043820614565866
Kim, E., & Klinger, R. (2018). Who feels what and why? Annotation of a literature corpus with semantic roles of emotions. In *COLING 2018—27th international conference on computational linguistics, proceedings*.
Kitchin, R. (2015). Making sense of smart cities: Addressing present shortcomings. *Cambridge Journal of Regions, Economy and Society, 8*(1), 131–136. https://doi.org/10.1093/cjres/rsu027
Kitchin, R., Cardullo, P., & Feliciantonio, C. D. (2019). Citizenship, justice, and the right to the Smart City. In *The right to the Smart City* (pp. 1–24). Emerald Group Publishing Ltd.. https://doi.org/10.1108/978-1-78769-139-120191001
Knebel, N. M. P., De Oliveira Fornasier, M., & Borges, G. S. (2022). Political economy of smart cities and the human rights: From corporative technocracy to sensibility. *Brazilian Journal of International Law, 19*(1), 258–273. https://doi.org/10.5102/rdi.v19i1.8096
Kumar, M., Raju, K. S., Kumar, D., Goyal, N., Verma, S., & Singh, A. (2021). An efficient framework using visual recognition for IoT based smart city surveillance. *Multimedia Tools and Applications, 80*(20), 31277. https://doi.org/10.1007/s11042-020-10471-x
Kusuma, A. T., & Supangkat, S. H. (2022). Metaverse fundamental technologies for smart city: A literature review. In *9th international conference on ICT for smart society, ICISS 2022*. https://doi.org/10.1109/ICISS55894.2022.9915079
Lee, J., Babcock, J., Pham, T. S., Bui, T. H., & Kang, M. (2022). Smart city as a social transition towards inclusive development through technology: A tale of four smart cities. *International Journal of Urban Sciences*. https://doi.org/10.1080/12265934.2022.2074076
Leitheiser, S., & Follmann, A. (2020). The social innovation–(re)politicisation nexus: Unlocking the political in actually existing smart city campaigns? The case of SmartCity Cologne, Germany. *Urban Studies, 57*(4), 894–915. https://doi.org/10.1177/0042098019869820
Levenda, A. M. (2019). Thinking critically about smart city experimentation: Entrepreneurialism and responsibilization in urban living labs. *Local Environment, 24*(7), 565–579. https://doi.org/10.1080/13549839.2019.1598957
Liu, W., Xu, Y., Fan, D., Li, Y., Shao, X. F., & Zheng, J. (2021). Alleviating corporate environmental pollution threats toward public health and safety: The role of smart city and artificial intelligence. *Safety Science, 143*, 105433. https://doi.org/10.1016/j.ssci.2021.105433

Lu, C. W., Huang, J. C., Chen, C., Shu, M. H., Hsu, C. W., & Tapas Bapu, B. R. (2021). An energy-efficient smart city for sustainable green tourism industry. *Sustainable Energy Technologies and Assessments, 47*, 101494. https://doi.org/10.1016/j.seta.2021.101494
Madyatmadja, E. D., Nindito, H., & Pristinella, D. (2020). Citizen behavior: The evaluation of complaint application that connected to smart city. *Advances in Science, Technology and Engineering Systems, 5*(4), 24–29. https://doi.org/10.25046/aj050403
Mahesa, R., Yudoko, G., & Anggoro, Y. (2019). Dataset on the sustainable smart city development in Indonesia. *Data in Brief, 25*, 104098. https://doi.org/10.1016/j.dib.2019.104098
Marciano, C. (2013). Unpacking a smart city model: The hegemony of ecological and information paradigms in urban space. *International Journal of Interdisciplinary Environmental Studies, 7* (3), 1–12. https://doi.org/10.18848/2329-1621/CGP/v07i03/53299
Marek, L., Campbell, M., & Bui, L. (2017). Shaking for innovation: The (re)building of a (smart) city in a post disaster environment. *Cities, 63*, 41–50. https://doi.org/10.1016/j.cities.2016.12.013
Martin, C. J., Evans, J., & Karvonen, A. (2018). Smart and sustainable? Five tensions in the visions and practices of the smart-sustainable city in Europe and North America. *Technological Forecasting and Social Change, 133*, 269–278. https://doi.org/10.1016/j.techfore.2018.01.005
McFarlane, C., & Söderström, O. (2017). On alternative smart cities: From a technology-intensive to a knowledge-intensive smart urbanism. *City, 21*(3–4), 312–328. https://doi.org/10.1080/13604813.2017.1327166
Nam, T., & Pardo, T. A. (2011). Conceptualizing smart city with dimensions of technology, people, and institutions. In *Proceedings of the 12th annual international digital government research conference: Digital Government innovation in challenging times* (pp. 282–291). https://doi.org/10.1145/2037556.2037602
Neirotti, P., De Marco, A., Cagliano, A. C., Mangano, G., & Scorrano, F. (2014). Current trends in smart city initiatives: Some stylised facts. *Cities, 38*, 25. https://doi.org/10.1016/j.cities.2013.12.010
Ngo, H. V., & Le, Q. (2021). Smart city: An approach from the view of smart urban governance. *International Journal of Sustainable Construction Engineering and Technology, 12*(1), https://doi.org/10.30880/ijscet.2021.12.01.029
Palmer, M., Kingsbury, P., & Gildea, D. (2005). The proposition bank: An annotated corpus of semantic roles. *Computational Linguistics, 31*(1), 71. https://doi.org/10.1162/0891201053630264
Partington, A. (2012). *Corpora: Corpus-based language learning, language processing and linguistics*. Edinburgh University Press.
Parycek, P., & Viale, G. (2017). *Drivers of smart governance: Towards to evidence-based policy-making: Vol. Part F1282* (H. C.C. & O. A., Eds.; pp. 564–565). Association for Computing Machinery. https://doi.org/10.1145/3085228.3085255
Paskaleva, K., Evans, J., & Watson, K. (2021). Co-producing smart cities: A quadruple helix approach to assessment. *European Urban and Regional Studies, 28*(4), 395. https://doi.org/10.1177/09697764211016037
Perkins, H. A. (2009). Out from the (green) shadow? Neoliberal hegemony through the market logic of shared urban environmental governance. *Political Geography, 28*(7), 395–405. https://doi.org/10.1016/j.polgeo.2009.09.007
Power Wogu, I. A., Njie, S. N. N., Katende, J. O., Ukagba, G. U., Edogiawerie, M. O., & Misra, S. (2020). The social media, politics of disinformation in established hegemonies, and the role of technological innovations in 21st century elections: The road map to US 2020 presidential elections. *International Journal of Electronic Government Research, 16*(3), 65. https://doi.org/10.4018/IJEGR.2020070104
Praharaj, S., & Han, H. (2019). Cutting through the clutter of smart city definitions: A reading into the smart city perceptions in India. *City, Culture and Society, 18*, 100289. https://doi.org/10.1016/j.ccs.2019.05.005
Pratama, A. B. (2018). Smart city narrative in Indonesia: Comparing policy documents in four cities. *Public Administration Issues, 2018*(6), 65–83. https://doi.org/10.17323/1999-5431-2018-0-6-65-83

Purnomo, F., & Prabowo, H. (2016). Smart city indicators: A systematic literature review. *Journal of Telecommunication, Electronic and Computer Engineering, 8*(3), 161–164.

Purwanto, E. A. (2018). Smart city as an upshot of bureaucratic reform in Indonesia. *International Journal of Electronic Government Research, 14*(3), 32–43. https://doi.org/10.4018/IJEGR.2018070103

Raaijen, T., & Daneva, M. (2017). Depicting the smarter cities of the future: A systematic literature review & field study. In *2017 Smart Cities symposium prague, SCSP 2017*. Institute of Electrical and Electronics Engineers Inc. https://doi.org/10.1109/SCSP.2017.7973352

Ruhlandt, R. W. S. (2018). The governance of smart cities: A systematic literature review. *Cities, 81*, 1–23. https://doi.org/10.1016/j.cities.2018.02.014

Sadowski, J., & Bendor, R. (2019). Selling smartness: Corporate narratives and the Smart City as a sociotechnical imaginary. *Science Technology and Human Values, 44*(3), 540. https://doi.org/10.1177/0162243918806061

Sáez-Martín, A., Haro-de-Rosario, A., & Caba-Perez, C. (2014). A vision of social media in the Spanish smartest cities. *Transforming Government: People, Process and Policy, 8*(4), 521–544. https://doi.org/10.1108/TG-03-2014-0010

Sanchez-Iborra, R., Bernal-Escobedo, L., & Santa, J. (2020). Eco-efficient mobility in smart city scenarios. *Sustainability (Switzerland), 12*(20). https://doi.org/10.3390/su12208443

Schuurman, D., Baccarne, B., De Marez, L., & Mechant, P. (2012). Smart ideas for smart cities: Investigating crowdsourcing for generating and selecting ideas for ICT innovation in a city context. *Journal of Theoretical and Applied Electronic Commerce Research, 7*(3), 11. https://doi.org/10.4067/S0718-18762012000300006

Scornavacca, E., Paolone, F., Za, S., & Martiniello, L. (2020). Investigating the entrepreneurial perspective in smart city studies. *International Entrepreneurship and Management Journal, 16*(4), 1197–1223. https://doi.org/10.1007/s11365-019-00630-4

Shelton, T., Zook, M., & Wiig, A. (2015). The "actually existing smart city.". *Cambridge Journal of Regions, Economy and Society, 8*(1), 13–25. https://doi.org/10.1093/cjres/rsu026

Shihab, M. R., & Hidayanto, A. N. (2021). Citizen adoption of eparticipation: A study of Jakarta smart city eparticipation initiatives. *Journal of Management Information and Decision Sciences, 24*(4), 1–12.

Shirazizadeh, M. (2019). Practical corpus linguistics: An introduction to corpus-based language analysis. *RELC Journal, 50*(2), 361. https://doi.org/10.1177/0033688217707631

Shokheh, M., Ebrahimi, M., & Yusoff, K. (2021). The role of Indonesian and Egyptians' NGOs in democratisation. *Geopolitics Quarterly, 17*(64), 289–306.

Sholeh, C., Sintaningrum, S., & Sugandi, Y. S. (2019). Formulation of innovation policy: Case of Bandung Smart City. *Jurnal Ilmu Sosial Dan Ilmu Politik, 22*(3), 10.22146/JSP.33698.

Söderström, O., Paasche, T., & Klauser, F. (2014). Smart cities as corporate storytelling. *City, 18*(3), 307. https://doi.org/10.1080/13604813.2014.906716

Sotillo, S. M., & Wang-Gempp, J. (2016). Using corpus linguistics to investigate class, ideology, and discursive practices in online political discussions. *Applied Corpus Linguistics*. https://doi.org/10.1163/9789004333772_007

Stagars, M. (2016). Open data in Southeast Asia: Towards economic prosperity, government transparency, and citizen participation in the ASEAN. In *Open data in Southeast Asia: Towards economic prosperity, Government transparency, and citizen participation in the ASEAN*. Springer International Publishing. https://doi.org/10.1007/978-3-319-32170-7

Stefanowitsch, A., & Gries, S. T. (2007). *Corpus-based approaches to metaphor and metonymy*. Mouton De Gruyter. https://journals.linguisticsociety.org/booknotices/?p=473

Strüver, A., Saltiel, R., Schlitz, N., Hohmann, B., Höflehner, T., & Grabher, B. (2021). A smart right to the city—Grounding corporate storytelling and questioning smart urbanism. *Sustainability (Switzerland), 13*(17). https://doi.org/10.3390/su13179590

Suryaningsih, W. E., & Chrisdanty, F. (2018). The role of the Surabaya City government in optimizing green open space. *Journal of Civil & Legal Sciences, 07*(02). https://doi.org/10.4172/2169-0170.1000248

Sweeting, D., de Alba-Ulloa, J., Pansera, M., & Marsh, A. (2022). Easier said than done? Involving citizens in the smart city. *Environment and Planning C: Politics and Space, 40*(6), 1365. https://doi.org/10.1177/23996544221080643

van Twist, A., Ruijer, E., & Meijer, A. (2023). Smart cities & citizen discontent: A systematic review of the literature. *Government Information Quarterly, 40*(2), 101799. https://doi.org/10.1016/j.giq.2022.101799

Vanolo, A. (2014). Smartmentality: The Smart City as disciplinary strategy. *Urban Studies, 51*(5), 883. https://doi.org/10.1177/0042098013494427

Visvizi, A., & Lytras, M. D. (2018). It's not a fad: Smart cities and smart villages research in European and global contexts. *Sustainability (Switzerland), 10*(8), 1–10. https://doi.org/10.3390/su10082727

Wiegand, V., & Mahlberg, M. (2019). *Corpus linguistics, context and culture*. Walter de Gruyter GmbH & Co KG. https://doi.org/10.1515/9783110489071

Yigitcanlar, T. (2015). Smart cities: An effective urban development and management model? *Australian Planner, 52*(1), 27. https://doi.org/10.1080/07293682.2015.1019752

Zhang, S. (2019). Public participation in the Geoweb era: Defining a typology for geo-participation in local governments. *Cities, 85*, 38–50. https://doi.org/10.1016/j.cities.2018.12.004

Zhang, J., Bates, J., & Abbott, P. (2022). State-steered smartmentality in Chinese smart urbanism. *Urban Studies., 59*, 2933. https://doi.org/10.1177/00420980211062888

Zhu, Y.-Q., & Alamsyah, N. (2022). Citizen empowerment and satisfaction with smart city app: Findings from Jakarta. *Technological Forecasting and Social Change, 174*, 121304. https://doi.org/10.1016/j.techfore.2021.121304

Zhu, W., Yan, R., & Song, Y. (2022). Analysing the impact of smart city service quality on citizen engagement in a public emergency. *Cities, 120*, 103439. https://doi.org/10.1016/j.cities.2021.103439

Anang Dwi Santoso is a lecturer of Public Administration, Faculty of Social and Political Sciences, Sriwijaya University. He received Bachelor of Public Administration from Universitas Brawijaya and holds Master of Public Administration from Universitas Gadjah Mada. His research interests are mainly about disaster management, disability studies, and e-government.

Januar Eko Aryansah is a lecturer of Public Administration, Faculty of Social and Political Sciences, Sriwijaya University, where he is now served as the Secretary of the Department. He received Bachelor of Political Sciences from the University of Bengkulu and Master of Science in Public Administration from Universitas Sriwijaya. His research interests are mostly about public policy.

Annada Nasyaya is a lecturer of Public Administration, Faculty of Social and Political Sciences, Sriwijaya University. She graduated from Magister of Public Administration, the Faculty of Social and Political Sriwijaya University in 2013. Since 2019, she has worked as a lecturer and researcher at the Department of Public Administration, Faculty of Social and Political Sciences, Sriwijaya University. Her research focuses on Development Administration, Strategic Management, and Gender.

Chapter 7
Artificial Intelligence-Based Metaverse as a Virtual Model of Smart Urbanism: Unveiling the Challenges for Cybersecurity

Pham Quang Huy and Vu Kien Phuc

Abstract Artificial intelligence (AI) is at the center of the Metaverse, a vast ecosystem adoption centered on Extended Reality technology. As the creation of the Metaverse is still in its early phases and many various technologies must be integrated in order for it to be realized, a number of cybersecurity concerns need to be resolved before the system can be used in reality for any significant purposes. A broad cyberattack surface will be produced by the fusion of several technologies, which surely leads to a number of brand-new cybersecurity problems. This study focuses on determining and assessing the challenges of cybersecurity in the AI-based Metaverse (AIM) model for smart urbanism. This study's explanatory sequential approach integrates qualitative and quantitative data for maximum advantage. Some of the methods used to gather data include closed-ended questionnaire survey and semi-structured interviews. In order to assess the survey data statistically, structural equation modeling is employed. The interview data is examined using content analysis to support and add to the findings from the questionnaire. In addition to potential directions for further research, recommendations based on the results include creating rules and best practices for the implementation AIM as a virtual model of smart urbanism.

Keywords Artificial intelligence · Challenges · Cybersecurity · Metaverse · Smart urbanism

P. Q. Huy (✉) · V. K. Phuc
School of Accounting, University of Economics Ho Chi Minh City (UEH), Ho Chi Minh City, Vietnam
e-mail: pquanghuy@ueh.edu.vn; phucvk@ueh.edu.vn

S. Majumdar et al. (eds.), *Smart Cities*, S.M.A.R.T. Environments,
https://doi.org/10.1007/978-3-031-59846-3_7

7.1 Introduction and Context

In order to address the sustainability challenges that cities are presently encountering, smart urbanism proposes an innovative approach that heavily relies on digital technologies (Marvin et al., 2016). Sustainability in urban planning is a hot topic in the field of urban and regional studies, and this approach is becoming recognized as essential for international urban policy (Mouton & Burns, 2021; Su & Fan, 2023). An urban technology infrastructure that integrates hardware, software, data, and digital services and applications is the backbone of smart cities (Mora et al., 2021). Research and development of the Metaverse has recently arisen as a significant trend in data-driven smart urbanism, specifically in relation to the construction of virtual cities built on artificial intelligence (AI) systems that are extraordinarily scalable (Allam et al., 2022; Bibri & Allam, 2022a, b). In the Metaverse, state-of-the-art technology is supreme, and there are solutions to many problems plaguing many areas of city life. The term "data-driven smart urbanism" is used to describe how people live in or interact with cities, according to Gregory et al. (2009) (Harvey, 2009). According to Kitchin (2014), smart cities aim to incorporate digital infrastructure and big data technologies strategically in order to create intelligent lifestyles, ecosystems, individuals, economies, and governments. These lifestyles are also said to be essential to the Metaverse (Allam et al., 2022).

Multiple scholarly studies have analyzed the adverse consequences and possible hazards of smart urbanism and the incorporation of socially disruptive technologies into daily life. These investigations have explored various aspects, such as technocratic governance, corporatization of governance, surveillance, privacy, control, and other related concerns (Bibri, 2021a, b). The Metaverse is poised to become a digital marketplace that will further amplify the influence and control of major tech conglomerates. The virtual worlds will be shaped by human users, leading to the consolidation of control and the exacerbation of inequality and social exclusion (Bibri & Allam, 2022b). The notion of the Metaverse and its underlying technologies are facing growing criticism, mistrust, and skepticism (Lee et al., 2021; Falchuk et al., 2018; Bibri & Allam, 2022a, b). In the contemporary world, cybersecurity has emerged as a crucial issue that necessitates safeguarding the computer network from potential threats (Komar et al., 2017; Alhayani et al., 2021). Organizations are increasingly relying on online technology to store financial and personal data, which has resulted in an increase in cyberattacks (Corallo et al., 2022). Cyberattacks also have a profound psychological effect on people, causing them to feel unhappy, tense, and stressed (Bada & Nurse, 2020).

More importantly, Al-Ayed (2022) noted that there are a number of worries and difficulties in the Metaverse with regard to cybersecurity using AI. The likelihood that AI will be used in cyberattacks is one of the biggest worries. AI has the potential to produce sophisticated malware and take advantage of security flaws. AI can be used to carry out automated cyberattacks far more quickly than a human attacker. This boosts the likelihood of an attacker's success by making it more difficult for defenders to keep up. The potential for AI to produce false or misleading information is another challenge. This could be used to trick people or create havoc. False information can also be used to skew decision-making in the financial markets. AI can also be used to find and gather private information about individuals. The data may be sold or used for marketing to third parties. Advocates for privacy are very concerned about this. Last but not least, AI may be used to create self-driving tools. These are several technologies that can select and interact with targets without requiring human involvement. This raises concerns about the potential for misuse and conflict escalation.

As a result, data-driven smart urbanism is far more safe, inclusive, and fair than the Metaverse when it comes to possible vulnerabilities and dangers (Bibri & Allam, 2022b). Given the negative impact on urban society and the numerous challenging issues brought up by Meta's social media platforms over the past 20 years, it is likely that the Metaverse will not be able to adequately justify investments in scientific and technological advancements, as these fields are often equated with societal progress (Bibri & Allam, 2022a). To mitigate the dangers of the Metaverse, Rosenberg (2022) proposes several unusual regulatory solutions, such as reducing the usage of synthetic personas, emotional analysis, virtual product placements, and surveillance. However, smart urbanization strategies and solutions are based on technological orthodoxies that are theoretically and empirically flawed. Accordingly, there is a lack of both theoretical grounding and empirical evidence to thoroughly assess the risks and unforeseen effects of smart urbanism‘s transformational processes on urban society's habits, practices, and institutions (Bibri, 2021a). In light of this, the present research seeks to identify and evaluate the cybersecurity threats to the AI-based Metaverse (AIM) smart urbanism model. The intriguing research questions raised in this study in order to achieve that goal are provided below:

RQ1. *What are the challenges of cybersecurity which are induced by the AIM for smart urbanism?*
RQ2. *How far does AIM impact these challenges of cybersecurity?*

Although the virtual world and the Metaverse were just introduced recently, there are still a lot of cybersecurity-related issues in the Metaverse that need to be resolved right now. Criminologists are worried that cybercriminals may take advantage of the lack of regulations around the Metaverse to launch attacks on its infrastructure, which stems from arguments regarding the concept's nature. Numerous security concerns that face Metaverse users are also faced by Internet users, including spam, malware assaults, data hacking, and privacy issues. The AIM can offer much better cybersecurity solutions by utilizing AI. Despite the usage of AI, security issues continue to be a significant challenge for the Metaverse (Pooyandeh et al., 2022). Unfortunately, there has been a paucity of the theoretical underpinning and empirics needed to conduct a comprehensive assessment of the hidden dangers and potential threats associated with the transformational processes of smart urbanism with regard to urban society's institutions, operations, and practices (Bibri, 2021a). In theoretical significance, the culmination of the research is to advance our understanding of cybersecurity issues related to the establishment of a Metaverse powered by AI for smart urbanism. Additionally, according to researchers' best knowledge, this may be the first academic paper to offer new and important insights on the remarkably distinctive effects of AIM implementation for smart urbanism on each cybersecurity challenge. Explanatory sequential design is used in these arguments. First, pertinent information is gathered from large-scale surveys and the literature on smart urbanization projects. Second, 23 expert interviews are carried out using a semi-structured process organized around the difficulties in implementing a Metaverse powered by AI for smart urbanism. In practical significance, the empirics offer managerial and operational pointers and suggests a roadmap, at the policy degree, regarding the AIM implementation for smart urbanism to practitioners with a holistic focus on the cybersecurity.

Aside from the introduction, the current research unfolds as follows. Theoretical basis and literature review are briefly covered in Sect. 7.2, while the development of hypotheses and the formulation of research models are covered in greater detail in Sect. 7.3. An extensive description of the research methodology used in this study is provided in Sect. 7.4. The specific outcomes of the explanatory sequential design are presented in detail in Sect. 7.5. Finally, Section 7.6 wraps up this investigation and describes the forthcoming orientations for upcoming academic works.

7.2 Literature Review and Theoretical Basis

Under the assumption of unchecked technological progress, Gurov and Konkova (2022) draw attention to the potential benefits and risks that the Metaverse might bring to mankind. When using AI to secure the Metaverse, there are a number of issues and difficulties to be aware of (Al-Ayed, 2022). The likelihood that AI will be utilized in cyberattacks and the prospect that it would be used to produce false or misleading information are the two biggest worries (Al-Ayed, 2022). The topic of user identification, for which biometric approaches are most frequently utilized, plays a significant role in the works provided in Pooyandeh et al. (2022)'s research on AIM cybersecurity and examines pertinent security concerns.

Artificial Intelligence-Based Metaverse Over the past few decades, AI has been extensively studied around the globe (Tang et al., 2021). Our culture has embraced AI due to its ability to assist various industries in tackling difficult problems and modernizing outdated procedures (Ali et al., 2023). In order to improve human decision-making, problem-solving skills, and technological innovation, AI systems use computers and other technologies to understand data (Mishra & Tripathi, 2021; Mustak et al., 2021). AI can develop a lot of content for the Metaverse on a wide scale and continuously because of how much content is produced in this digital universe (Pooyandeh et al., 2022).

The development of the Metaverse, a three-dimensional virtual world in which physical distance has no effect on its bounds, is the subsequent step in the digital transformation process (Abbate et al., 2022). A combination of the words "universe" and "meta" and the phrase "Metaverse" means "abstraction," "virtuality," "beyond," or transcendence" (Lee & Kim, 2022). The Metaverse is an imaginary three-dimensional virtual environment that provides virtual reality experiences as an alternative to the real world, as stated by Mehta et al. (2023). According to Tlili et al. (2022), state-of-the-art technologies such as blockchain, edge computing, AI, and VR/AR will find new opportunities in the Metaverse. Metaverse implementations have emerged across various industries, driven by the promise of better industrial services and a more sustainable society (Tlili et al., 2023).

Smart Urbanism Academics generally agree that "smart urbanism" hinges on the concept of an ever-expanding web of interconnected digital sensors and other devices embedded in urban fabric (Kitchin, 2014). For the purpose of managing and modifying urban government and urban lifestyles, smart urbanism turns cities and residents into functional datasets (Marvin et al., 2016). The infrastructure, operational, service-oriented, and governmental elements of smart urbanism are those that can be digitally upgraded, retrofitted, or even "twinned" (Tomko & Winter, 2019).

Cybersecurity A cyberattack is defined by Bokhari and Myeong (2023) as an organized effort to cause harm to non-combatant opponents through the targeting of computer networks, relevant data, programs, and electronic information. Any member of society can fall victim to cyberattacks including phishing, denial of service, viruses, and ransomware (Weaver et al., 2022). This kind of assault has the potential to disable or destroy devices, as well as provide the attacker authority over, the ability to delete or block the devices, and the ability to steal their data (Pooyandeh et al., 2022). Cybersecurity is defined as a set of practices and policies aimed at protecting information systems, networks, applications, and data against intrusion, damage, or malicious attack (Jaipong et al., 2022).

7.3 Research Model Development and Hypothesis Formulation

Metaverse depictions of future data-driven smart cities will show their unique characteristics. Urbanism, as it pertains to city services and city administration, describes the unique aspects of city life (Bridge, 2009). Data-driven smart urbanism and platform urbanism built on analytics and AI are great at accommodating these features (Bibri & Allam, 2022b). With this merger comes the inevitable datafication, algorithmization, and platformization of all things social, including the social organization that emerges from these interactions and activities (Bibri & Allam, 2022a).

There are several well-established social media, sharing economy, and e-governance platforms in data-driven smart cities, and the broad adoption of new networked digital technologies that are now part of everyday life all lend credence to the push to build and actualize the Metaverse through platformization (Bibri & Allam, 2022a). Thanks to hyper-connectivity, all the actors in this network can freely exchange data, products, services, resources, and values with one another.

Both the original intent of the data's generation and the many subsequent uses for the repackaged data are completely disregarded, and the importance of getting the consent of users who care about the data is also ignored. There will be more effects and risks brought about by digital and computational technologies, but this is an example of the Metaverse vision in action, with the end objective of virtually changing people's lives, livelihoods, and opportunities in urban society (Bibri & Allam, 2022a). A clear breach of the data minimization principle exists when data is not limited to what is necessary to complete certain tasks and when it is used exclusively for its intended purpose (Zuboff, 2019). Digital and computationally mediated interactions, communications, and activities are commonplace, and unique identifiers are being used more and more to access virtual services. As a result, users in the Metaverse will almost certainly leave traces and have other traces captured about them as they go about their daily lives (Bibri & Allam, 2022a). To make matters even more difficult, each of the multiple virtual worlds that will be created in the Metaverse is expected to have its own unique set of laws that will control how users behave and what they can do (Lee et al., 2021). Based on the explanations demonstrated above, the speculative hypotheses for the current investigation are posited as follows:

Hypothesis 1 (H1). *AIM induces in an impact on SC in a significant and negative manner.*

Hypothesis 2 (H2). *AIM induces in an impact on RC in a significant and negative manner.*

Hypothesis 3 (H3). *AIM induces in an impact on IC in a significant and negative manner.*

Hypothesis 4 (H3). *AIM induces in an impact on CC in a significant and negative manner.*

Hypothesis 5 (H3). *AIM induces in an impact on AC in a significant and negative manner.*

All the hypotheses in this research have been indicated in the proposed model (Fig. 7.1) as seen below.

7.4 Materials and Methods

As the predominant purpose of the current research is to give rise to insights that has application in the real world and to construct theory from practice, a mixed-method approach is employed. As stated by Creswell and Creswell (2018), the mixed-method approach not only helps minimizing the weaknesses of both quantitative and qualitative methods but also enables integrating their advantages. Researchers can obtain a better grasp of the phenomena they are studying through the mixed-methods technique, which permits cross-validation and verification of findings in a single study (Creswell & Creswell, 2018; Alsharari & Aljohani, 2023). The questionnaire is initially distributed to participants in a quantitative study, after which

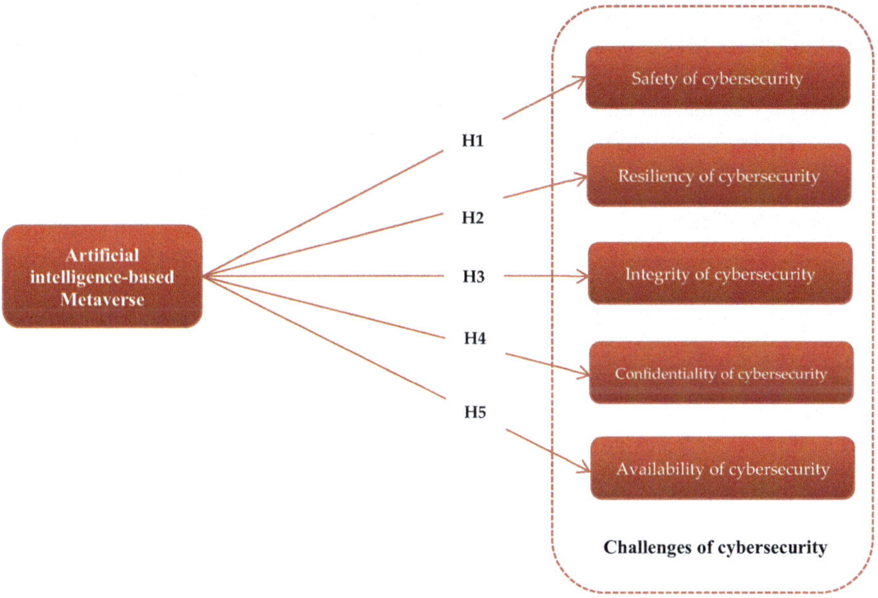

Fig. 7.1 Hypothesized proffered model

the data is analyzed quantitatively. The qualitative study goes on to provide additional support for the quantitative research findings.

7.4.1 Quantitative Stage

In order to put the proposed theoretical model to the test, this study employed a quantitative approach, which provides trustworthy statistics of precision for analyzing the survey data. The researchers in this study use a cross-sectional design since it allows them to look at people's attitudes right now (Kesmodel, 2018). A quantitative approach based on structured questionnaires is utilized to verify the relationships in the proposed model (Selem et al., 2022). All of the constructions' study metrics are taken from earlier works, with linguistic modifications made to fit the specifics of the current investigation.

After being initially produced in English, the questionnaire is validated using a back-translation method (Wilson, 2014). To confirm their language equivalently, the scales are first translated into Vietnamese and then back into English. The first version of the survey is sent to four professors for them to check for clarity and offer suggestions for how to make it better. The difficult to understand scales are finally removed. In order to examine the validity of the questions being utilized and to gather a sample of potential responses from respondents, the structured questionnaire has undergone a preliminary pilot test (Kant Pal et al., 2021). Research methods are evaluated for validity and reliability using Cronbach's alpha, as stated by Hair et al. (2012). The results show that all of the items have Cronbach's alpha values greater than 0.70. The internal consistency of a trustworthy scale is also shown by Cronbach's alpha scores that fall within the range of 0.50–0.70 (Hinton et al., 2014). According to the results of the pilot test, all items should be retained for the future analysis.

Artificial Intelligence-Based Metaverse Implementation The AIM is supposed to generate several benefits in smart urbanism applications namely better team work and collaboration, effective testing and training, business opportunities, digital representations and physical objects, greater human connection, and innovation in entertainment services. As such, the criteria applied to evaluate better team work and collaboration, effective testing and training, business opportunities, digital representations and physical objects, greater human connection, and innovation in entertainment services are based on the work of Yaqoob et al. (2023) as a point of reference.

Challenges of Cybersecurity In this research, the challenges for cybersecurity comprise safety of cybersecurity, resiliency of cybersecurity, integrity of cybersecurity, confidentiality of cybersecurity, and availability of cybersecurity. As such, the criteria applied to evaluate safety of cybersecurity, resiliency of cybersecurity, integrity of cybersecurity, confidentiality of cybersecurity, and availability of cybersecurity are based on the work of Arpaci and Sevinc (2022) as a point of reference.

The results from public sector entities, such as public hospitals, public universities, and public schools, are gathered using a standardized survey design. The participants work as accountants and information technology personnel for the organizations in the interim. The participants are contacted by phone in order to inform them about the study, assure their privacy, and assess their interest in participating before the formal interviews take place. The printed survey questionnaires are handed out to the intended respondents.

This study uses a non-random sampling instrument because non-random sampling strategies frequently contained very small sample sizes that sought to understand the richness and depth of the phenomenon being studied (Fossey et al., 2002). Alternatively, researchers can seek out and include informants who may narrow the scope of their questions while also providing detailed information pertinent to the study's aim (Parker & Northcott, 2016). On the other hand, it is smart to choose a sample that meets really practical needs, including being close by, being available within a certain time frame, or having a strong willingness to take part (Etikan et al., 2016). Therefore, this research makes use of a hybrid of convenience and snowball sampling methods. Ten replies are required for every case or questionnaire issue, following the advice of Kline (2011). Finally, 800 questionnaires are distributed to these organizations between December 2022 and September 2023 using a convenience sample methodology. This method is preferred since it is less expensive, quicker, and simpler for responders to complete (Mahmud et al., 2022; Jean-Noël et al., 2020). Following screening and examination of questionnaires, 612 cases

Table 7.1 Demographic information

Demographic profile	Usable responses	Weight (%)
Gender		
Male	401	65.52
Female	211	34.48
Age		
Below 25	14	2.29
25–under 35	84	13.73
35–under 45	392	64.05
45–under 55	117	19.12
Over 55	5	0.81
Experience (years)		
Below 10	29	4.74
10–below 20	461	75.33
20–below 30	95	15.52
Over 30	27	4.41
Education		
Undergraduate	491	80.23
Postgraduate	121	19.77

remained for analysis, representing a data loss rate of 23.5%. An exhaustive synopsis of the survey's demographic data is presented in Table 7.1.

7.4.2 Qualitative Stage

Building on the perspectives of Edmondson and McManus (2007), qualitative technique is acceptable since it may be used to fill in gaps in the literature and develop theories.

This instance makes use of a semi-structured interview with subject matter experts. According to Levitt et al. (2018), qualitative methods are useful for studying experiences since they collect data through the use of natural language, which leads to theory. According to Saunders et al. (2019), the semi-structured interview is able to examine each expert's unique experiences and views while also guaranteeing that all interviewees are asked the same range of questions and levels of exploration. In cases when the respondent possesses relevant expertise, semi-structured interviews can also be employed (McIntosh & Morse, 2015). During this stage, semi-structured interviews on telephones are used due to their many benefits, such as allowing participants to focus on the research subject and form conclusions on perceived causality.

The importance of including individuals with background in cybersecurity, AI, and the Metaverse highlights the usefulness of using purposive sampling for the interview sample. Researchers can proactively choose participants who are both knowledgeable about the topic at hand and willing to participate in the study by employing a technique known as purposive sampling (Thottoli et al., 2019; Thottoli,

2021). Triangulation—the process of cross-checking participant responses across participants with similar organizational and industry affiliations—is made easier and more effective by using the snowballing sampling technique (Etikan et al., 2016). Nonetheless, the results have limited external validity because they can only be extended to the sampled group or to those with similar characteristics (Andrade, 2021). By assuring the qualitative data gathering technique is rigorous, the validity and dependability of the research findings can be ameliorated and enhanced. After determining that theoretic saturation has been attained, 23 semi-structured interviews with experts are undertaken (Saunders et al., 2019).

Because of the importance of the panel of experts and the need for careful management in light of the reflection on the group decision's quality, 23 experts have agreed to participate in the current study's qualitative phase. The demographic breakdown of the expert group is depicted in Table 7.2.

Thematic analysis is used to examine the interview transcripts. The qualitative data has been organized and categorized based on the themes and patterns that emerge from the participant replies (Alsharari & Aljohani, 2023). The six-step thematic analysis procedure followed by Braun and Clarke (2019) consists of getting to know the data, coding it, creating themes, reviewing them, defining and identifying them, and finally, writing stuff up.

Table 7.2 Profile of panel of expert

Item	Frequency	Valid (%)	Item	Frequency	Valid (%)
Gender			*Organization*		
Female	6	26.09	Public university	6	26.09
Male	19	82.61	Private university	1	4.35
			Treasury	1	4.35
Experience (years)			Department of Finance	1	4.35
Under 10	8	34.78	Public hospital	2	8.70
10–under 20	13	56.52	Public administration center	1	4.35
20–under 30	2	8.70	Commercial Bank	3	13.04
			National Bank	1	4.35
			Tax Department	1	4.35
			Department of Natural Resources and Environment	2	8.70
			Department of Construction	2	8.70
			Department of Transport	2	8.70
Career			*Education*		
Vice president	2	8.70	Master's degree	21	91.30
Deputy	2	8.70	PhD degree	2	8.70
Official	15	65.22			
University lecturer	4	17.39			

7.5 Inferential Statistics

7.5.1 *Quantitative Stage*

Measurement Model for Confirmatory Factor Analysis The CFA is carried out to evaluate the measurement model (Nye, 2022). It is set up through utilizing suggested benchmark of "goodness of fit" measures proposed by Bagozzi and Yi (2012). Building on the statistical results in Fig. 7.2, the measurement model fit indexes for the CFA are fluctuated within the acceptable threshold. More concretely, chi-square to degree of freedom (χ^2/df) = 1.700; goodness of fit index (GFI) = 0.931, comparative fit index (CFI) = 0.965, Tucker–Lewis index (TLI) = 0.958, and root mean square error of approximation (RMSEA) = 0.034 (Bagozzi & Yi, 2012; Byrne, 2013).

Construct Validity and Reliability Evaluation The reliability of the constructs is evaluated using Cronbach's alpha and composite reliability (CR), with a recommended value more than 0.70 (Garson, 2015). Hair et al. (2017) suggest that the 0.70 literature cutoffs should be used as the accepted threshold for Cronbach's alpha and CR values. The study by Preko et al. (2023) examines convergent validity by calculating the average variance extracted (AVE) and factor loading for each measuring item and construct. Values for the average variance extracted (AVE) and item loadings both exceed 0.50. All of the measures have excellent content and convergent validity, as shown by Lathabhavan and Griffiths (2023). Table 7.3 displays the characteristics of these outputs, which highlight the convergent validity, internal consistency, and reliability of the measurement model.

Correlations and Discriminant Validity According to Hew and Syed Abdul Kadir (2016), the discriminant validity criterion is validated, and the AVE scores are found to be significantly higher than the maximum shared variance, indicating that the AVE is robust. Based on the statistical results shown in Table 7.4, it is reasonable to assume that discriminant validity is not an issue.

Full Structural Equation Modeling and Hypotheses Testing Building on the statistical results in Fig. 7.3, the structural model fit indexes are fluctuated within the acceptable threshold. More concretely, chi-square to degree of freedom (χ^2/df) = 1.757; goodness of fit index (GFI) = 0.922, comparative fit index (CFI) = 0.958, Tucker–Lewis index (TLI) = 0.955, and root mean square error of approximation (RMSEA) = 0.035 (Bagozzi & Yi, 2012; Byrne, 2013).

The mathematical results on the respective β values for each hypothesis in Table 7.5 highlight markedly negative interconnections between AIM and challenges for cybersecurity. More concretely, AIM induces a significant and negative

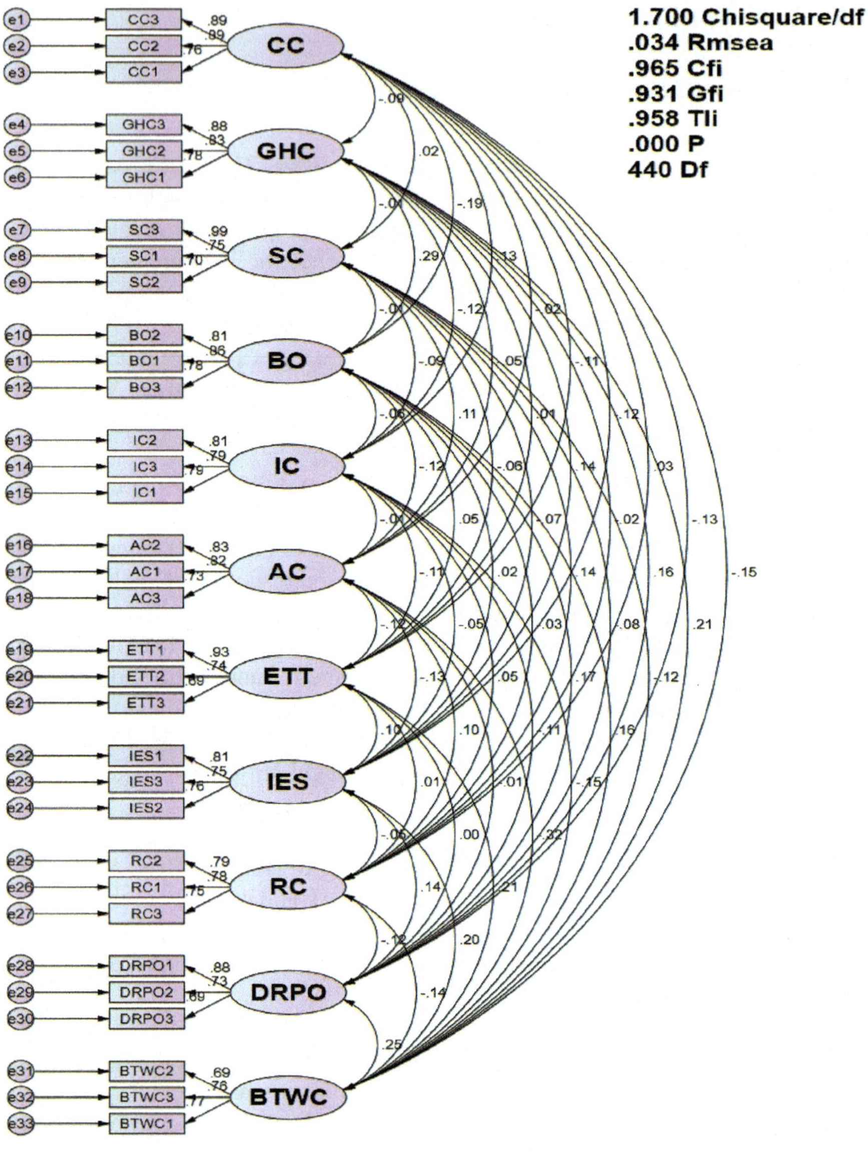

Fig. 7.2 CFA results

impact on AC ($\beta = -0.345$; critical ratio = −4.913, $p < 0.001$), CC ($\beta = -0.273$; critical ratio = −4.266, $p < 0.001$), IC ($\beta = -0.220$; critical ratio = −3.468, $p < 0.001$), RC ($\beta = -0.195$; critical ratio = −3.092; $p < 0.01$), and SC ($\beta = -0.164$; critical ratio = −2.882; $p < 0.01$). Thus, H1–H5 are buttressed.

Table 7.3 Results and summary of the measurement model assessment

Construct	Item acronyms	Convergent validity		Construct reliability		Inference
		Factor loadings ranges	AVE	Cronbach's alpha	Composite reliability	
Artificial intelligence-based Metaverse	**AIM**					
Better team work and collaboration	BTWC	0.725–0.761	0.551	0.781	0.786	Retained
Effective testing and training	ETT	0.664–0.971	0.632	0.823	0.835	Retained
Business opportunities	BO	0.777–0.851	0.666	0.852	0.856	Retained
Digital representations and physical objects	DRPO	0.668–0.890	0.594	0.805	0.813	Retained
Greater human connection	GHC	0.770–0.910	0.692	0.869	0.871	Retained
Innovation in entertainment services	IES	0.748–0.818	0.599	0.815	0.817	Retained
Safety of cybersecurity	**SC**	0.696–0.989	0.678	0.851	0.860	Retained
Resiliency of cybersecurity	**RC**	0.763–0.783	0.595	0.813	0.815	Retained
Integrity of cybersecurity	**IC**	0.769–0.822	0.632	0.835	0.837	Retained
Confidentiality of cybersecurity	**CC**	0.753–0.911	0.720	0.882	0.885	Retained
Availability of cybersecurity	**AC**	0.732–0.830	0.634	0.837	0.838	Retained

7.5.2 *Qualitative Stage*

Semi-structured interviews are conducted in person after scheduled appointments or over the phone at a mutually suitable time. The interviews took place between June 2023 and September 2023 and were planned around workload and shifting work schedules. Between 30 and 40 min were allotted for each interview. The principal investigator performed each interview for consistency's sake. The notes were meticulously documented for each interview, with all identifying information removed to facilitate analysis. Examples that would support and act as illustrations of reporting openness were chosen once the final themes had been decided upon. Thus, Table 7.6 presents the qualitative results.

Table 7.4 Results and summary of discriminant validity using Fornell–Larcker process

	CC	GHC	SC	BO	IC	AC	ETT	IES	RC	DRPO	BTWC
CC	**1**										
GHC	−0.086	**1**									
SC	0.016	−0.007	**1**								
BO	−0.190	0.291	−0.009	**1**							
IC	0.129	−0.115	−0.087	−0.065	**1**						
AC	−0.016	−0.051	0.113	−0.116	−0.009	**1**					
ETT	−0.114	0.010	−0.064	0.050	−0.110	−0.125	**1**				
IES	−0.119	0.136	−0.067	0.024	−0.048	−0.134	0.104	**1**			
RC	0.032	−0.016	0.138	−0.034	0.050	0.103	0.011	−0.047	**1**		
DRPO	−0.133	0.160	−0.076	0.168	−0.114	−0.011	−0.003	0.140	−0.121	**1**	
BTWC	−0.149	0.206	−0.120	0.157	−0.147	−0.325	0.211	0.204	−0.136	0.252	**1**

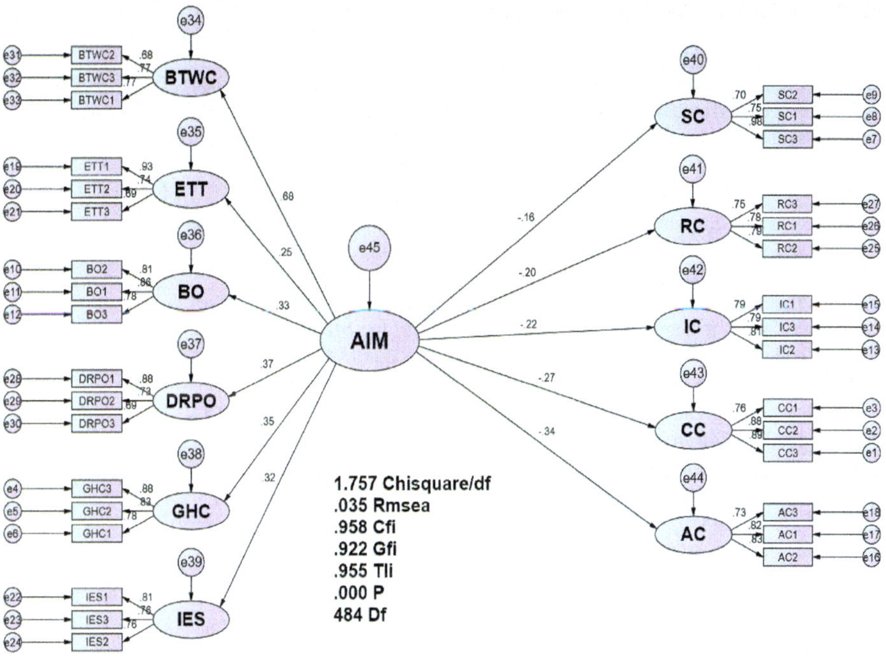

Fig. 7.3 SEM results

Table 7.5 Results and summary of hypotheses' acceptance

Hypothesis no.	Hypothesized path			Estimate	S.E.	C.R.	Status
H1	AIM	→	SC	−0.164***	0.193	−2.882	Buttressed
H2	AIM	→	RC	−0.195***	0.200	−3.092	Buttressed
H3	AIM	→	IC	−0.220**	0.151	−3.468	Buttressed
H4	AIM	→	CC	−0.273***	0.200	−4.266	Buttressed
H5	AIM	→	AC	−0.345*	0.183	−4.913	Buttressed

Notes: *$p < 0.05$; **$p < 0.01$; ***$p < 0.001$

7.6 Final Deliberations

7.6.1 Policy Implications

This viewpoint chapter provides practitioners and policymakers with valuable ideas that can help them better understand and approach the difficulties posed by the AIM with a more nuanced and realistic attitude. It is asserted that a thorough and strong cybersecurity architecture is necessary for smart urbanization in order to protect their integrity and direct their response to any threats. In order to ensure privacy, resiliency, confidentiality, availability, and integrity, strong cybersecurity platforms should be built and implemented for smart urbanization. A more secure digital

Table 7.6 Summary of qualitative results

Exemplar quotation of qualitative results	
…in my opinion…I think…the risk of cyberattacks in the Metaverse …has grown exceedingly difficult to mitigate…so I think… all organizations must think about cybersecurity …and privacy issues…this is because the Metaverse …is built in a cyber environment … to provide users with the necessary services …in a secure and effective manner…	*P01, Official, Tax Department*
…there are a number of possible attacks in the Metaverse…they are attacks against the system, …attacks against the network, …attacks against the database, …and attacks against devices…so I think… it is necessary to improve security methods …and rules to take into account the properties of the Metaverse… to secure the virtual environment of the Metaverse from cyberattacks and cybercrime…I think… encryption should be used …to retain sensitive data securely …to reduce the effects of unauthorized access.	*P06, Lecturer, Public university*
… I think …the greatest cyberthreat in the Metaverse is data privacy…So … users who are unaware of their immediate surroundings while using the Metaverse …may run the danger of bodily harm… Users' personal data… including details about their online interactions and activities …may be gathered and stored by the Metaverse.	*P14, Official, Department of Construction*
…I think…each community must include characteristics …that promote security and privacy…and… the ongoing problem of data security in the Metaverse is a direct result of the daily occurrence of millions of cyberattacks….and I think…the platform's potential to affect data integrity poses a risk to Metaverse users.	*P17, Employee, Commercial bank*
…I think…the possibility of virtual marketplaces in the metaverse… where users can trade …or acquire virtual goods …raises the possibility of theft by hackers… if it enables users to stay anonymous or share private information with others …the metaverse may be used for cyberbullying or harassment.	*P20, Deputy head of division, Department of Finance*
…As you know …Urban societies are going through a large-scale digital transition …that has been expedited and amplified …thanks to the powerful convergence of artificial intelligence. However, …Artificial intelligence has both positive and negative consequences…with the latter being a detrimental one …that speeds up and intensifies the initial phase of cyberattacks…	*P23, Official, Department of Transport*

environment will come from this. Policymakers and urban planners should build up a robust and secure digital environment, as well as design specific strategies to encourage the adoption of advanced security measures and provide guidelines for responding to possible attacks. Alternatively, the government must develop and implement successful cybersecurity awareness initiatives to raise public knowledge of how people should protect their data and privacy. Accordingly, governmental organizations can provide individuals with the cybersecurity practices, knowledge, and tools they need to recognize, stop, and react to cyberattacks. In addition to the incentives for partaking in cyber-awareness programs, the assistance for training and education initiatives may be included. More notably, it is also important to emphasize the development of public–private partnerships in order to improve comprehensive cyberinfrastructure.

7.6.2 Limitations and Future Scopes

Like any statistically based studies, there are a number of limitations to the results that should be acknowledged and taken into account in subsequent studies. First off, because all of the samples came from Vietnam, additional cross-regional study is needed to confirm the findings before they can be applied broadly. Future studies can also enhance the outcomes by building on the current research findings and using a larger sample size in nations with diverse technological infrastructures. The second constraint relates to the results' generalizability because this study's foundation is a non-probabilistic sampling technique. The chosen sociodemographic characteristics are not used to distribute the respondents. Furthermore, the majority of the users have expertise with the technology, which can lead to selection bias. Therefore, data from individuals with various backgrounds and experiences might be included in future studies.

Acknowledgments This chapter was funded by University of Economics Ho Chi Minh City (UEH) in Vietnam.

References

Abbate, S., Centobelli, P., Cerchione, R., Oropallo, E., & Riccio, E. (2022). A first bibliometric literature review on Metaverse. In *2022 IEEE technology and engineering management conference (TEMSCON EUROPE)* (pp. 254–260). IEEE.

Al-Ayed, F. (2022). Contemporary cybersecurity challenges in Metaverse using MetMetav. In *2022 international conference on computational science and computational intelligence (CSCI)* (pp. 1–4). IEEE.

Alhayani, B., Jasim Mohammed, H., Zeghaiton Chaloob, I., & Saleh Ahmed, J. (2021). Effectiveness of artificial intelligence techniques against cyber security risks apply of IT industry. *Materials Today: Proceedings, 531*, 1–6. https://doi.org/10.1016/j.matpr.2021.02.531

Ali, S., Abuhmed, T., El-Sappagh, S., Muhammad, K., Alonso-Moral, J. M., Confalonieri, R., Guidotti, R., Ser, J. D., Díaz-Rodríguez, N., & Herrera, F. (2023). Explainable artificial intelligence (XAI): What we know and what is left to attain trustworthy artificial intelligence. *Information Fusion, 99*, 1–52. https://doi.org/10.1016/j.inffus.2023.101805

Allam, Z., Sharifi, A., Bibri, S. E., Jones, D. S., & Krogstie, J. (2022). The Metaverse as a virtual form of smart cities: Opportunities and challenges for environmental, economic, and social sustainability in urban futures. *Smart Cities, 5*, 771–801. https://doi.org/10.3390/smartcities5030040

Alsharari, N. M., & Aljohani, M. S. (2023). The benchmarking implementation and management control process as influenced by interplay of environmental and cultural factors: Institutional and contingency perspectives. *Benchmarking: An International Journal*, 1–22. https://doi.org/10.1108/BIJ-11-2022-0733

Andrade, C. (2021). The inconvenient truth about convenience and purposive samples. *Indian Journal of Psychological Medicine, 43*(1), 86–88. https://doi.org/10.1177/0253717620977000

Arpaci, I., & Sevinc, K. (2022). Development of the cybersecurity scale (CS-S): Evidence of validity and reliability. *Information Development, 38*(2), 218–226. https://doi.org/10.1177/0266666921997512

Bada, M., & Nurse, J. R. C. (2020). The social and psychological impact of cyberattacks. In *Emerging cyber threats and cognitive vulnerabilities* (pp. 73–92). https://doi.org/10.1016/b978-0-12-816203-3.00004-6

Bagozzi, R., & Yi, Y. (2012). Specification, evaluation and interpretation of structural equation models. *Journal of Academic Marketing Science, 40*(1), 8–34. https://doi.org/10.1007/s11747-011-0278-x

Bibri, S. E. (2021a). Data-driven smart sustainable cities of the future: An evidence synthesis approach to a comprehensive state-of-the-art literature review. *Sustainable Futures, 3*, 1–23. https://doi.org/10.1016/j.sftr.2021.100047

Bibri, S. E. (2021b). Data-driven smart eco-cities and sustainable integrated districts: A best-evidence synthesis approach to an extensive literature review. *European Journal of Futures Research, 9*, 1–43. https://doi.org/10.1186/s40309-021-00181-4

Bibri, S. E., & Allam, Z. (2022a). The Metaverse as a virtual form of data-driven smart urbanism: On post-pandemic governance through the prism of the logic of surveillance capitalism. *Smart Cities, 5*, 715–727. https://doi.org/10.3390/smartcities5020037

Bibri, S. E., & Allam, Z. (2022b). The Metaverse as a virtual form of data-driven smart cities: The ethics of the hyper-connectivity, datafication, algorithmization, and platformization of urban society. *Computational Urban Science, 2*(22), 1–22. https://doi.org/10.1007/s43762-022-00050-1

Bokhari, S. A. A., & Myeong, S. (2023). The influence of artificial intelligence on E-governance and cybersecurity in smart cities: A stakeholder's perspective. *IEEE Access, 11*, 69783–69797. https://doi.org/10.1109/ACCESS.2023.3293480

Braun, V., & Clarke, V. (2019). Reflecting on reflexive thematic analysis. *Qualitative Research in Sport, Exercise and Health, 11*(4), 589–597. https://doi.org/10.1080/2159676x.2019.1628806

Bridge, G. (2009). Urbanism. In R. Kitchin & N. Thrift (Eds.), *International encyclopedia of human geography* (pp. 106–111). Elsevier.

Byrne, B. M. (2013). *Structural equation modeling with AMOS: Basic concepts, applications, and programming*. Routledge.

Corallo, A., Lazoi, M., Lezzi, M., & Luperto, A. (2022). Cybersecurity awareness in the context of the industrial Internet of Things: A systematic literature review. *Computers in Industry, 137*, 1–16. https://doi.org/10.1016/j.compind.2022.103614

Creswell, J. W., & Creswell, J. D. (2018). *Research design: Qualitative, quantitative, and mixed methods approaches*. Sage.

Edmondson, A., & McManus, S. (2007). Methodological fit in management field research. *Academy of Management Review, 32*, 1155–1179. https://doi.org/10.5465/AMR.2007.26586086

Etikan, I., Musa, S. A., & Alkassim, R. S. (2016). Comparison of convenience sampling and purposive sampling. *American Journal of Theoretical and Applied Statistics, 5*, 1–4. https://doi.org/10.11648/j.ajtas.20160501.11

Falchuk, B., Loeb, S., & Neff, R. (2018). The social Metaverse: Battle for privacy. *IEEE Technology and Society Magazine, 37*(2), 52–61. https://doi.org/10.1109/mts.2018.2826060

Fossey, E., Harvey, C., Mcdermott, F., & Davidson, L. (2002). Understanding and evaluating qualitative research. *Australian & New Zealand Journal of Psychiatry, 36*(6), 717–732. https://doi.org/10.1046/j.1440-1614.2002.01100.x

Garson, G. D. (2015). *Structural equation modeling*. Statistical Associates Publishers.

Gregory, D., Johnston, R., & Pratt, G. (2009). *Dictionary of human geography*. Wiley-Blackwell.

Gurov, O., & Konkova, T. (2022). Metaverses for human or human for Metaverses. *Artificial Societies, 17*(1), 1–11. https://doi.org/10.18254/S207751800019011-1

Hair, J. F., Sarstedt, M., Ringle, C. M., & Mena, J. A. (2012). An assessment of the use of partial least squares structural equation modeling in marketing research. *Journal of the Academy of Marketing Science, 40*(3), 414–433. https://doi.org/10.1007/s11747-011-0261-6

Hair, J. F., Hult, G. T. M., Ringle, C. M., & Sarstedt, M. (2017). *A primer on partial least squares structural equation modeling (PLS-SEM)* (2nd ed.). Sage.

Harvey, D. (2009). *Social justice and the city*. Edward Arnold.

Hew, T.-S., & Syed Abdul Kadir, S. L. (2016). Understanding cloud-based VLE from the SDT and CET perspectives: Development and validation of a measurement instrument. *Computers & Education, 101*, 132–149. https://doi.org/10.1016/j.compedu.2016.06.004

Hinton, P. R., McMurray, I., & Brownlow, C. (2014). *SPSS explained* (2nd ed.). Routledge/Taylor & Francis Group.

Jaipong, P., Siripipattanakul, S., Sriboonruang, P., & Sitthipon, T. (2022). A review of Metaverse and cybersecurity in the digital era. *International Journal of Computing Sciences Research*, 1–8. https://doi.org/10.25147/ijcsr.2017.001.1.122

Jean-Noël, B., Gongbing, B., & Ali, Z. (2020). Exploring the factors influencing the adoption of supply chain finance in supply chain effectiveness: Evidence from manufacturing firms. *Journal of Business & Industrial Marketing, 36*(5), 706–716. https://doi.org/10.1108/JBIM-01-2020-0047

Kant Pal, S., Mukherjee, S., Baral, M. M., & Aggarwal, S. (2021). Problems of big data adoption in the healthcare industries. *Asia Pacific Journal of Health Management, 16*(4), 282–287.

Kesmodel, U. S. (2018). Cross-sectional studies—What are they good for? *Acta Obstetricia et Gynecologica Scandinavica, 97*(4), 388–393. https://doi.org/10.1111/aogs.13331

Kitchin, R. (2014). The real-time city? Big data and smart urbanism. *GeoJournal, 79*(1), 1–14. https://doi.org/10.1007/s10708-013-9516-8

Kline, R. B. (2011). Convergence of structural equation modeling and multilevel modeling. In M. Williams & W. P. Vogt (Eds.), *Handbook of methodological innovation* (pp. 562–589). Sage.

Komar, M., Kochan, V., Dubchak, L., Sachenko, A., Golovko, V., Bezobrazov, S., & Romanets, I. (2017). High performance adaptive system for cyber-attacks detection. In *2017 9th IEEE international conference on intelligent data acquisition and advanced computing systems: Technology and applications (IDAACS), 2* (pp. 853–858). IEEE.

Lathabhavan, R., & Griffiths, M. D. (2023). Antecedents and job outcomes from a self-efficacy perspective while working from home among professionals during the COVID-19 pandemic. *International Journal of Manpower*, 1–29. https://doi.org/10.1108/IJM-04-2022-0185

Lee, U.-K., & Kim, H. (2022). UTAUT in Metaverse: An "Ifland" case. *Journal of Theoretical and Applied Electronic Commerce Research, 17*, 613–635. https://doi.org/10.3390/jtaer17020032

Lee, L. H., Braud, T., Zhou, P., Wang, L., Xu, D., Lin, Z., Kumar, A., Bermejo, C., & Hui, P. (2021). All one needs to know about Metaverse: A complete survey on technological singularity, virtual ecosystem, and research agenda. *Journal of Latex Class Files, 14*(8), 1–15. https://doi.org/10.13140/RG.2.2.11200.05124/8

Levitt, H. M., Bamberg, M., Creswell, J. W., Frost, D. M., Josselson, R., & Suárez-Orozco, C. (2018). Journal article reporting standards for qualitative primary, qualitative meta-analytic, and mixed methods research in psychology: The APA Publications and Communications Board task force report. *American Psychologist, 73*(1), 26–46. https://doi.org/10.1037/amp0000151

Mahmud, A., Ding, D., Hasan, M., Ali, Z., & Amin, M. B. (2022). Employee psychological reactions to microcorporate social responsibility and societal behavior: A structural equation modeling analysis. *Current Psychology*, 1–15. https://doi.org/10.1007/s12144-022-02898-2

Marvin, S., Luque-Ayala, A., & McFarlane, C. (Eds.). (2016). *Smart urbanism—Utopian vision or false dawn?* Routledge.

McIntosh, M. J., & Morse, J. M. (2015). Situating and constructing diversity in semi-structured interviews. *Global Qualitative Nursing Research, 2*, 1–12. https://doi.org/10.1177/2333393615597674

Mehta, M., Pancholi, G., & Saxena, D. A. (2023). Metaverse changing realm of the business world: A bibliometric snapshot. *Journal of Management Development, 42*(5), 373–387. https://doi.org/10.1108/JMD-01-2023-0006

Mishra, S., & Tripathi, A. R. (2021). AI business model: An integrative business approach. *Journal of Innovation and Entrepreneurship, 10*(1), 1–21. https://doi.org/10.1186/s13731-021-00157-5

Mora, L., Deakin, M., Zhang, X., Batty, M., De Jong, M., Santi, P., & Appio, F. P. (2021). Assembling sustainable smart city transitions: An interdisciplinary theoretical perspective. *Journal of Urban Technology, 28*(1–2), 1–27. https://doi.org/10.1080/10630732.2020.1834831

Mouton, M., & Burns, R. (2021). (Digital) neo-colonialism in the smart city. *Regional Studies, 55*(12), 1890–1901. https://doi.org/10.1080/00343404.2021.1915974

Mustak, M., Salminen, J., Plé, L., & Wirtz, J. (2021). Artificial intelligence in marketing: Topic modeling, scientometric analysis, and research agenda. *Journal of Business Research, 124*, 389–404. https://doi.org/10.1016/j.jbusres.2020.10.044

Nye, C. D. (2022). Reviewer resources: Confirmatory factor analysis. *Organizational Research Methods*, 1–21. https://doi.org/10.1177/10944281221120541

Parker, L. D., & Northcott, D. (2016). Qualitative generalising in accounting research: Concepts and strategies. *Accounting, Auditing & Accountability Journal, 29*(6), 1100–1131. https://doi.org/10.1108/aaaj-04-2015-2026

Pooyandeh, M., Han, K.-J., & Sohn, I. (2022). Cybersecurity in the AI-based Metaverse: A survey. *Applied Sciences, 12*(12993), 1–23. https://doi.org/10.3390/app122412993

Preko, A., Amoako, G. K., Dzogbenuku, R. K., & Kosiba, J. (2023). Digital tourism experience for tourist site revisit: An empirical view from Ghana. *Journal of Hospitality and Tourism Insights, 6*(2), 779–796. https://doi.org/10.1108/JHTI-10-2021-0294

Rosenberg, L. (2022). *Regulation of the Metaverse: A roadmap*. Conference: 6th international conference on virtual and augmented reality simulations (ICVARS 2022), Brisbane, Australia.

Saunders, M. N. K., Lewis, P., & Thornhill, A. (2019). *Research methods for business students* (8th ed.). Pearson.

Selem, K. M., Ahmad, M. S., Belwal, R., & AlKayid, K. (2022). Fear of COVID-19, hotel employee outcomes and workplace health and safety management practices: Evidence from Sharm El-Sheikh, Egypt. *Tourism and Hospitality Research*, 1–16. https://doi.org/10.1177/14673584221119374

Su, Y., & Fan, D. (2023). Smart cities and sustainable development. *Regional Studies, 57*(4), 722–738. https://doi.org/10.1080/00343404.2022.2106360

Tang, K.-Y., Chang, C.-Y., & Hwang, G.-J. (2021). Trends in artificial intelligence-supported e-learning: A systematic review and co-citation network analysis (1998–2019). *Interactive Learning Environments, 31*(4), 2134–2152. https://doi.org/10.1080/10494820.2021.1875001

Thottoli, M. M. (2021). Knowledge and use of accounting software: Evidence from Oman. *Journal of Industry – University Collaboration, 3*(1), 2–14. https://doi.org/10.1108/JIUC-04-2020-0005

Thottoli, M. M., Thomas, K. V., & Ahmed, E. R. (2019). Adoption of audit software by audit firms: A qualitative study. *Journal of Information and Computational Science, 9*(9), 768–776.

Tlili, A., Huang, R., Shehata, B., Liu, D., Zhao, J., Metwally, A. H. S., Wang, H., Denden, M., Bozkurt, A., Lee, L.-H., Beyoglu, D., Altinay, F., Sharma, R. C., Altinay, Z., Li, Z., Liu, J., Ahmad, F., Hu, Y., Salha, S., Abed, M., & Burgos, D. (2022). Is Metaverse in education a blessing or a curse: A combined content and bibliometric analysis. *Smart Learning Environments, 24*, 1–31. https://doi.org/10.1186/s40561-022-00205-x

Tlili, A., Huang, R., & Kinshuk. (2023). Metaverse for climbing the ladder toward 'Industry 5.0' and 'Society 5.0'? *The Service Industries Journal, 43*(3–4), 260–287. https://doi.org/10.1080/02642069.2023.2178644

Tomko, M., & Winter, S. (2019). Beyond digital twins—A commentary. *Environment and Planning B: Urban Analytics and City Science, 46*(2), 395–399. https://doi.org/10.1177/2399808318816992

Weaver, G. A., Feddersen, B., Marla, L., Wei, D., Rose, A., & Van Moer, M. (2022). Estimating economic losses from cyber-attacks on shipping ports: An optimization-based approach. *Transportation Research Part C: Emerging Technologies, 137*, 1–29. https://doi.org/10.1016/j.trc.2021.103423

Wilson, V. (2014). Research methods: Triangulation. *Evidence Based Library and Information Practice, 9*, 74–75. https://doi.org/10.18438/B8WW3X

Yaqoob, I., Salah, K., Jayaraman, R., & Omar, M. (2023). Metaverse applications in smart cities: Enabling technologies, opportunities, challenges, and future directions. *Internet of Things, 23*, 1–19. https://doi.org/10.1016/j.iot.2023.100884

Zuboff, S. (2019). *The age of surveillance capitalism: The fight for a human future at the new frontier of power*. Profile Books.

Pham Quang Huy is an Advanced Lecturer in the School of Accounting, University of Economics Ho Chi Minh City, Vietnam. He also earned the Certified Public Accountant and has strong expertise in accounting and management in public sector. Prior to this chapter, he has finished publishing numerous peer-reviewed journals and created a significant contribution to the management in public sector through several scientific research projects in Vietnam. He has recently worked on the project targeting at exploring the new management application in performance enhancement in Vietnam public sector organizations under the support of state budget. His current studies consist of performance measurement and management in public sector, internal control, and quality of financial statement information in public education sector, as well as the impact of technological initiatives on the orientations and operations in public sector.

Vu Kien Phuc is a lecturer in School of Accounting at University of Economics Ho Chi Minh City with Vinh Long Campus in Vietnam. Her research interests are mainly concentrated around accounting and management in private and public sector. She is always being the best teacher in students' heart in her teaching courses. She also completed several articles as well as studies for international conferences. Her present studies include measuring and managing performance in public sector, accounting practices in private and public sector. She has many papers published in the reputed journals with the research mainstreams of green finance and accounting, innovation, auditing, public sector organizations, and related others.

Chapter 8
Digital Twin: A Transformative Tool for Smart Cities

Rahul Joshi and Rhythma Badola

Abstract The concept of digital twins (DTs) has attracted considerable attention from scholars owing to recent breakthroughs in information and communication technology (ICT) since it can potentially bring revolutionary changes in several sectors. The term "Digital twin" pertains to the digital portrayal of a tangible object, which can replicate its physical characteristics via platforms and real-time data exchange. Digital twins, in conjunction with the Internet of Things, fifth-generation wireless networks, blockchain, collaborative computing, simulation, and artificial intelligence technologies, provide significant prospects for the reformation of the existing urban government paradigm toward smart cities. Digital twins are widely regarded as a novel and crucial foundation for developing contemporary smart city infrastructure. Smart cities use the integration of Big Data and the Internet of Things (IoT) to enhance the quality of life for their residents. Due to its intricate nature, implementing these technologies poses a challenging endeavor, necessitating the ongoing acquisition, consolidation, and examination of data. Therefore, adopting a systematic strategy that focuses on digital transformation is essential to address urban challenges effectively. Substantial endeavors are being made to construct city information models that facilitate encoding urban entities and their interconnections that help in decision-making processes. The successful management of diverse and abundant information necessitates using a shared knowledge foundation strengthened by extensive vocabularies and ontologies. This chapter analyzes and interprets the significance of digital twin technology within the context of smart cities. It investigates the potentially transformative impact of digital twin technology on the construction sector, highlighting its significance as a game-changing technological advancement. Furthermore, it examines several viewpoints about the function of digital twins inside this sector and describes how the digital twin might contribute to advancing innovative city development.

R. Joshi (✉)
School of Media Studies & Humanities, Manav Rachna International Institute of Research & Studies, Faridabad, Haryana, India

R. Badola
Patanjali India Limited, Haridwar, Uttarakhand, India

S. Majumdar et al. (eds.), *Smart Cities*, S.M.A.R.T. Environments,
https://doi.org/10.1007/978-3-031-59846-3_8

Keywords Blockchain · Digital twin · Development · IoT · Smart cities · Transformation · Urban

8.1 Introduction

Undoubtedly, digital solutions have the potential to provide substantial benefits to an organization, benefits that were previously unattainable before the emergence of interconnected and intelligent technology. Recently, there has been a notable interest in the digital twin concept, which refers to a digital representation of a physical item or process that closely mirrors its real-time state. The purpose of a digital twin is to enhance company performance by enabling optimization strategies. Businesses faced challenges in accessing and using the digital twin, along with the extensive data it handles, mainly owing to constraints in digital technology capabilities and the high costs associated with processing, storage, and bandwidth. However, this situation has changed in recent times. Nevertheless, these challenges have significantly decreased in recent times. The convergence of information technology (IT) and operations technology (OT) has been facilitated by substantial cost reductions and advancements in power and capabilities. These developments have resulted in exponential transformations, empowering executives to effectively integrate IT and OT to create and use a digital twin. The significance of the digital twin and its relevance for organizations deserve investigation. The concept of a digital twin enables organizations to establish a comprehensive digital representation of their goods, including the whole lifecycle spanning from the initial design and development stages to the conclusion of the product's lifecycle. Consequently, this may facilitate their comprehension of the product's design, the system responsible for its creation, and the practical use of the product in real-world scenarios.

Leng and Jiang (2019) argued, in the traditional methodology, designers use computer-aided simulation and engineering tools to facilitate the design process, forecast the lifecycle, and execute diverse physical testing methodologies. The optimization of design is undertaken to maximize performance and reduce design costs. However, this technique has limitations regarding tolerances, tactics, configuration interactions, and planning. Nevertheless, the advancement of the computer industry, characterized by the integration of artificial intelligence, improved processing speed, increased algorithms, and the continuous growth of computational power, has significantly impacted several aspects of the goods and production line. According to Wan et al. (2017), one notable outcome is the emergence of the digital twin concept, which facilitates real-time control and digitization. A digital twin may represent a physical thing, process, or system. By integrating data and intelligence that capture the characteristics, context, and dynamics of a physical system, an interface is created that enables the monitoring of historical and current operations and the ability to make predictions about future outcomes (Grieves, 2014).

Hence, Constante (2018) argued, "The concept of a digital twin involves the integration of artificial intelligence, software analytics, and machine learning data to develop digital simulation models that dynamically adapt and evolve in response to changes occurring in their corresponding physical counterparts." This system offers simultaneous real-time tracking and updates from several sources. The process involves generating digital representations of items to imitate their behavior. The virtual models can comprehend the condition of tangible items through sensing data, enabling them to make estimations and conduct analyses on the ever-changing dynamics. Tao et al. (2018) stated that the digital twin concept can potentially optimize the whole manufacturing process.

8.2 History of Digital Twin

Despite the significant surge in the popularity of DT technology in recent years, it is essential to note that the underlying principle is only partially novel. The concept was first conceived in 2002 at the University of Michigan by Michael Grieves in collaboration with Product Lifecycle Management (PLM) (Grieves, 2016). The model being examined has three discrete components: the physical world, the digital space, and an intermediary technology that enables the exchange of data and knowledge between these two domains. The framework under consideration has been referred to as the "Mirrored Spaces Model" in prior scholarly works (Grieves, 2005). David Gelernter introduced the "Mirror Worlds" notion in 1991, whereby software models imitate reality based on data obtained from the actual world (Gelernter, 1993). Framing et al. (2003) proposed "an agent-based architecture where each product item has a corresponding 'virtual counterpart' or agent associated with it." This proposal addressed the inefficiencies associated with the paper-based transfer of production information in the context of Product Lifecycle Management. Grieves suggested a conceptual model that underwent a name modification from the initial designation of the "Mirrored Spaces Model" to the revised appellation of the "Information Mirroring Model." They argued, "The model highlights the significance of a bidirectional connection mechanism between two places, as well as the existence of several virtual spaces for a single physical area." These virtual spaces allow for exploring different concepts or designs (Grieves, 2009). During the period in question, the actual implementation of DT was hindered by many technological constraints, including limited computer power, inadequate connection between devices and the Internet, challenges related to data storage and administration, and the underdevelopment of machine algorithms.

Shafto et al. (2012) stated, the term "Digital Twin" was first used in NASA's preliminary edition of the technical roadmap in 2010. Within the NASA roadmaps, the acronym DT was further denoted as "Virtual Digital Fleet Leader." The National Aeronautics and Space Administration (NASA) was the pioneering organization responsible for formulating the first concept of DT. This concept was "an integrated simulation of a vehicle or system that incorporates multiple physics, scales, and probabilities. It uses the most accurate physical models, sensor data updates, historical fleet information, and other relevant factors to replicate its airborne counterpart's operational lifespan accurately." Rosen et al. (2015) argued, although the first reference to DT can be traced back to the 2010 roadmap, it is worth noting that NASA had previously used a similar idea during the Apollo program. This included the construction of two identical spaceships designed to serve as mirror images of one another. Subsequently, the United States Air Force used DT technology, mirroring the approach taken by NASA, to facilitate its aircraft's design, maintenance, and predictive capabilities (Tuegel, 2012). The objective was to use digital twin technology to replicate the aircraft's physical and mechanical characteristics to predict fatigue or structural fractures. This approach effectively extends the aircraft's remaining operational lifespan. In their publications, Gockel et al. (2012) introduced the concept of DT specifically for aircraft, referring to it as the "Airframe Digital Twin" (ADT). They argued, "The ADT is a computational model designed to oversee the whole lifespan of an aircraft effectively. In addition to its use in monitoring, DT has been suggested as a viable approach for sustainable space exploration and developing future generations of aerospace vehicles" (Fig. 8.1).

8.3 Developing the DT Concept

The idea of digital twin has three distinct and significant components. The first crucial component pertains to tangible goods, namely actual objects existing in physical space. The second crucial component is including virtual items, namely those that exist inside a virtual environment. The third significant component consists of connections, namely the linkage of information and data, which integrates tangible and intangible items (Grieves, 2014). These connections may also be referred to as links. There exist a connection for the transmission of data and a separate connection for the transmission of information. The flow of data takes place from the actual world into the virtual world, whereas the flow of information takes place in the other direction, from the virtual world into the actual world and the virtual sub-spaces. Grieves and Vickers (2017) described digital twin in three following types:

(a) *Digital twin prototype*: This is the document that depicts a typical physical artifact. Digital twin prototype bears significant responsibility. The provided content encompasses the necessary information to depict and create a tangible iteration that emulates or duplicates the digital counterpart. The information comprises various elements, including essential items, a fully annotated 3D model, various materials accompanied by their specified specifications and corresponding invoices, bills for different services rendered, and bills for disposal and related processes.
(b) *Digital twin instance*: In this process, an individual digital twin is established to connect with a real-world product over its entire lifecycle, providing a comprehensive description of this object. Based on its application, the DTI digital twin can encompass various sets of information, such as a fully annotated 3D model incorporating Geometric Dimensioning and Tolerancing (GD&T). This model

Fig. 8.1 The timeline of historical innovations in digital twin

comprehensively describes the physical instance's geometry, including all its constituent components. Additionally, the digital twin includes a bill of materials comprising a comprehensive list of historical and current components. Another bill included in the bill of process contains a comprehensive account of the activities carried out during the production of this physical instance.

(c) *Digital twin environment*: This integrated and transdisciplinary physics application ecosystem aims to operate on a digital twin for various objectives. These many rationales include predictive and interrogative aspects:

1. *Predictive*: The use of digital twins, as implied by its name, primarily serves the goal of prediction. The prediction may be conducted not just for future behavior but also for the future functioning of the tangible object itself. During the prototype phase, predictions will be made about the product's behavior and the elements that exhibit variations in their low and high tolerances. During the prototyping phase, it is necessary to predict the behavior of the planned product

and its components, considering the variations within their high and low tolerances.

2. *Interrogative*: This situation would apply to circumstances where individuals may be interrogated on their present and previous experiences. Various examples of goods may provide associated data, which can be used to forecast future conditions.

8.4 Applications Involved in DT Technology

Digital twinning is a revolutionary technology that involves creating virtual copies of physical assets, systems, or processes. It has gained significant attention in various industries due to its potential to provide timely and valuable information, streamline processes, reduce costs, and enhance decision-making. This detailed section examines the many uses of DT technology across several domains.

Manufacturing and Industry 4.0 The industrial sector has been at the forefront of embracing DT technology as a crucial component of the Industry 4.0 revolution. Digital twins generate virtual representations of industrial processes, allowing the continuous real-time monitoring and analysis of these operations. This practice contributes to optimizing production efficiency, waste reduction, and improvement of quality control. Manufacturers can effectively anticipate machine failures and strategically plan maintenance activities using digital twins to monitor equipment in real time. This proactive approach enables them to save both downtime and associated expenses.

Aerospace and Automotive Engineering DT plays a pivotal role in several high-tech sectors, exhibiting numerous significant applications. Aerospace engineers use digital replicas, known as digital twins, to assess the aerodynamic properties, structural soundness, and overall performance of aircraft before the construction of tangible prototypes. This practice yields significant time and resource savings. Automotive manufacturers use computer simulations to conduct crash tests and analyze vehicles' performance, eliminating the need for extensive physical testing.

Power Sector The power sector is undergoing significant transformation due to the use of DT. Digital twins are used by operators to monitor and enhance power plant operations in real time, hence enhancing energy production efficiency and minimizing periods of inactivity. Digital twins are crucial in managing renewable energy sources, including wind and solar farms, by optimizing their performance and seamlessly integrating them into the existing power grid infrastructure.

Environmental Monitoring and Conservation DT plays a crucial role in facilitating environmental preservation and conservation endeavors. Researchers develop computerized replicas of ecosystems to effectively monitor alterations, investigate species' behavior, and forecast the ramifications of environmental shifts. Digital twins play a vital position in the simulation of climatic trends, hence facilitating climate change research and enhancing catastrophe preparation efforts.

Supply Chain Management Organizations use digital twins to monitor inventory levels in real time, hence mitigating surplus stock and enhancing the efficiency of the supply chain. In addition, digital twins are instrumental in enhancing operational efficiency by facilitating route optimization, truck monitoring, and demand forecasting. This technology assures the timely and efficient delivery of items to their intended locations, streamlining logistics processes, and increasing productivity.

Building and Facility Management Digital twins (DTs) can effectively monitor building systems like heating, cooling, and lighting to optimize energy consumption and ensure occupant comfort. Additionally, when utilized for predictive maintenance, DTs have proven to reduce equipment downtime and extend the operational lifespan of assets.

Retail and Customer Experience Retailers use digital twins to enhance the efficiency of shop layouts in order to optimize client flow and improve product positioning. The use of real-time inventory monitoring systems is a preventive measure against stockouts and overstock problems.

Defense and Military Applications DTs can replicate intricate mission situations, furnishing military personnel with realistic training scenarios. Digital twins are computer-based models replicating military equipment and vehicles' operational characteristics and behaviors in many scenarios.

Internet of Things (IoT) and Sensor Networks Digital twins provide a virtual representation of real-world data, facilitating sensor information analysis and use. The integration of Internet of Things (IoT) data into digital twins has the potential to facilitate predictive maintenance and improve decision-making strategies.

8.5 Overview of Digital Twin in Smart Cities

The idea of smart cities has played a vital role in facilitating the digital transformation of urban areas. A smart city is a city that employs contemporary digital technology to enhance the provision of municipal services, the development of infrastructure, and the overall well-being of its residents. Nevertheless, a more expansive interpretation of the concept incorporates a socio-technical framework and examines smart cities through the lenses of both economic and environmental considerations. The digital twin has inherent operational capabilities that render it well-suited for implementation in various applications. For it to operate effectively, the system needs the availability of data.

Digital twin is a technological advancement that has the potential to facilitate the planning and development processes associated with smart cities. The urban planners engaged in developing a smart city use digital twin technology to assess many hypothetical situations and make informed decisions based on the outcomes. Digital twin technology spans several application domains and sectors within the framework of smart cities. One such application is the management of traffic, which may be effectively facilitated via the implementation of digital twin systems. Furthermore, the use of DT technology extends to several areas within the smart city framework, including but not limited to construction and healthcare. As a result, the digital twin can provide extensive information about the domains and industries in which it is implemented, hence offering potential use. Ruohomäki et al. (2018) argued, the acquired data may be disseminated throughout several municipal departments, therefore becoming advantageous. The foundation of a digital twin lies in integrating sensor data with a city model. The concept of digital twin has been implemented and used in several practical applications, with advantageous outcomes for each of them. One notable instance of using digital twin technology is seen in Bridgestone Company's tyre development process (Shah & Suthar, 2022). This exemplifies the practical application of digital twin technology, which has shown its efficacy in several other domains.

In a digital twin city, data about the operational state of infrastructure, the allocation of municipal resources, and the movement of people, goods, and vehicles would be gathered using sensors, cameras, and other digital subsystems. The integration of technologies such as 5G, which enables the transmission of data to the cloud and facilitates connectivity with the local administration, has the potential to enhance the city's overall efficiency. The implementation of digital twin cities is expected to revolutionize urban planning, management, and services. It is a critical step toward the creation of smart cities, enabling comprehensive data visualization and integration of intelligent systems in urban planning and service provision.

8.6 Applications of DT in Smart Cities

Digital twins have a multitude of applications that can be utilized to organize, manage, and improve urban areas, ranging from mobility, sustainability, energy efficiency, waste management, security monitoring, to infrastructure maintenance (Vohra, 2022).

The following section highlights and examines specific applications of digital twin technology.

8.6.1 Managing the Flow of Traffic

Traffic congestion is a prevalent issue that may manifest on many routes inside urban areas, mainly due to the increasing volume of cars on the roadways. Dealing with the ever-challenging task of managing this intensifying problem has posed a growing challenge for traffic authorities. Technology may provide significant support in tackling this challenging situation, with digital twins as a possible remedy.

Digital twins are an indispensable tool for traffic management in smart cities, providing numerous benefits:

1. *Identification of Vulnerabilities*: Digital twins have exceptional capabilities in identifying specific traffic network locations prone to congestion or operational inefficiencies. The proactive identification of areas needing improvement facilitates the implementation of specific strategies for enhancement.
2. *Utilization of DTs in Testing Solutions*: Digital twins provide a significant opportunity to conduct experiments and evaluate prospective traffic solutions before implementing them in the physical realm. This mitigates the potential hazards linked to unverified operations.
3. *Flow of traffic*: Traffic flow optimization may be achieved by using digital twins, which provide valuable insights. These insights can be used to fine-tune and optimize traffic flow, creating smoother and more efficient transportation networks.
4. *Management becomes easy*: Digital twin technology enables traffic departments to optimize their operations and decision-making procedures, enhancing the effectiveness and adaptability of traffic management in response to dynamic situations.

8.6.2 Construction

Efficient building operations are crucial for advancing diverse structures, particularly in smart cities. DT technology has emerged as an exceptional tool in optimizing various stages of construction, from the initial design process to project management. In the design process, digital twins play a critical role. By utilizing simulation tools, engineers and architects can assess the potential consequences of design modifications and make well-informed decisions. This approach streamlines the design process and ensures that final designs align with the project's objectives. This technology offers significant benefits throughout the building process. Real-time monitoring of project progress allows project managers to closely track ongoing activities. Moreover, remote monitoring capabilities make it possible to promptly identify and resolve any deviations from intended procedures, ensuring a smooth and efficient project administration. Proactively predicting and detecting faults or difficulties before they physically manifest is another advantage of digital twins. This approach reduces disruptions and lowers the likelihood of expensive rework, facilitating a more streamlined construction process. Digital twins can gather vast amounts of data about various aspects of the building. This data can optimize different processes, strengthen safety protocols, and enhance the project's overall performance.

8.6.3 Systemic Health Monitoring

DT technology represents a viable approach to monitor structural integrity. Digital twins enable real-time data transfer from various structures to a digital twin representation. This representation continuously updates itself, receives real-time changes in data, and monitors the structural integrity of objects. It enables the convenient practice of remote monitoring while also providing valuable insights into prospective structural challenges and health-related considerations. Implementing digital twin facilitates predictive maintenance strategies for structures, offering a viable and advantageous approach. In conclusion, incorporating digital twin technology into smart urban environments has shown significant efficacy in monitoring the structural integrity of various structures. Digital twins streamline monitoring the well-being of diverse infrastructures and offer immediate and valuable insights into their current state. They also enable the real-time identification of maintenance and upgrading requirements in buildings from a distant location, assisting in mitigating safety problems arising from structural deficiencies. Integrating digital twins into building and maintenance methods of smart cities serves to preserve the structural integrity of critical infrastructure, augmenting both safety and durability.

8.6.4 Healthcare

Integrating healthcare into a digital twin architecture within smart cities presents exciting opportunities to improve public health, healthcare delivery, and overall well-being. By replicating hospital layouts and patient flows, digital twins can facilitate the construction of healthcare facilities that are optimized for efficiency, ensuring that hospitals are structured efficiently, conveniently accessible, and equipped to manage an inflow of patients during times of crisis. Digital twins may also support the development and assessment of healthcare facilities that prioritize infection control measures, protecting both patients and healthcare personnel during worldwide health emergencies, such as pandemics. Digital twins can generate virtual representations of patients, including their medical records, genetic data, and up-to-date health metrics. This facilitates healthcare practitioners in customizing treatment plans and actions according to the specific requirements of each patient. Real-time monitoring of patients may be achieved by using wearable devices and sensors, enabling the timely identification of health conditions, and enhancing overall patient results. It also offers hospitals and clinics the ability to optimize resource allocation, including staff schedules, bed usage, and administration of medical equipment. This approach guarantees the adequate provision of healthcare services and reduces the duration of waiting periods. Maintaining a consistent supply of pharmaceuticals, medical equipment, and personal protective gear is also paramount, and digital twins can monitor and forecast inventory levels, enabling the timely acquisition of goods and services while anticipating any interruptions in the supply chain. Digital twins facilitate the implementation of telemedicine by offering a safe and immersive environment that enables virtual consultations between patients and healthcare providers. In addition, DT-enabled health kiosks can provide essential health evaluations in public locations such as airports or shopping malls.

8.6.5 Efficient Drainage Systems

DT technology has emerged as a viable method to optimize the functionality of drainage systems. Water system administrators, including water managers, often encounter water quality and flood protection challenges. The emergence of digital twins for water systems has generated a burgeoning interest in their development. According to Bartos & Kerkez (2021), "The purpose of DT is to provide detailed representations of surface water systems. Integrating sensor data and online models is executed smoothly, comprehensively understanding the system's dynamics, and facilitating accurate control." One significant development in this field is the authors' introduction of "pipedream," an end-to-end simulation engine. Pipedream is a real-time tool for modeling and estimating the condition of natural and urban drainage networks. The engine merges two entities. The first component is a novel hydraulic solver that employs the principles of the "one-dimensional Saint Venant equations." The second approach utilizes a "Kalman filtering technique" to incorporate observed data and update the hydraulic conditions. The authors have shown the effectiveness of their simulation engine in accurately predicting and estimating hydraulic conditions by using sensor data obtained from real-world watersheds. The benefits of this technology are wide-ranging and substantial. The Pipedream system facilitates the prompt detection of flood occurrences, providing timely alerts to minimize harm. Maintenance and remediation are crucial in improving the comprehension of infrastructure upkeep and repair requirements, enabling proactive management.

8.6.6 Power Grid

To maintain a stable and uninterrupted energy supply, the integration of digital twins is crucial. "Hitachi Energy's IdentiQ is an outstanding digital twin solution for high-voltage direct current (HVDC) and power quality applications. Its primary objective is to promote flexibility, sustainability, and security in the global energy system, thereby accelerating the transition towards a carbon-neutral future"

(Taqaddas, 2023). This example showcases the advantages of utilizing digital twin systems in power networks. Smart grids enable the seamless integration of renewable energy sources, such as solar panels and wind turbines. These systems facilitate bidirectional power flow, allowing customers to sell any surplus energy back to the grid. This feature serves to foster the promotion of energy sustainability. Smart cities use energy storage technology, such as batteries, to store surplus energy during times of low demand and then discharge it during periods of high demand. This practice alleviates pressure on the electrical grid and fosters the grid infrastructure's stability. Smart grids have the potential to improve invoicing accuracy and boost customer service.

8.6.7 Environmental Monitoring

In order to ensure the continued success of smart cities, it is crucial to incorporate environmental monitoring. By utilizing a digital twin framework, the effectiveness of environmental monitoring can be further enhanced. DT technology enables cities to create real-time simulations of their ecosystems, which allows for well-informed decision-making, mitigation of environmental consequences, and overall improvement of citizens' quality of life. One significant application of ecological monitoring in smart cities is real-time air-quality monitoring. By utilizing data from various sensors and sources, digital twins can provide accurate and reliable representations of air quality across urban areas. This enables municipal authorities to identify areas with high levels of pollution, anticipate changes in air quality, and implement strategies to combat air pollution. By taking proactive measures to address air-quality concerns, urban areas can improve public health outcomes and reduce healthcare costs. Digital twins also play a crucial role in monitoring and managing the quality of water supply in urban areas. By combining data from sensors located in water treatment facilities, distribution networks, and natural water sources, urban areas can ensure the provision of safe drinking water, quickly detect pollution incidents, and improve water treatment processes. Additionally, digital twins facilitate water resource management, including leak detection, demand forecasting, and efficient distribution, which contributes to water conservation in urban areas. Effective trash management is another crucial component of environmental monitoring in smart cities. Digital twins can monitor and manage the amount of waste produced, optimize waste collection routes, and monitor landfill capacity. Metropolitan regions can make substantial strides toward sustainability by implementing waste reduction, recycling, and composting initiatives to mitigate environmental effects. This organization's platform offers modeling and simulation capabilities for assessing the effects of climate change on urban environments, including increased temperatures, rising sea levels, and severe weather events. Urban areas can increase their resilience to climate change impacts and minimize risks by utilizing digital twins to evaluate vulnerabilities and formulate adaptation strategies. As the importance of conserving biodiversity within urban settings becomes increasingly recognized, digital twins can be used to effectively map green areas, monitor animal populations, and evaluate the impact of urban development on nearby ecosystems. This information can serve as a guiding framework for the strategic development and effective administration of urban parks and green infrastructure, promoting ecological sustainability within cities.

8.7 Digital Twin and India's Urban Transformation

According to the United Nations, India is home to 1.4 billion people, accounting for roughly 17.7% of the global population. As of 2023, 35% of the population resided in urban areas. The International Monetary Fund (IMF) predicts that India's economy will grow by 6.1% in 2023 and around 6.5% in 2024. The Indian government has placed great emphasis on digital transformation, particularly in urban departments. Through various initiatives such as the Smart City Mission, AMRUT Scheme, Swachh Bharat Mission-Urban, HRIDAY, and Pradhan Mantri Awas Yojana-Urban, significant funds have been allocated and effective implementation has been demonstrated by diverse departments. The "location" factor is a pivotal element in urban initiatives and is recognized by all stakeholders. Spatial analysis is a valuable resource that reveals concealed geographic connections, and it contributes to every stage of urban infrastructure projects, ranging from planning and designing to constructing, operating, and maintaining. The digital twin technology in India is now in its early developmental phase, exhibiting restricted use within specific sectors. Nevertheless, many noteworthy ongoing efforts exemplify the potential for broader implementation.

8.7.1 India's Smart Cities: An Overview of Urban Change

The Indian Government has initiated a selection process for 20 smart cities out of the 100 proposed across different states and union territories of India in the fiscal year 2017–2018. To achieve the goals of the innovative city concept, it is crucial to have a dependable and secure communication channel. The current communication systems in smart cities require improvement (Mishra & Chakraborty, 2020). Therefore, it is imperative to enhance the existing service infrastructure using artificial intelligence and soft computing methodologies. The primary objective of enhancing the quality of life in a smart city is to establish connectivity among various communication and information technologies. For instance, integrating sensors and the Internet of Things (IoTs) can enhance the comfort and safety of inhabitants. A notable example is creating a communication infrastructure within residential settings that enables the integration of telephony and other communication technologies, such as personal computers, through the Internet of Things (IoT) in a smart city.

8.7.1.1 Successful Stories of Smart Cities in India

Pune: The Leader in Urban Evolution Using Digital Twins
Pune, a bustling city in Maharashtra, has been recognized for its innovative use of digital twin technology to drive urban revitalization. The city, renowned for its rich cultural heritage and thriving IT industry, faces a host of urban challenges, including traffic congestion, inadequate public transport, and insufficient waste management infrastructure. In response, the Pune Municipal Corporation (PMC) launched a comprehensive smart city initiative, with the primary objective of improving traffic management, urban planning, and environmental monitoring. To begin, a vast amount of data were gathered from various sources, such as IoT sensors, traffic cameras, waste management systems, and environmental monitoring stations. In collaboration with technology partners, this data was used to create an accurate digital replica of the city, known as a digital twin. The digital twin utilized real-time data feeds and advanced simulation and modeling methods to precisely replicate the functioning of the urban environment. PMC established a monitoring center equipped with data analytics capabilities and implemented digital twin technology to enable real-time monitoring of urban operations by municipal authorities, including traffic management, garbage disposal, and environmental monitoring. Pune has fostered public engagement by providing real-time data through mobile applications and online platforms, enabling citizens to report concerns, receive information, and participate in the decision-making process of the city. The use of real-time traffic data and simulations has enabled the city to optimize traffic signals, redirect traffic flow during periods of congestion, and ultimately reduce journey durations, resulting in less stress for citizens. DT technology has enabled data-driven approaches to urban planning, enabling simulation of the potential effects of development projects on aspects such as traffic, utilities, and the environment, promoting sustainable growth (Saranya & Amutha, 2023). Overall, the digital twin platform has increased citizens' engagement in urban area, resulting in greater participation in the municipal decision-making process.

Amaravati: The Greenfield Smart City Powered by DT
Amaravati has been proposed as the capital city of Andhra Pradesh, India. This innovative smart city initiative prioritizes comprehensive urban infrastructure development using digital twins. The goal is to create a sophisticated metropolis that values sustainability and aesthetic appeal. The digital twin provides a detailed coverage of the entire city, allowing for effective planning and coordination. Amaravati's development approach is systematic and coordinated. The city's infrastructure, transportation, and public spaces are designed to create a harmonious urban environment. The digital twin infrastructure also plays a crucial role in electronic government, ensuring openness, effectiveness, and availability in the provision of public services. Citizens can easily access information and interact with the government. The digital twin technology ensures effective coordination and planning, while also serving as a fundamental infrastructure for electronic government

(Augustine, 2020). The city's preparation for emergency situations through catastrophe simulations has made it a safe and efficient place for its citizens.

Jaipur: Developing an Integrated Command and Control Center
Jaipur is renowned for its rich cultural and historic heritage. The city is making significant strides toward transforming itself into a modern, sustainable urban center via a range of smart city initiatives. These programs focus on different aspects of urban development and service provision to the citizens. To achieve this vision, Jaipur has implemented digital twin technology to enhance urban planning, infrastructure management, and overall governance. One of the most significant challenges faced by Jaipur is traffic congestion, especially in its historic urban areas. As Jaipur experiences rapid urbanization, meticulous planning is essential to accommodate its growing population and development. Planners and politicians can use the digital twin to visually represent the potential effects of new projects, make informed decisions based on data, and ensure that urban growth aligns with the city's long-term vision. The state of Rajasthan, of which Jaipur is a part, is prone to natural disasters, particularly floods. According to Mylonas et al. (2021), digital twins play a crucial role in enhancing disaster preparedness by simulating various catastrophic situations. Simulations enable municipal authorities to formulate efficient evacuation strategies, distribute resources appropriately, and synchronize emergency response efforts. Digital twins provide a framework for continuously monitoring infrastructure conditions and real-time performance. Predictive maintenance, enabled by data obtained from the digital twin, allows maintenance teams to proactively detect problems, optimize resource allocation, and minimize equipment downtime. Digital twins have the potential to enhance citizen participation by providing real-time information and interactive platforms, thereby granting people enhanced access. Jaipur residents can use digital technologies to report various concerns, avail themselves of public services, and remain updated on the progress and advancements inside the city.

Bhubaneswar: Digital Twin for Better Urban Amenities
Bhubaneswar, also known as the "Temple City of India," is famous for its rich cultural heritage and rapid urban growth. The city has undergone significant expansion and development, resulting in various challenges such as increased traffic congestion, environmental concerns, and the need for improved urban amenities. Traffic congestion is a major issue in Bhubaneswar, especially during peak hours. Digital twins are used to model and simulate traffic patterns, enabling the optimization of traffic signal timings, planning road extensions, and more effective traffic management. Real-time traffic data from sensors, cameras, and GPS devices are incorporated into the digital twin to enhance decision-making. With the expansion of Bhubaneswar, careful urban planning is required. Kodieswari et al. (2023) argued, digital twins can generate a virtual representation of the city's current infrastructure and potential growth initiatives, enabling urban planners and authorities to visualize the potential effects of future developments, make informed decisions based on data, and ensure that infrastructure investments align with the city's long-term

vision. Establishing a two-way communication channel promotes transparency and encourages citizens' active participation in municipal administration.

Indore: A Visionary State for Innovative Planning
Indore, situated in the Indian state of Madhya Pradesh, has been at the forefront of implementing innovative plans that emphasize digital twin technology for urban development and administration. One of the significant issues in Indore is traffic congestion, especially during peak hours. Digital twins are used to model and simulate traffic patterns, which helps in optimizing traffic signal timings, planning road extensions, and managing traffic flow more effectively. Real-time traffic data from sensors, cameras, and GPS devices feed into the digital twin, enabling better decision-making. Efficient waste management and sanitation are critical to maintaining a healthy and eco-friendly urban environment. Digital twins monitor garbage collection, routing, and disposal procedures, optimizing routes in real time, resulting in decreased fuel consumption and emissions, contributing to cleaner streets. In the context of Indore's ongoing growth, urban planning plays a crucial role in achieving sustainable development. Cheela et al. (2021) stated that, digital twins can generate a virtual replication of a city, including its current infrastructure and potential growth endeavors. This capability allows urban planners and authorities to visually comprehend the consequences of new projects, make choices based on empirical evidence, and verify that infrastructure developments follow the city's long-term vision. Indore's strategic use of digital twins aims to foster a more livable, sustainable, and technologically advanced urban environment. However, sustained efforts and financial investments will be necessary to fully harness digital twins' capabilities in the process of Indore's smart city metamorphosis.

8.8 Challenges Ahead in Smart Cities

Botín-Sanabria et al. (2022) argued, incorporating digital twins into smart city infrastructure offers numerous benefits, such as enhanced urban planning, infrastructure management, and personalized citizen services. Nevertheless, it is important to acknowledge the inherent complexities and challenges associated with this trend.

Within the context of digital twin smart cities, several significant issues are present, which are detailed below:

Integrating High-Quality Data Accurate data is an essential ingredient in the creation of smart cities. This data can be sourced from a variety of channels, including IoT sensors, security cameras, social media, and public records. However, harmonizing data from multiple sources can be time-consuming, and it is important that it is both accurate and reliable. Outdated or inaccurate data can result in substandard simulations and erroneous decisions. Hence, maintaining the accuracy of data is a perpetual challenge.

Confidentiality and Safety Digital twin systems can gather and evaluate substantial quantities of personal information. Given the potential dangers posed by cyberattacks and data breaches, safeguarding data privacy and security is indispensable. Maintaining the security of the digital twin platform is imperative in preventing cyberattacks. Additionally, as digital twins become increasingly integrated into urban infrastructure, they become appealing targets for malicious actors. In order to enhance privacy protection, digital twin systems may use methods such as data anonymization and de-identification. These methodologies guarantee eliminating or concealing personally identifiable information (PII) from datasets, mitigating the potential for data breaches or infringements on privacy.

Scalability As cities grow and change, it is crucial that digital twins are scalable to support the integration of new areas, infrastructure, and services. The primary obstacle is managing the vast amounts of data and complexity that come with expansion. To achieve comprehensive urban planning, there must be effective communication and data-sharing capabilities across digital twin systems within and between cities. However, achieving interoperability is challenging due to the various data formats and protocols used.

Data Governance Data governance is crucial in implementing digital twin systems that include several stakeholders, such as government agencies, commercial firms, and individuals. Establishing unambiguous data ownership and access rights is essential to ensure effective management and control of the data involved. The issue of data ownership, access, and monetization arising from sensor-generated data might give rise to complex governance frameworks.

Digital Divide The digital divide is a significant problem in terms of achieving equal access to technology and the advantages of digital twin systems. Marginalized populations may need more access to cellphones, laptops, or high-speed Internet, impeding their capacity to participate actively in digital twin platforms. Promoting digital literacy and mitigating gaps in access are crucial factors in the development of inclusive smart cities that cater to the needs of all inhabitants.

Mitigating Algorithmic Bias The success of digital twin systems heavily relies on sophisticated algorithms that process data and make decisions. However, a major obstacle that must be overcome is algorithmic bias, which could lead to unfair outcomes. To address this issue, advanced machine learning techniques, fairness-aware algorithms, and ongoing monitoring are required to prevent unintentional bias against certain communities or groups.

Regulatory and Legal Frameworks Effectively adhering to a multitude of rules across various levels, be it local, national, or international, can prove to be a significant technical challenge. For this reason, advanced compliance monitoring systems must be developed and deployed, continuously monitoring, and analyzing changes to legal obligations and evaluating the digital twin's adherence to these dynamic rules. Navigating the complexities of data residency laws and jurisdictional borders can be a daunting task, but technical solutions such as data partitioning, dispersed data storage, and geofencing systems can be employed to ensure compliance with relevant regulations, keeping data within established legal boundaries. Farsi et al. (2020) stated that, as the regulatory framework is in a constant state of evolution to incorporate technological advancements, it is imperative for smart cities to ensure that their digital twin systems comply with the ever-increasing legal requirements. This may involve updating the system, modifying algorithms, and revising policies from a technical standpoint.

8.9 Innovative Cloud-Based Methodology for Improving People's Life Quality in Smart Cities

Cloud computing refers to using distant server-based internetworks to store, manage, and process data, as opposed to relying on a local server or external discs. In the current age characterized by digitalization and virtualization, cloud computing has emerged as a prominent trend for the maintenance and deployment of software. Prominent multinational corporations and sectors, like IBM, Google, Amazon, and Microsoft, have adopted the cloud computing paradigm for data management and processing. The advent of virtualization has sparked a paradigm shift in contemporary technology. Virtual storage in conjunction with virtual operating systems has seen rapid growth in recent years, to the extent that their presence is now considered commonplace. To a certain degree, the dominance of the virtual realm has superseded that of the physical realm, and this phenomenon also applies to the notion of cloud computing. One notable characteristic of cloud computing is the absence of resource ownership for its users, who incur charges according to a per-user framework. The facilitation of customer services is made possible with online on-demand service delivery (Moser, 2001).

Cloud computing consists of three levels of services: infrastructure as a service (IaaS), platform as a service (PaaS), and software as a service (SaaS). The Software as a Service (SaaS) paradigm involves hosting applications by a service provider or vendor, who then provides the required services to clients over a wireless or wired network. Typically, this approach provides clients access to business software capability at a reduced cost while simultaneously enabling them to reap the same benefits without considering the associated complexities of program installation, licensing, and support management (Gaur et al., 2015). The Software as a Service provider can either host the software applications on a private server or deploy them on a cloud computing infrastructure controlled by external entities such as Amazon or Google. Incorporating cloud computing with the "pay-as-you-go" model allows application service providers to reduce their investment in infrastructure services. This integration also enables IoT-based cloud computing systems to prioritize delivering enhanced customer service. Furthermore, it is essential to acknowledge, identify, and rectify the security concerns associated with the Software as a Service paradigm in cloud computing prior to its adoption as a service (Lee et al., 2014). Platform as a Service (PaaS) refers to a collection of interconnected services that facilitate the development and deployment of software applications on cloud computing infrastructures. Therefore, this technique does not rely on singleton sets. Platform as a Service can effectively handle and oversee user subscriptions, security measures, resource metering, and the dissemination of additional services. The Platform as a Service assumes a vital role inside cloud computing systems as it facilitates migrating bespoke software development to cloud-based environments. According to the National Institute of Standards and Technology (NIST), Platform as a Service (PaaS) is the capacity for consumers to deploy applications, either produced by themselves or purchased from other sources, into the cloud infrastructure. Mell and Grance (2011) argued, "These applications are developed using programming languages and tools supported by the PaaS provider. Several popular cloud systems based on Platform as a Service (PaaS) include Google App Engine (GAE), Microsoft Windows Azure (MWA), and Ground Operating Systems (GOS)."

The Internet protocols for Infrastructure as a Service (IaaS) govern various computer resources, including processing and storage capabilities. Infrastructure as a Service (IaaS) can partition, allocate, and dynamically adjust the existing resources to create impromptu systems that align with customers' specific needs. This is a precise depiction of Infrastructure as a Service (IaaS). Moreover, within the context of cloud computing, clients using Platform as a Service may be regarded as developers of applications. These customers are responsible for designing, developing, and implementing software products tailored explicitly for deployment in a cloud-based environment. "The individuals responsible for uploading programs to the cloud and configuring, monitoring, and managing deployed apps inside the cloud are also considered integral members of the Platform-as-a-Service framework. Therefore, the compensation for PaaS providers may be determined based on several factors, including the number of PaaS users, the volume of processing activities, the storage capacity, the length of platform utilization, and the number of network resources required by PaaS applications" (Manvi & Shyam, 2014). Several significant drawbacks of Platform as a Service include concerns over data security, restricted flexibility, the potential for customer lock-in (where clients are constrained in their options due to vendor dependencies), and challenges related to integration. Integrating Platform-as-a-Service (PaaS) offerings with various cloud infrastructures and other applications can result in an unforeseen and unregulated escalation in the intricacy of cloud computing systems (Yousif et al., 2015; Alam et al., 2016). Therefore, researchers in cloud computing need to consider developing and maintaining a flawless communication system based on cloud technology in the following years.

8.10 Conclusion and Further Scope of Study

The smart city idea is now undergoing a transformative shift toward greater environmental sustainability. Consequently, it is entering a new phase characterized by integrating artificial intelligence technologies, sensor networks, and information and communication technologies (ICT) to enable the smartification of many aspects of the city. Moreover, using semantic modeling techniques, it is possible to tackle various personalized services inside a smart city setting effectively. Our objective is to concentrate on the critical aspects of the smart city ecosystem. The urban environment plays a crucial role in addressing the impacts of climate change and promoting the preservation of natural resources. Urbanization is a burgeoning and rapidly expanding phenomenon on a global scale. The development of intelligent systems and their interaction with other artificial intelligence systems is necessary to enhance service provision, address population growth pressures, and mitigate the consequences of global warming on smart cities.

The design, operation, and administration of smart cities should be approached at the system level, primarily prioritizing environmental sustainability and development. Conventional industries and value chains are transforming in response to evolving times and emerging commercial prospects. Moreover, novel business models have been initiated by emerging players, contributing to the organic growth of the environment. Conversely, revolutionary technologies and paradigm changes are significantly influencing the environment. Hence, urbanization has led to a significant transformation in the urban landscape, resulting in a dire scarcity of clean and unpolluted air inside metropolitan areas.

The subsequent endeavors of this chapter include doing empirical investigations on the concepts above that have been deliberated. The research may include identifying several categories of real-time heterogeneous information and developing a practical semantic knowledge framework for integrating sensor data from smart cities for the goal of analysis. The upcoming design of the smart city may include data interoperability and scalability concerns.

References

Alam, A. F., Soltanian, A., Yangui, S., Salahuddin, M. A., Glitho, R., & Elbiaze, H. (2016, June). A cloud platform-as-a-service for multimedia conferencing service provisioning. In *2016 IEEE symposium on computers and communication (ISCC)* (pp. 289–294). IEEE.

Augustine, P. (2020). The industry use cases for the digital twin idea. *Advances in Computers, 117*(1), 79–105.

Bartos, M., & Kerkez, B. (2021). Pipedream: An interactive digital twin model for natural and urban drainage systems. *Environmental Modelling & Software, 144*, 105120.

Botín-Sanabria, D. M., Mihaita, A. S., Peimbert-García, R. E., Ramírez-Moreno, M. A., Ramírez-Mendoza, R. A., & Lozoya-Santos, J. D. J. (2022). Digital twin technology challenges and applications: A comprehensive review. *Remote Sensing, 14*(6), 1335.

Cheela, V. S., Ranjan, V. P., Goel, S., John, M., & Dubey, B. (2021). Pathways to sustainable waste management in Indian smart cities. *Journal of Urban Management, 10*(4), 419–429.

Constante, T. (2018). Contributions for a simulation framework for designing and evaluation manufacturing systems [Doctoral dissertation]. https://hdl.handle.net/10216/114229

Farsi, M., Daneshkhah, A., Hosseinian-Far, A., & Jahankhani, H. (Eds.). (2020). *Digital twin technologies and smart cities*. Springer.

Främling, K., Holmström, J., Ala-Risku, T., & Kärkkäinen, M. (2003). Product agents for handling information about physical objects. *Report of Laboratory of Information Processing Science Series B, TKO-B, 153*(03).

Gaur, A., Scotney, B., Parr, G., & McClean, S. (2015). Smart city architecture and its applications based on IoT. *Procedia Computer Science, 52*, 1089–1094.

Gelernter, D. (1993). *Mirror worlds: Or the day software puts the universe in a shoebox... How it will happen and what it will mean*. Oxford University Press.

Gockel, B., Tudor, A., Brandyberry, M., Penmetsa, R., & Tuegel, E. (2012). Challenges with structural life forecasting using realistic mission profiles. In *53rd AIAA/ASME/ASCE/AHS/ASC structures, structural dynamics and materials conference 20th AIAA/ASME/AHS adaptive structures conference 14th AIAA* (p. 1813).

Grieves, M. W. (2005). Product lifecycle management: The new paradigm for enterprises. *International Journal of Product Development, 2*(1–2), 71–84.

Grieves, M. (2009). Back to the future: Product lifecycle management and the virtualization of product information. In *Product realization: A comprehensive approach* (pp. 1–13).

Grieves, M. (2014). Digital twin: Manufacturing excellence through virtual factory replication. *White paper, 1*(2014), 1–7.

Grieves, M. (2016). Origins of the Digital Twin Concept. https://doi.org/10.13140/RG.2.2.26367.61609

Grieves, M., & Vickers, J. (2017). Digital twin: Mitigating unpredictable, undesirable emergent behavior in complex systems. In *Transdisciplinary perspectives on complex systems: New findings and approaches* (pp. 85–113).

Kodieswari, A., Sabarmathi, K. R., Kavitha, A., & Gayathiri, N. R. (2023). Digital twins and cyber-physical system architecture for smart factory. In *Digital twin for smart manufacturing* (pp. 267–289). Academic Press.

Lee, J. H., Hancock, M. G., & Hu, M. C. (2014). Towards an effective framework for building smart cities: Lessons from Seoul and San Francisco. *Technological Forecasting and Social Change, 89*, 80–99.

Leng, J., & Jiang, P. (2019). Dynamic scheduling in RFID-driven discrete manufacturing system by using multi-layer network metrics as heuristic information. *Journal of Intelligent Manufacturing, 30*, 979–994.

Manvi, S. S., & Shyam, G. K. (2014). Resource management for infrastructure as a service (IaaS) in cloud computing: A survey. *Journal of Network and Computer Applications, 41*, 424–440.

Mell, P., & Grance, T. (2011). Reports on Computer Systems Technology. National Institute of Standards and Technology Special Publication 800-145. https://nvlpubs.nist.gov/nistpubs/legacy/sp/nistspecialpublication800-145.pdf

Mishra, K. N., & Chakraborty, C. (2020). A novel approach toward enhancing the quality of life in smart cities using clouds and IoT-based technologies. In *Digital twin technologies and smart cities* (pp. 19–35).

Moser, M. A. (2001). What is smart about the smart communities movement. *EJournal, 10*(1), 1–11.

Mylonas, G., Kalogeras, A., Kalogeras, G., Anagnostopoulos, C., Alexakos, C., & Muñoz, L. (2021). Digital twins from smart manufacturing to smart cities: A survey. *IEEE Access, 9*, 143222–143249.

Rosen, R., Von Wichert, G., Lo, G., & Bettenhausen, K. D. (2015). About the importance of autonomy and digital twins for the future of manufacturing. *Ifac-Papersonline, 48*(3), 567–572.

Ruohomäki, T., Airaksinen, E., Huuska, P., Kesäniemi, O., Martikka, M., & Suomisto, J. (2018, September). Smart city platform enabling digital twin. In *2018 international conference on intelligent systems (IS)* (pp. 155–161). IEEE.

Saranya, M., & Amutha, B. (2023, August). An insight into federated learning based digital twin applications in Smart City. In *2023 5th international conference on inventive research in computing applications (ICIRCA)* (pp. 1295–1302). IEEE.

Shafto, M., Conroy, M., Doyle, R., Glaessgen, E., Kemp, C., LeMoigne, J., & Wang, L. (2012). Modeling, simulation, information technology & processing roadmap. *National Aeronautics and Space Administration, 32*(2012), 1–38.

Shah, V., & Suthar, A. (2022). Possibilities with digital twin. In *Digital twin technology: Fundamentals and applications* (pp. 219–232).

Tao, F., Cheng, J., Qi, Q., Zhang, M., Zhang, H., & Sui, F. (2018). Digital twin-driven product design, manufacturing and service with big data. *The International Journal of Advanced Manufacturing Technology, 94*, 3563–3576.

Taqaddas, A. E. (2023). Innovative leadership comparison of ion beam applications, SA Belgium against its competitors. *Journal of Clinical Oncology Reports, 2*(1), 01.

Tuegel, E. (2012, April). The airframe digital twin: Some challenges to realization. In *53rd AIAA/ASME/ASCE/AHS/ASC structures, structural dynamics and materials conference 20th AIAA/ASME/AHS adaptive structures conference 14th AIAA* (p. 1812).

Vohra, M. (2022). Digital twin in smart cities: Application and benefits. In *Digital twin technology: Fundamentals and applications* (pp. 159–172).

Wan, J., Tang, S., Li, D., Wang, S., Liu, C., Abbas, H., & Vasilakos, A. V. (2017). A manufacturing big data solution for active preventive maintenance. *IEEE Transactions on Industrial Informatics, 13*(4), 2039–2047.

Yousif, A., Farouk, M., & Bashir, M. B. (2015, April). A cloud-based framework for platform as a service. In *2015 International conference on cloud computing (ICCC)* (pp. 1–5). IEEE.

Rahul Joshi is an assistant professor of journalism and mass communication at Faculty of Mass Media & Humanities, MRIIRS (Deemed to be University), Faridabad, Haryana. He has completed his Ph.D (media studies) from MRIIRS, Faridabad. He has done his masters in Journalism & Mass Communication from GGSIP University. He was a gold medalist in MA (MC) in GGSIP University. He has 11 years of academic and industry experience. His specializations are in radio journalism and television production. He has published over 15 research papers in different reputed international and national journals. Before stepping into academics, he was Anchor cum Assistant Producer in various National Hindi and English News Channels like ANI, India News, News X, APN News, News 18 Network, and Zee News. He has also been associated with World's biggest Radio Broadcasting House, All India Radio as a Radio Presenter and Production Executive for the five consecutive years. His teaching journey started with a 27-month stint at the Media school of DME, Noida, affiliated to Guru Gobind Singh Indraprastha University, Delhi, as an Assistant Professor. His areas of research interest are radio production, TV production, and digital media. His teaching expertise also lies in radio and TV production.

Rhythma Badola has been an ayurvedic advisor (Doctor) at Patanjali India Limited for the past 5 years. She is a qualified B.A.M.S. (Bachelor of Ayurvedic Medicine & Surgery) and B.P.T. (Bachelor of Physiotherapy) with good professional knowledge in all aspects of ayurveda. She has attended many workshops, conferences, and seminars on ayurveda in different states.

Part III
Sustainable Urban Ecosystems: Integrating Circular Economy and Planning in Smart Cities

Chapter 9
Smart Sustainable City Initiatives and the Circular Economy: Exploring Upcycling Organizations in Cape Town

Sara Calvo and Andres Morales

Abstract The increasing interest in the circular economy highlights its potential in addressing the environmental, economic, and social issues inherent in traditional linear production and consumption systems. Simultaneously, the rise of smart cities represents a pivotal shift in urban development, utilizing technology and innovation to improve urban life and promote sustainability. This chapter examines the role of the circular economy in the sustainable evolution of smart cities. It investigates the complex interplay between circular economy practices and the development of smart cities, using a case study of upcycling initiatives in Cape Town, South Africa, as an example. The study uncovers several key themes, including innovation, cultural value preservation, the importance of local networks, environmental awareness, and the generation of social and economic benefits.

Keywords Circular economy · Sustainable smart cities, upcycling · South Africa

9.1 Introduction

The twenty-first century's rapid urbanization has introduced a slew of challenges, from dwindling resources to environmental harm and a surge in waste production. In response, the notion of "smart cities"has gained traction as a potential remedy to these urban issues. These cities distinguish themselves by employing technology and data-driven methods to enrich residents' lives while curbing detrimental environmental effects. Nevertheless, achieving sustainable development within smart cities demands more than just technological advancements; it calls for a

S. Calvo (✉)
The Business School, The Open University, Milton Keynes, UK
e-mail: sara.calvo-martinez@open.ac.uk

A. Morales
Universidad Internacional de la Rioja, Logroño, Spain
e-mail: andres.morales@unir.net

S. Majumdar et al. (eds.), *Smart Cities*, S.M.A.R.T. Environments,
https://doi.org/10.1007/978-3-031-59846-3_9

comprehensive approach that integrates circular economy principles. This chapter delves into the role of the circular economy in bolstering sustainable development within smart cities.

Smart cities represent urban regions that harness technology and data to enhance various facets of city life, spanning transportation, energy efficiency, waste management, and citizen involvement. On the flip side, sustainable development encompasses a multifaceted concept aiming to satisfy current needs without jeopardizing future generations' ability to meet their own requirements. In the smart city context, sustainable development involves discovering innovative solutions to urban dilemmas while minimizing adverse environmental and social impacts. The circular economy stands as a framework designed to uncouple economic growth from resource consumption by advocating for efficient resource use, reuse, and recycling. Departing from the traditional linear economy's "take-make-dispose" model, the circular economy envisions a system where products and materials undergo reuse, repair, or recycling to prolong their lifespan (Ellen MacArthur Foundation, 2013).

The global pursuit of sustainable development has led to increasing attention on innovative approaches to resource management and waste reduction. Among these approaches, upcycling, a process that transforms discarded materials into new, higher-value products, has gained prominence. Upcycling social enterprises play a pivotal role in advancing sustainable development by not only reducing waste but also creating economic opportunities and promoting social inclusion. This chapter explores upcycling social enterprises and their significant contribution to sustainable development. Upcycling social enterprises contribute significantly to waste reduction and resource efficiency. By diverting waste materials from landfills and incineration, they help reduce the environmental burden of waste disposal. Simultaneously, they minimize the extraction of new raw materials, thereby conserving natural resources and decreasing the carbon footprint of production processes. This chapter aims to address the following research question: Does the circular economy contribute to the sustainable development of smart cities? It does this using case study examples of upcycling organizations in Cape Town, South Africa.

9.2 Circular Economy's Role in Enhancing Sustainable Development in Smart Cities

The concept of sustainable development, centered on the idea of meeting current needs without jeopardizing future generations' abilities to meet their own, has gained immense importance amid swift urban growth and environmental hurdles. Smart cities, propelled by technological innovations and data-centric methods, have emerged as a promising avenue to fulfill sustainable development objectives in urban settings. This essay delves into the intricate interplay between sustainable development and smart cities, highlighting their potential to craft urban environments that are ecologically responsible, socially inclusive, and economically vibrant. Academic references supporting this exploration offer evidence of the transformative impact of smart city endeavors on sustainable development.

The concept of sustainable development gained prominence through the Brundtland Report, which defined it as development meeting present needs without compromising future generations' capacities to meet their own needs. This principle underscores the imperative of addressing environmental, social, and economic dimensions concurrently to secure long-term well-being.

Smart cities signify a shift in urban planning and development paradigms. These cities harness cutting-edge technologies like the Internet of Things (IoT), data analytics, and artificial intelligence to elevate the efficiency and quality of urban services. Encompassing domains such as transportation, energy, waste management, and healthcare, smart city initiatives aspire to create urban environments that are not just more livable but also sustainable (Anthopoulos & Fitsilis, 2018).

9.2.1 Sustainable Development in Smart Cities

Sustainable development is a global imperative, and smart cities offer a promising path toward its realization. Academic research supports the positive impact of smart city initiatives on environmental sustainability, social inclusion, and economic prosperity. Smart cities leverage technology and data to optimize resource utilization, promote equitable access, and foster innovation, all of which contribute to the holistic vision of sustainable development. However, challenges related to privacy, digital inequality, and resource constraints must be addressed to ensure that the benefits of smart cities are accessible to all. As the world continues to urbanize, the symbiotic relationship between sustainable development and smart cities will become increasingly vital in shaping a more prosperous and sustainable future for urban populations.

9.2.1.1 Environmental Sustainability

Smart cities prioritize environmental sustainability by reducing energy consumption, greenhouse gas emissions, and resource waste. For instance, energy-efficient building designs, smart grid systems, and renewable energy integration contribute to reduced carbon footprints (Cugurullo et al., 2019). These initiatives align with the sustainable development goal of combating climate change. A study by Cugurullo et al. (2019) investigated the environmental impact of smart city initiatives in Barcelona. The research found that smart energy systems and IoT-enabled urban planning led to significant reductions in energy consumption and greenhouse gas emissions. Such findings underscore the potential of smart cities to contribute to environmental sustainability.

9.2.1.2 Social Inclusion

Inclusive development is a core component of sustainable development, emphasizing equitable access to opportunities, services, and resources. Smart cities incorporate digital inclusion strategies to bridge the digital divide, ensuring that technological advancements benefit all residents, regardless of socio-economic status (Nam & Pardo, 2011). Additionally, smart transportation systems promote accessible mobility for all, reducing social disparities. Nam and Pardo (2011) conducted research on the digital divide in smart cities. Their study revealed that targeted interventions, such as public Wi-Fi networks and digital literacy programs, can bridge the digital divide and promote social inclusion. Access to digital services enhances opportunities for marginalized communities.

9.2.1.3 Economic Prosperity

Sustainable development encompasses economic prosperity as a means to improve living standards. Smart city investments stimulate economic growth by fostering innovation and entrepreneurship. The development of technology clusters, business incubators, and coworking spaces in smart cities encourages economic diversification (Komninos, 2013a, b). These economic activities generate jobs and increase overall urban prosperity. Komninos (2013a, b) examined the economic development impact of smart city strategies in several European cities. The research highlighted how smart city investments can attract businesses and stimulate innovation, leading to job creation and economic growth. These findings affirm the role of smart cities in fostering economic prosperity.

Despite their potential, smart cities encounter several challenges and considerations in achieving sustainable development:

1. Privacy and Security Issues: The extensive data collected in smart cities raises concerns regarding privacy and security. Safeguarding citizens' data and implementing strong cybersecurity measures are crucial to uphold public trust (Batty et al., 2019).
2. Digital Disparities: Despite the manifold advantages offered by smart city technologies, disparities in digital access persist. Ensuring fair access to digital services and bridging the digital gap are imperative for fostering inclusive development (Townsend, 2013).
3. Resource Constraints: Implementing smart city initiatives requires significant financial and technological resources. Smaller cities or those in developing regions may face challenges in adopting these technologies.

9.2.2 Circular Economy's Contribution to Sustainable Development in Smart Cities

The circular economy represents a fundamental shift in how we manage resources and materials, aligning perfectly with the goals of sustainable development within smart cities. Academic research provides compelling evidence of the circular economy's positive impact on resource efficiency, waste reduction, economic growth, and social inclusion in smart city contexts. By optimizing resource use, minimizing waste generation, and promoting the sharing economy, smart cities embrace circular economy principles to create urban environments that are more sustainable, resilient, and inclusive. As smart cities continue to evolve and expand, the integration of circular practices will remain essential in shaping a more sustainable and prosperous future for urban communities.

9.2.3 Principles of the Circular Economy

The circular economy marks a departure from the conventional linear economic model of "take-make-dispose" to one that advocates for the regeneration of resources and materials. Essential principles within the circular economy encompass:

Resource Efficiency: Maximizing the value extracted from resources through recycling, reusing, and remanufacturing (Ellen MacArthur Foundation, 2013).

Waste Reduction: Minimizing waste generation by designing products and processes for longevity and disassembly.

Sharing and Collaborative Consumption: Promoting shared use of products and services to minimize resource consumption.

9.2.3.1 Resource Efficiency

Smart cities leverage technology and data-driven approaches to optimize resource use. For example, sensor networks and data analytics are used to manage energy consumption more efficiently, reducing waste and emissions (Anthopoulos & Fitsilis, 2018). Circular economy principles align with these efforts by emphasizing the importance of prolonging the lifespan of resources and materials, ultimately reducing the need for extraction and production. A study by Anthopoulos and Fitsilis (2018) investigated the role of smart city technologies in resource optimization. The research showed that smart energy management systems, driven by data analytics, led to significant improvements in resource efficiency and energy conservation. Such findings illustrate the synergy between smart cities and circular economy principles in maximizing resource value.

9.2.3.2 Waste Reduction

Effective waste management is a critical component of sustainability in smart cities. Circular practices, such as recycling, upcycling, and waste-to-energy technologies, enable cities to divert materials from landfills and reduce waste volumes (Buenrostro-Delgado et al., 2017). Waste reduction not only conserves resources but also mitigates environmental pollution. Buenrostro-Delgado et al. (2017) conducted research on sensor-based waste collection systems in smart cities, focusing on Barcelona, Spain. Their study demonstrated that such systems improved waste management efficiency, reduced waste generation, and aligned with circular economy principles by optimizing resource use.

9.2.3.3 Sharing Economy

The sharing economy is an integral part of both the circular economy and smart cities. Platforms that facilitate collaborative consumption, such as ride-sharing and peer-to-peer rentals, promote resource efficiency and reduce individual ownership (Hamari et al., 2016). This leads to reduced resource consumption and congestion while supporting economic growth. Research by Hamari et al. (2016) explored the environmental and economic impacts of the sharing economy. Their findings highlighted the potential for collaborative consumption platforms in smart cities to reduce resource consumption, lower environmental footprints, and support economic growth—a testament to the circular economy‘s role in smart city sustainability.

9.2.3.4 Economic Growth and Job Creation

Circular economy practices within smart cities stimulate economic growth by fostering innovation and entrepreneurship (Bocken et al., 2016). The development of circular business models, such as product-as-a-service or remanufacturing, generates new economic opportunities (Korhonen et al., 2018). Furthermore, circular economy initiatives often lead to the creation of green jobs, particularly in sectors like recycling and refurbishment (Rizos et al., 2016). Economic growth is a central pillar of sustainable development, as it improves living standards and contributes to poverty reduction.

9.2.3.5 Social Inclusion

In smart cities, circular economy practices can enhance social inclusion by providing job opportunities and improving access to affordable goods and services. For instance, recycling and waste management initiatives create employment opportunities for marginalized communities (Lombardi et al., 2012). Additionally, sharing and collaborative consumption platforms can reduce the cost of living, making essential services more accessibles. Social inclusion is a key dimension of sustainable development, promoting equality and reducing disparities within urban populations.

9.3 Upcycling: An Eco-friendly Innovation and the Role of Upcycling Organizations

Upcycling is a powerful approach to resource management that contributes significantly to sustainability and waste reduction. By transforming waste materials into valuable and functional products, upcycling aligns with the principles of the circular economy and encourages responsible consumption. Upcycling organizations play a pivotal role in promoting and facilitating this practice by offering education, resources, and support to individuals and businesses. As the world seeks innovative solutions to environmental challenges, upcycling emerges as an effective strategy that empowers individuals and communities to make a positive impact on the planet, all while fostering economic opportunities and creativity. The academic references included in this essay underscore the multidisciplinary relevance of upcycling in the context of sustainability and resource management.

Upcycling is a multifaceted process with the primary aim of reusing discarded or waste materials in a way that enhances their value, functionality, or aesthetic appeal (Chabowski, 2019). Unlike traditional recycling, which often involves breaking down materials to their base components, upcycling seeks to maintain the original integrity of the materials while giving them new life (Mont, 2020). This concept can be applied to various materials and products, from repurposing old furniture to creating fashion pieces from discarded textiles.

One of the fundamental principles of upcycling is repurposing. This involves finding new uses for items that might otherwise be discarded (Leal Filho et al., 2020). For example, old wine barrels can be turned into unique tables, or discarded car parts can be transformed into artistic sculptures. By emphasizing creativity and innovative thinking, upcycling fosters a mindset of sustainability and responsible consumption (Moore & Hunt, 2018).

9.3.1 The Significance of Upcycling in Sustainability

The practice of upcycling holds immense significance in the context of sustainability. It aligns with the principles of the circular economy, where resources are used efficiently, and waste is minimized (Geissdoerfer et al., 2017). Several key reasons highlight the importance of upcycling in promoting sustainability:

Resource Conservation: Upcycling reduces the need for new raw materials, which can help preserve natural resources and reduce the environmental impact of resource extraction and manufacturing (Bocken et al., 2016).

Waste Reduction: By diverting materials from landfills and incinerators, upcycling directly contributes to waste reduction, minimizing the adverse environmental effects of landfilling, such as soil and water contamination (Filimonau et al., 2019).

Energy Savings: Upcycling often requires less energy compared to recycling or manufacturing new products from raw materials, leading to reduced energy consumption and greenhouse gas emissions (Zamani et al., 2021).

Stimulating Creativity: Upcycling encourages creative thinking and innovative problem-solving, nurturing sustainable design practices and eco-friendly solutions (Stevenson & Parnell, 2020).

Economic Opportunities: The upcycling industry can generate economic opportunities by supporting local artisans and businesses that specialize in creating upcycled products, thus contributing to job creation and community development (Blomsma & Brennan, 2017).

Consumer Education: Through the creation and promotion of upcycled products, upcycling raises awareness among consumers about the environmental impacts of their choices, encouraging responsible consumption and informed decision-making (Jandric et al., 2017).

9.3.2 Upcycling Organizations and Their Role

To advance the practice of upcycling, various organizations have emerged, serving as catalysts for change and fostering a more sustainable and circular economy. These upcycling organizations play an integral role in promoting upcycling by providing essential resources, expertise, and platforms for collaboration. Their contributions can be categorized into several key areas:

Education and Resources: Upcycling organizations offer workshops, training programs, and resources to educate individuals and businesses about the principles and techniques of upcycling. These initiatives help people develop the skills necessary to transform waste materials into valuable products (Klodawsky et al., 2021).

Fostering Collaboration: Many upcycling organizations serve as intermediaries, facilitating collaborations between designers, artisans, craftspeople, and manufacturers. These collaborations can lead to the development of unique and marketable upcycled items, promoting circularity (De Jesus et al., 2018).

Supporting Artisans and Entrepreneurs: Upcycling organizations often provide guidance and support to artisans and entrepreneurs interested in establishing businesses centered around upcycled products. This includes assistance with business development, marketing, and access to markets (Palmer et al., 2020).

Advocacy and Awareness: These organizations engage in advocacy efforts and public awareness campaigns to highlight the environmental and social benefits of upcycling. Through these initiatives, they encourage consumers to choose upcycled products over conventional ones.

Showcasing Success Stories: Upcycling organizations frequently showcase success stories and case studies to inspire others to adopt upcycling practices. These narratives demonstrate the potential for upcycling to transform waste into valuable resources and encourage innovation.

Certification and Standards: Some upcycling organizations develop certification standards to ensure the authenticity and quality of upcycled products. These standards help build trust among consumers and promote the growth of the upcycling industry (Gregson et al., 2015).

9.3.3 *Examples of Upcycling Organizations*

Several upcycling organizations have emerged globally, each contributing uniquely to the promotion and advancement of upcycling. Here are some notable examples:

Precious Plastic (Global): Precious Plastic is an open-source project that provides blueprints and resources for building small-scale plastic recycling machines. It empowers communities to transform plastic waste into valuable products, addressing the issue of plastic pollution.

TerraCycle (Global): TerraCycle specializes in recycling and upcycling hard-to-recycle materials, such as cigarette butts, used pens, and plastic waste. They collaborate with brands and individuals to collect and repurpose these materials, diverting them from landfills.

Remake (USA): Remake is a nonprofit organization that focuses on transforming the fashion industry. They educate consumers about the environmental and social impacts of fast fashion and promote sustainable and upcycled clothing alternatives.

Precious Forests (Australia): Precious Forests upcycles waste timber and discarded materials to create unique furniture and homewares. They prioritize sustainability by using locally sourced and reclaimed materials.

9.4 Smart Cities in South Africa: Navigating the Path to Sustainable Urban Development

As urbanization continues to reshape the global landscape, cities worldwide are increasingly embracing smart city initiatives to tackle the numerous challenges stemming from rapid urban expansion. South Africa, with its distinctive mix of urban diversity and developmental requirements, has emerged as an ideal setting for implementing smart city concepts. Smart cities offer a promising strategy for confronting the intricate challenges spurred by rapid urbanization within South Africa. While the journey toward becoming a smart city is not devoid of obstacles, the potential advantages—such as economic growth, sustainability enhancements, and an elevated quality of life—render it a worthwhile pursuit.

Several factors are driving the development of smart cities in South Africa:

Urbanization: South Africa is experiencing rapid urbanization, with a significant portion of its population living in cities. The growth of urban centers creates a pressing need for smart solutions to address issues like congestion, resource management, and service delivery.

Economic Development: Smart cities can attract investment and spur economic growth. South Africa aims to position itself as a technological hub, and smart cities can play a pivotal role in achieving this goal (Chinguno & Mbohwa, 2019).

Sustainability: Concerns about environmental sustainability, energy efficiency, and resource conservation drive the adoption of smart technologies. South Africa faces energy challenges, and smart cities can contribute to more sustainable energy use.

Inclusion and Equity: Smart city initiatives can address social inequalities and enhance the quality of life for all citizens. In South Africa, where historical disparities persist, this is a crucial consideration.

9.4.1 Key Components of Smart Cities in South Africa

To transform South African cities into smart cities, several key components must be integrated:

Digital Infrastructure: Building the necessary digital infrastructure, including high-speed Internet access, is fundamental to enabling smart city solutions (Komninos, 2013a, b).

IoT and Data Analytics: Utilizing IoT devices to collect data and applying analytics to derive insights can optimize urban services such as traffic management and waste collection.

Transportation: Developing smart transportation systems can reduce traffic congestion and emissions while improving mobility and connectivity (Dijkstra et al., 2015).

Energy Management: Implementing energy-efficient solutions, renewable energy sources, and grid management can enhance sustainability and resilience.

Governance and Citizen Engagement: Smart city governance involves citizen participation, transparency, and responsiveness. Engaging citizens in decision-making processes is crucial for the success of smart city initiatives (Van Zyl et al., 2021).

9.4.2 Challenges Faced by Smart Cities in South Africa

Despite the potential benefits, smart cities in South Africa face several challenges:

Infrastructure Gaps: Existing infrastructure deficiencies, especially in rural areas, pose challenges for the deployment of digital solutions.
Financial Constraints: Funding smart city projects can be a significant hurdle, as many municipalities in South Africa have limited resources.
Digital Divide: Bridging the digital divide and ensuring equitable access to smart city services are crucial but challenging (Van Zyl et al., 2021).
Data Privacy and Security: The collection and utilization of vast amounts of data raise concerns about privacy and cybersecurity.

9.4.3 Opportunities for Sustainable Urban Development

While challenges exist, the development of smart cities in South Africa presents numerous opportunities:

Job Creation: Smart city initiatives can create jobs in technology, infrastructure development, and service provision.
Improved Quality of Life: Enhanced services, efficient transportation, and better resource management can improve the quality of life for urban residents (Chinguno & Mbohwa, 2019).
Economic Growth: Smart cities can attract foreign investment and stimulate local economies.
Environmental Sustainability: By optimizing resource use and reducing emissions, smart cities contribute to environmental sustainability.

9.4.4 The Circular Economy and Upcycling in South Africa

South Africa, a country with diverse socio-economic and environmental challenges, has recognized the potential of a circular economy to address issues such as resource scarcity, unemployment, and waste management (Department of Environment, Forestry and Fisheries, 2019). Key components of the circular economy in South Africa include:

Waste Management: South Africa faces significant waste management challenges, with many landfills nearing capacity. The circular economy promotes waste reduction, recycling, and upcycling to divert materials from landfills (Majozi et al., 2019).

Resource Efficiency: Resource efficiency is a cornerstone of the circular economy. South Africa's abundant natural resources make it imperative to use them efficiently and sustainably (Mafini et al., 2019).

Job Creation: A circular economy can generate employment opportunities, particularly in sectors such as recycling and upcycling (Rivas & Zorpas, 2019).

Upcycling, a central component of the circular economy, involves transforming waste or discarded materials into products of higher value or quality (Ghosh et al., 2018). This practice not only reduces waste but also promotes creativity and innovation while conserving resources. In the South African context, upcycling offers several benefits:

Waste Reduction: Upcycling diverts materials from landfills, reducing the environmental burden of waste disposal (Nhamo et al., 2019).

Resource Conservation: Upcycling reduces the demand for new raw materials, preserving natural resources such as timber, minerals, and energy (Mkhize & Pretorius, 2020).

Economic Opportunities: Upcycling enterprises can create job opportunities, especially in marginalized communities (Majozi et al., 2019).

Local Innovation: Upcycling encourages local innovation and the development of sustainable design solutions.

9.4.4.1 Current Initiatives and Challenges

South Africa has made strides in embracing the circular economy and upcycling:

National Waste Management Strategy: South Africa's National Waste Management Strategy includes provisions for waste minimization, recycling, and extended producer responsibility, aligning with circular economy principles (Department of Environment, Forestry and Fisheries, 2019).

Social Enterprises: Social enterprises and community-based initiatives have emerged in South Africa, focusing on upcycling waste materials into marketable products (Nhamo et al., 2019).

Government Support: The South African government has shown interest in supporting circular economy initiatives, including funding research and development in the recycling and upcycling sectors.

Despite these positive developments, challenges persist:

Infrastructure: Inadequate recycling and upcycling infrastructure in certain regions hinders progress (Nhamo et al., 2019).

Education and Awareness: Public awareness about the benefits of upcycling and the circular economy remains limited (Rivas & Zorpas, 2019).

Regulatory Framework: The lack of comprehensive policies and regulations for waste management and recycling can impede growth in the circular economy (Department of Environment, Forestry and Fisheries, 2019).

A circular economy, with upcycling as a critical component, holds immense potential for South Africa to address pressing socio-economic and environmental challenges. Academic references demonstrate that upcycling can reduce waste, conserve resources, create jobs, and promote local innovation. However, it is essential to overcome existing challenges, including infrastructure limitations and regulatory gaps. As South Africa continues to develop and refine its approach to the circular economy and upcycling, it can move closer to achieving sustainable development, while simultaneously addressing environmental and social concerns. The following section will look at the methodology used to explore upcycling organizations in Cape Town.

9.5 Methodology

9.5.1 Context: Cape Town

Cape Town, the legislative capital of South Africa and a city renowned for its natural beauty, has been at the forefront of sustainable development efforts, smart city initiatives, and circular economy practices. With its unique geographical and environmental challenges, Cape Town has set an example for urban centers worldwide by integrating sustainability into its urban planning. This section explores how Cape Town has embraced sustainable development, smart city technologies, and the circular economy, drawing from academic references to highlight its innovative approach. Cape Town's journey toward sustainable development has been shaped by a commitment to environmental conservation, social inclusion, and economic growth (Sarova et al., 2019).

The city has implemented various conservation measures to protect its biodiversity and manage water resources, exemplified by the Cape Town Water Fund, which focuses on sustainable water management (Herrmann & Snyman, 2019). Cape Town has strived to address socio-economic disparities through inclusive policies, such as affordable housing initiatives and equitable access to services (De Lannoy et al., 2016). Sustainable tourism, renewable energy projects, and green technology innovation have contributed to Cape Town's economic growth while minimizing environmental impacts (Van der Merwe & Kotzé, 2018).

Cape Town has emerged as a leader in the adoption of smart city technologies to enhance urban efficiency and sustainability (Van Zyl et al., 2021). The city has invested in intelligent transportation systems, including traffic management and public transport integration, to reduce congestion and promote eco-friendly commuting (Kotzé & Van der Merwe, 2019). Cape Town's commitment to digital governance and citizen engagement has improved service delivery and transparency (Benjamin & Kyobe, 2019). Smart energy management, incorporating renewable sources and demand-side management, has been vital in reducing the city's carbon footprint (Ejechi et al., 2020).

9.5.1.1 Circular Economy Practices in Cape Town

Cape Town has actively embraced the principles of the circular economy to reduce waste, conserve resources, and foster sustainable growth (Majozi et al., 2019). Academic references highlight the following circular economy initiatives:

Waste Management: The city has implemented comprehensive waste reduction and recycling programs, diverting materials from landfills and promoting upcycling and reuse (Nhamo et al., 2019).

Resource Efficiency: Cape Town has prioritized resource efficiency in industries such as agriculture and manufacturing, emphasizing sustainable practices and responsible consumption (Sarova et al., 2019).

Innovation Ecosystem: The city's innovation ecosystem supports circular economy startups and green enterprises, encouraging creative solutions to environmental challenges.

Cape Town's commitment to sustainable development, smart city initiatives, and the circular economy demonstrates its dedication to building a resilient, inclusive, and environmentally responsible urban center. Academic references underscore the city's innovative approaches to conservation, technology integration, and waste reduction, setting an example for cities globally. As Cape Town continues to evolve and adapt to changing environmental and social dynamics, its experiences and practices offer valuable insights and inspiration for urban areas seeking to foster sustainable development and create smarter, more circular cities.

9.5.1.2 Upcycling Initiatives in Cape Town

Cape Town's upcycling initiatives have taken root in various sectors, from art and fashion to community development and education. Several key aspects of upcycling in Cape Town include:

Art and Design: Cape Town boasts a vibrant art and design scene that has integrated upcycling into its creative process. Local artists and designers transform discarded materials into stunning artworks, sculptures, and fashion pieces. The Woodstock Exchange, for example, is a hub for upcycled art and design studios, where talented individuals breathe new life into reclaimed materials (Christians, 2019).

Fashion and Textiles: Upcycling has found a home in Cape Town's fashion industry. Designers and fashion brands are reimagining clothing, using discarded textiles and materials to create unique and sustainable fashion lines. These upcycled fashion items reflect the city's commitment to ethical and eco-friendly clothing (Moloisi, 2020).

Community Development: Upcycling initiatives in Cape Town often have a strong community development component. Organizations such as "Hout Bay Recycling for Good" empower local communities to collect and upcycle waste materials, creating employment opportunities while promoting environmental stewardship (Southey-Swartz, 2020).

Education and Awareness: Several educational institutions and nonprofit organizations in Cape Town are dedicated to educating the public about the importance of upcycling. They offer workshops, training programs, and awareness campaigns that teach individuals and businesses how to incorporate upcycling practices into their daily lives (Leal Filho et al., 2020).

Sustainable Tourism: Cape Town's tourism industry has embraced the concept of sustainable and responsible tourism. Visitors can explore upcycled art galleries, attend eco-friendly craft markets, and learn about the city's upcycling initiatives as part of their travel experience (Cape Town Tourism, 2021).

9.5.1.3 Organizations and Impact

Cape Town is home to various organizations and enterprises dedicated to upcycling practices. These entities play a pivotal role in driving the upcycling movement and its impact on the city:

The Recycle Swop Shop: This community-based organization in Cape Town encourages children to collect recyclable materials and exchange them for goods like toys, clothing, and school supplies. By engaging young citizens, the organization instills the values of recycling and upcycling from an early age (The Recycle Swop Shop, n.d.).

Red Bull Amaphiko: A social entrepreneurship program in Cape Town, Red Bull Amaphiko, supports upcycling and sustainable business ventures. It provides mentorship and resources to entrepreneurs who aim to make a positive social and environmental impact (Red Bull Amaphiko, n.d.).

Greenpop: Focusing on environmental education and reforestation, Greenpop encourages tree planting and community engagement through upcycling workshops. It promotes the use of recycled materials for creating art installations and functional items (Greenpop, n.d.).

The impact of upcycling practices in Cape Town extends beyond environmental benefits. It fosters a sense of community, creativity, and sustainability among its residents. By providing employment opportunities, promoting environmental awareness, and reducing waste, upcycling initiatives contribute to a more inclusive and responsible city.

Cape Town's upcycling practices exemplify its commitment to sustainable living, creativity, and community development. From the art studios of Woodstock Exchange to community-based initiatives like the Recycle Swop Shop, the city embraces the ethos of transforming waste into innovation. These upcycling efforts not only divert materials from landfills but also inspire a new generation of eco-conscious citizens. Cape Town serves as a model for cities worldwide, demonstrating how upcycling can be an integral part of a city's identity, fostering sustainable living, and promoting positive social and environmental change.

9.5.2 Data Collection and Analysis

Video research is a valuable method for exploring and understanding case study organizations in depth. By capturing real-life interactions and behaviors, it provides multidimensional, contextual, and temporally rich data. To ensure the effectiveness of video research, researchers must address ethical considerations, use appropriate data collection techniques, apply rigorous analysis methods, and consider the challenges associated with interpretation and data storage. As an increasingly accessible and sophisticated method, video research offers a dynamic approach to studying and documenting case study organizations in contemporary research (Knoblauch et al., 2019).

This study employs a multiple case study design, which allows for an in-depth exploration of upcycling organizations and their contributions to Sustainable Development smart cities in South Africa. Multiple case studies are a robust qualitative research method that enables the examination of complex phenomena within their real-world context. Table 9.1 shows the characteristics of the three selected organizations for this study.

Data collection involves a combination of methods, including semistructured interviews, document analysis, and direct observations. Semistructured interviews were conducted with key stakeholders within each organization, including founders, employees, and beneficiaries. For each organization, 5 interviews were taken with the cofounder/leader of the project and the employees. The interviews lasted between 60 and 90 min each. Document analysis included the examination of

Table 9.1 Characteristics of the organizations selected for the study

Name	Sustainable impact	Socio-economic impact	Link to website and video case studies
Case 1: *Sealand (South Africa)*	*Designing fashion*: A company specializing in crafting bags using plastic retrieved from the ocean	Hired individuals facing financial hardship in Cape Town	https://www.youtube.com/watch?v=AASEOi3FYJs https://sealandgear.com
Case 2: *Hands of honour (South Africa)*	*Industrial design*: This company involves repurposing outdated inventory and recyclable materials	Hired individuals experiencing homelessness	http://handsofhonour.co.za https://www.youtube.com/watch?v=oLFlJuS3sA8
Case 3: *H18 (South Africa)*	*Fashion design:* A company that focused on the production of cotton crochet products	Employed disadvantaged women (refugees)	https://www.youtube.com/watch?v=JqEl07_VaLs&t=129s https://h18.co.za

reports, publications, and organizational documents to gather historical and contextual data. Direct observations were conducted to gain a firsthand understanding of the daily operations and practices of each organization. Informed consent was obtained from all participants, and their anonymity and confidentiality were preserved. The study was adhered to ethical guidelines and principles of research involving human subjects, as outlined by relevant institutional review boards and ethical committees.

9.5.2.1 Data Analysis

Data analysis follows a thematic approach, which involves coding and categorizing qualitative data to identify recurring themes and patterns. This thematic coding was performed using NVIVO. Following the individual case analyses, a cross-case analysis was conducted to identify commonalities and differences among the cases. This step enabled the extraction of overarching themes related to the contributions of upcycling organizations in South Africa. It also facilitated the development of a comprehensive understanding of the role of these enterprises in sustainable development. To ensure the validity and reliability of the study, multiple data sources, including interviews, documents, and observations, were triangulated. Member checking, where participants reviewed and validate the findings, were also employed to enhance the study's trustworthiness. Additionally, the research team engaged in continuous discussions and peer debriefing to minimize bias and enhance the credibility of the analysis.

9.6 Presentation of Findings

This section outlines the five themes derived from the data analysis conducted through NVIVO. These themes encompass: (1) innovation, (2) preservation of cultural values, (3) fostering social and economic impact, (4) environmental awareness, and (5) cultivation of local networks.

9.6.1 Innovation

The interview responses underscored the paramount importance of innovation for the participants. They firmly believed that innovation would enhance the appeal of their offerings to customers. One respondent emphasized the significance of retrieving waste from the sea and converting it into exquisite products, as illustrated in the following excerpt:

Our aim is to repurpose waste from the sea, contributing to its preservation while crafting functional products for our clientele. Additionally, we differentiate ourselves by offering customers the opportunity to repair their purchases—a unique and distinctive service (Interviewee 1).

Moreover, there was substantial evidence indicating the vital role of an innovation lab focused on addressing market needs. This was deemed a cornerstone for their success, as exemplified in the following statements:

We maintain an innovation lab dedicated to consistently exploring distinctive and practical products tailored for our customers. Innovation lies at the heart of our company's ethos (Interviewee 5).

9.6.2 Preservation of Cultural Values

In numerous instances, participants highlighted the significant role of cultural values as a driving force behind their upcycling enterprises. This was vividly demonstrated in excerpts from the interview transcripts. For instance, a pivotal interviewee overseeing an upcycling organization elaborated:

This product encapsulates the essence of vibrant South African attire, traditionally worn by women (Interviewee 10).

The research findings further indicate that some respondents place emphasis on the aesthetic aspects of their office and building spaces. According to these interviewees, this focus has the potential to evoke pleasure and positive value when appreciated or experienced in an aesthetic manner. This sentiment is articulated in the following statements:

I believe that in doing so, we help preserve the essence of the region, keeping its traditional lifestyle visible. Additionally, from an aesthetic perspective, it becomes an attraction for customers (Interviewee 12).

9.6.3 Fostering Social and Economic Impact

The results revealed that economic advantages held significance for respondents when evaluating upcycling enterprises. This aligns with prior research indicating the significance of generating economic impact and cost savings in contemplating such approaches. An illustration of this perspective is evident in the following quote where a respondent emphasized the efficacy of this initiative as a profitable avenue:

A lot of people are attracted to products that have been reused and therefore this can bring us economic benefits (Interviewee 4).

In addition, respondents reported that this type of dynamics helped them save money as avoiding buying new materials was very useful for their economy.

We save a lot of money like this. Otherwise, if we have to buy new materials, it costs a lot of money, we are lucky people appreciate it (Interviewee 7).

In several cases, interviewees reported that it was good to contribute to the local economy of families that did not have many financial resources. Therefore, the social impact of these three upcycling organizations was pointed as they all employed local people from the community.

We are employing homeless people from the Cape flats one of the most disadvantaged areas of the city and where we have our facilities (Interviewee 11).

With this organisation, we are employing disadvantaged women from the local community, we are helping them by providing a job and supporting their families (Interviewee 5).

We are employing people from the local community... our intention is to empower locals and create wealth in the neighborhood (Interviewee 2).

9.6.4 Environmental Awareness

The research findings have demonstrated that these endeavors were primarily motivated by the desire to yield positive impacts on the environment, aiming to contribute to its preservation and conservation. Respondents expressed enthusiasm about their strategies to champion environmental values and their commitment to persist in these efforts. As highlighted by one respondent:

The main purpose of our organisation is to get a charming, special environment considering nature, this is extremely important for us (Interviewee 5).

The term "upcycling" was emphasized by some of the respondents relating this to the importance of reusing and restoring materials.

Yes, of course, many junks in houses and companies now we have restored and are visible. They are things that were not thrown away but that at the time seemed old because new furniture arrived. Although many things were thrown away and it is a shame. Above all, we have used old objects from these places and create it beautiful products (Interviewee 6).

It is an opportunity to recover the old, that no longer had a use, and give it one. I am entertained by restoring it. We spend time making them pretty. This way, we save money. It gives us pride (Interviewee 6).

9.6.5 Cultivation of Local Networks

The significance of local networks in fostering the development of these initiatives was highlighted in the interview responses. Both neighbors and relatives were underscored as pivotal in accessing necessary products, as evidenced in the quotes provided. This aspect is particularly intriguing since prior studies on this subject have not emphasized the importance of local networks.

> *Some neighbors have given me materials, people that are not using them anymore (Interviewee 5).*

> *Some things local fishermen find they bring it to us (Interviewee 1).*

> *We got them from people in the area who no longer wanted them. We have not bought anything (Interviewee 8).*

9.7 Conclusions

In this chapter, we redefine "upcycling" as a holistic approach that transforms underused materials or objects into more valuable ones. This work makes a substantial addition to the current body of knowledge by expanding the conversation on the circular economy, the function of upcycling entities, and their impact on sustainable development and the advancement of smart cities. To answer the central question—how does the circular economy aid sustainable growth in smart cities?—our findings strongly support a positive correlation.

We present evidence of various factors that foster upcycling in this realm, including local community networks, cultural impacts, the generation of social and economic benefits, awareness of environmental issues, and the safeguarding of cultural heritage. These results are in line with prior research indicating motivations for upcycling, such as ecological concerns, economic savings, personal fulfillment, and an appreciation for the unique aesthetic of upcycled items. Furthermore, our research underscores the critical role of upcycling in preserving urban cultural traditions, with innovation identified as a key catalyst, echoing earlier findings.

A notable discovery, not extensively covered in existing literature but revealed in our study, is the vital role of community networks in promoting upcycling practices. Participants often cited sourcing materials from their local community, which represents a significant addition to existing research.

In concluding, we recognize the limitations of our study and suggest avenues for future research. While this investigation provides insights into the contribution of upcycling organizations to sustainable development in smart cities, its applicability is limited due to its concentrated focus on specific instances within one city, incorporating only three case studies. The unique socio-cultural dynamics of each city play a crucial role in shaping motivations and goals. Therefore, more extensive, comparative research involving a wider array of interviews is essential for a deeper understanding. Future work could benefit from examining upcycling organizations in different parts of Cape Town and South Africa to better understand their impacts and influences. Furthermore, cross-regional studies, including upcycling initiatives in various African countries, could provide valuable insights into the differences in resources and infrastructure, enhancing our comprehension of these practices.

Acknowledgement This project was partly funded by the British Council to support upcycling initiatives in South Africa. The project was conducted by Minca Ventures Ltd. and Hands of Honour.

References

Anthopoulos, L., & Fitsilis, P. (2018). De las ciudades inteligentes a las cognitivas. *Investigación de sistemas cognitivos, 52*, 372–382.

Batty, M., Axhausen, K. W., Giannotti, F., Pozdnoukhov, A., Bazzani, A., Wachowicz, M., et al. (2019). Ciudades inteligentes del futuro. *Temas especiales de la revista europea de física, 214*(1), 481–518.

Benjamin, T. D., & Kyobe, M. E. (2019). A Smart City framework for South African municipalities: The case of Cape Town. In *proceedings of the international conference on information technology & systems* (pp. 1013–1021).

Blomsma, F., & Brennan, G. (2017). The emergence of upcycling: A new fashion economy. *Fashion Theory, 21*(4), 499–522.

Bocken, N. M., De Pauw, I., Bakker, C., & Van Der Grinten, B. (2016). Product design and business model strategies for a circular economy. *Journal of Industrial and Production Engineering, 33*(5), 308–320.

Buenrostro-Delgado, O., Cortés-Berrueco, L. E., & Ponce-Cueto, E. (2017). Smart waste collection system for efficient urban waste management. *IEEE Access, 5*, 23926–23938.

Cape Town Tourism. (2021). *Responsible Tourism*. [Online] Available at: https://www.capetown.travel/responsible-tourism/. Accessed 4 Oct 2021.

Chabowski, B. R. (2019). El reciclaje como fenómeno de los mercados emergentes. *Revista Internacional de Investigación en Gestión y Mercados, 12*(1), 1–14.

Chinguno, C. N., & Mbohwa, C. (2019). Smart cities: A path to economic growth in South Africa. *South African Journal of Industrial Engineering, 30*(2), 92–106.

Christians, T. (2019). Upcycling in woodstock: Fashion and furniture finds new life in Cape Town. *IOL News*. [Online] Available at: https://www.iol.co.za/weekend-argus/news/upcycling-in-woodstock-fashion-and-furniture-finds-new-life-in-cape-town-38987896. Accessed 4 Oct 2021.

Cugurullo, F., Tojo, N., & Codagnone, C. (2019). Ciudades inteligentes: Definiciones, dimensiones, desempeño e iniciativas. *Revista de tecnología urbana, 26*(1), 3–17.

De Jesus, A., Diaz, C., Fisk, L. J., & Mont, O. (2018). Upcycling: From recycling to creating value. *Sustainability, 10*(1), 209.

De Lannoy, A., Giesecke, S., & Wilkins, N. (2016). Evaluating spatial equity in Cape Town: A multidimensional approach. *The South African Geographical Journal, 98*(3), 448–471.

Department of Environment, Forestry and Fisheries. (2019). *National waste management strategy: implementation plan 2019–2020*. Government of South Africa.

Dijkstra, L., Garcilazo, E., & McCann, P. (2015). The economic performance of European cities and City regions: Myths and realities. *European Planning Studies, 23*(9), 1673–1685.

Ejechi, O. A., Stellenbosch, R., & Van Heerden, J. (2020). Sustainable and smart cities in South Africa: A conceptual study of the role of digital Technologies in Reducing Carbon Emissions. *Energies, 13*(3), 652.

Ellen MacArthur Foundation. (2013). *Towards the circular economy*. Retrieved from https://www.ellenmacarthurfoundation.org/assets/downloads/publications/Ellen-MacArthur-Foundation-Towards-the-Circular-Economy-vol.1.pdf

Filimonau, V., Dehe, B., Filimonau, V. Y., & Mika, M. (2019). Motivating sustainable waste management in hotels: The role of upcycling. *Journal of Sustainable Tourism, 27*(11), 1194–1210.

Geissdoerfer, M., Savaget, P., Bocken, N. M., & Hultink, E. J. (2017). The circular economy—A new sustainability paradigm? *Journal of Cleaner Production, 143*, 757–768.

Ghosh, A., Raturi, A., & Bhatia, A. (2018). Upcycling: A sustainable way of living. *Journal of Cleaner Production, 192*, 411–421.

Greenpop. (n.d.). *About us*. [Online] Available at: https://greenpop.org/about-us/. Accessed 4 Oct 2021.

Gregson, N., Crang, M., Fuller, S., & Holmes, H. (2015). Interrogating the circular economy: The moral economy of resource recovery in the EU. *Economy and Society, 44*(2), 218–243.

Hamari, J., Sjöklint, M., & Ukkonen, A. (2016). The sharing economy: Why people participate in collaborative consumption. *Journal of the Association for Information Science and Technology, 67*(9), 2047–2059.

Herrmann, T. M., & Snyman, H. N. (2019). Urban water governance in Cape Town, South Africa: A review of the City's water management. *Water, 11*(11), 2383.

Jandric, P., Knox, J., Besley, T., Ryberg, T., Suoranta, J., & Hayes, S. (2017). Postdigital science and education. *Educational Philosophy and Theory, 49*(6), 597–602.

Klodawsky, F., Siltanen, J., & Andrew, C. (2021). Crafting a more Sustainable City: Social enterprise, upcycling, and green jobs. *Cities, 108*, 103056.

Knoblauch, H., Schnettler, B., Raab, J., & Soellner, R. (2019). Video analysis: Methodology and methods. *Qualitative Social Research, 20*(4), 1–15.

Komninos, N. (2013a). Modelos de negocio de ciudades inteligentes: desafíos y oportunidades en el caso de las redes inteligentes. *Revista de tecnología urbana, 20*(2), 3–15.

Komninos, N. (2013b). *Intelligent cities and globalisation of innovation networks*. Routledge.

Korhonen, J., Honkasalo, A., & Seppälä, J. (2018). Circular economy: The concept and its limitations. *Ecological Economics, 143*, 37–46.

Kotzé, P., & Van der Merwe, T. (2019). Smart Mobility for Smart Cities: A case study of public transport in Cape Town, South Africa. In *Proceedings of the 2019 international conference on computational science and computational intelligence* (pp. 767–772).

Leal Filho, W., Salvia, A. L., Brandli, L. L., Günther, D., & Molthan-Hill, P. (2020). Sustainable development and upcycling. In *Handbook of sustainability science and research* (pp. 1–15). Springer.

Lombardi, P., Giordano, S., Farouh, H., & Yousef, W. (2012). Modelling the smart city performance. *Innovation: The European Journal of Social Science Research, 25*(2), 137–149.

Mafini, C., Letsoalo, T., & Sipamla, R. (2019). The circular economy and waste diversion in South Africa: A review. *Resources, 8*(1), 10.

Majozi, T., Boje, C., & Annegarn, H. (2019). Challenges in shaping South Africa's waste management policy. *Journal of Environmental Planning and Management, 62*(9), 1606–1622.

Mkhize, T. G., & Pretorius, J. H. C. (2020). Assessing the uptake of the circular economy concept in South Africa. *Sustainability, 12*(11), 4487.

Moloisi, O. (2020). Upcycling in fashion: Sustainability and ethical practices in the south African fashion industry. *Sustainability, 12*(21), 9095.

Mont, O. (2020). Upcycling as an emerging business model. *Journal of Industrial Ecology, 24*(3), 525–527.

Moore, J., & Hunt, D. V. (2018). Upcycling and the circular economy: The influence of product understanding on consumer evaluation. *Sustainability, 10*(8), 2702.

Nam, T. Pardo, TA (2011). Ciudad inteligente como innovación urbana: centrándose en la gestión, las políticas y el contexto. En Actas de la Quinta Conferencia Internacional sobre Teoría y Práctica de la Gobernanza Electrónica (págs. 185–194).

Nhamo, G., Phd, K., & Maringa, K. (2019). Moving towards a circular economy: The role of upcycling in waste management in South Africa. *Resources, 8*(4), 203.

Palmer, M. S., Rickard, J., & Robbins, D. (2020). Upcycling and value creation: The case of a bicycle cooperative. *Business Strategy and the Environment, 29*(8), 2883–2896.

Red Bull Amaphiko. (n.d.). *About Amaphiko*. [Online] Available at: https://www.redbullamaphiko.com/about. Accessed 4 Oct 2021.

Rivas, J. L., & Zorpas, A. A. (2019). Circular economy: A new strategic vision for environmental sustainability in South Africa. *Sustainability, 11*(18), 4960.

Rizos, V., Behrens, A., Kafyeke, T., Hirschnitz-Garbers, M., & Ioannou, A. (2016). *The circular economy: Barriers and opportunities for SMEs*. Retrieved from https://publications.europa.eu/en/publication-detail/-/publication/82b3b3b2-8145-11e6-9bca-01aa75ed71a1

Sarova, S., Sarova, M., & Kinyua, R. (2019). Sustainable urbanization and regional development in Cape Town, South Africa. *European Journal of Sustainable Development, 8*(1), 267–276.

Southey-Swartz, I. (2020). Hout Bay recycling for good: A model for upcycling and community empowerment in South Africa. *Resources, 9*(6), 69.

Stevenson, F., & Parnell, R. (2020). Sustainable creativity through upcycling: An exploration of the role of creative agencies. *Journal of Business Research, 120*, 457–464.

The Recycle Swop Shop. (n.d.). *About us*. [Online] Available at: https://www.recycleswopshop.com/aboutus. Accessed 4 Oct 2021.

Townsend, A. M. (2013). *Ciudades inteligentes: big data, hackers cívicos y la búsqueda de una nueva utopía*. W. W. Norton & Company.

Van der Merwe, L. E., & Kotzé, P. (2018). Promoting sustainable tourism development: A case study of Cape Town, South Africa. *Sustainability, 10*(8), 2900.

Van Zyl, A., Bagula, A., & Petit, M. (2021). Smart City governance framework for South African municipalities.

Zamani, B., Yousefi, M., & Khademalrasoul, A. (2021). Energy-efficient upcycling: A case study in the construction industry. *Journal of Cleaner Production, 290*, 125639.

Dr. Sara Calvo (PhD) is a Senior Lecturer in Entrepreneurship at the Department of Public Leadership & Social Enterprise, where she brings academic experience from posts held in the UK and Spain. Prior to joining the Open University, she was a Senior Lecturer at Middlesex University. She is a certified Social Enterprise Business Consultant and a certified Coach by the International Coaching Federation. She is the cofounder of the social enterprise, Minca Ventures, an organization that works to create impact and sustainability projects where they carry out consulting, coaching, and training work. She is also the founder of the Buen Vivir Institute, a social enterprise company that offers online education courses and coaching services for people and organizations. Creator of the Buen Vivir method, she has developed the self-knowledge program called "Build your Good Living."

Dr. Andres Morales (PhD) is a Lecturer in International Development at Universidad International de La Rioja. He is a researcher in the field of social enterprise and innovation, international development, indigenous communities, innovation in education, sustainability, sustainable enterprises, and visual research. He is the cofounder of Minca Ventures Ltd. and has a YouTube channel with a podcast "Impact Story channel" where he shows impact case studies across the globe.

Chapter 10
xEMS: Control Towers of Optimal Energy Flow Management for Smart City Realization

Bo Jie, Jumpei Baba, and Akiko Kumada

Abstract Smart city is an initiative to solve various problems and creates new value, including comfort and convenience, by utilizing advanced technologies to improve the efficiency and sophistication of urban and regional functions and services. In order to realize smart city, the introduction of renewable energy, stable and efficient energy supply and utilization technologies are indispensable. To achieve the carbon neutrality, the introduction of huge volumes of renewable energy, the spread of EVs, and the acceleration of electrification have created unprecedented challenges for energy, especially power systems. Various energy management systems can effectively solve these problems at each level of the system.

This chapter unfolds the details of each EMS (Energy management system), focusing on the various issues that result from the realization of smart city. For example, GEMS (Grid EMS) is the energy management of the power infrastructure, primarily aimed at maintaining the power quality (voltage, frequency, stability, etc.) of the power grid. BEMS (Build EMS) is a system for energy management of air conditioning, lighting, motive power, etc. in offices, buildings, stores, etc. In addition to visualizing the overall energy consumption of a building, BEMS can support optimal energy operation by controlling air conditioning, lighting, and other equipment in the building. FEMS (Factory EMS) is a system that achieves optimization in conjunction with manufacturing schedules by rationalizing energy consumption by monitoring the energy utilization and operating status of production facilities at factories, and by improving energy intensity by performing predictive maintenance of production and facilities. HEMS (House EMS) primarily aims at reducing energy

B. Jie (✉) · A. Kumada
Department of Electrical Engineering and Information Systems Graduate School of Engineering, The University of Tokyo, Tokyo, Japan
e-mail: kaiha-jiebo@g.ecc.u-tokyo.ac.jp

J. Baba
Department of Advanced Energy Graduate School of Frontier Sciences, The University of Tokyo, Tokyo, Japan

S. Majumdar et al. (eds.), *Smart Cities*, S.M.A.R.T. Environments,
https://doi.org/10.1007/978-3-031-59846-3_10

consumption and electricity bills. It can streamline the consumption of electricity, helping to reduce CO2 emissions. And CEMS (Community EMS) is a technology for ecology and residential comfort.

Keywords Smart city · Carbon neutrality · Energy management system (EMS) · Electricity power system · Energy system

10.1 Introduction

The concept of a smart city entails the utilization of sophisticated technologies to enhance the effectiveness and sophistication of metropolitan and regional operations and services, therefore addressing a range of issues and generating novel value, such as convenience and comfort. In order to realize smart city, the introduction of renewable energy, stable and efficient energy supply and utilization technologies are indispensable. To achieve the carbon neutrality, the introduction of huge volumes of renewable energy, the spread of electric vehicles (EVs), and the acceleration of electrification have created unprecedented challenges for energy, especially power systems. Various energy management systems can effectively solve these problems at each level of the system. For elucidating this, researchers have put a lot of effort into this topic.

The possibility of cities achieving smartness without simultaneously attaining sustainability was examined (Yigitcanlar et al., 2019). The findings indicate significant obstacles that challenges confronting Smart Cities in the pursuit of sustainable societies. (a) Smart city policies exhibit a pronounced focus on technology-centric approaches. (b) Smart city practices entail intricate dynamics and intricacies. (c) Smart city concepts are often formulated in an improvised and non-systematic way. The notion of smart cities has attracted significant interest from scholars in many disciplines, such as computer science and engineering, owing to its pertinence to multiple actors and the accompanying advantages and obstacles in its execution. The research undertaken by Ismagilova et al. (2019) explores several aspects of smart cities, encompassing smart transportation, intelligent living, smart circumstances, smart citizens, intelligent governance, and intelligent architecture, as well as the corresponding technologies and concepts. *Choi.* put forward a proposition for a distributed architectural design based on hierarchy that effectively combines the advantages of hierarchy combined with distributed architectures (Choi, 2019). This article presents a comprehensive overview of the potential applications of this framework in the context of smart city settings, focusing on selected upcoming technologies. *Nguyen et al.* implemented a smart solar microgrid for disaster protection purposes in the Dawu neighborhood, located in Pingtung County, Kaohsiung City (Nguyen et al., 2020). The installation site is situated at an elevation of 1000 m above sea level. Backpropagation neural networks are employed for the purpose of forecasting complimentary grid power. This is achieved by the utilization of many inputs, including the load power, battery voltage, battery charging/discharging current, as well as the voltage and current readings obtained from the solar panel.

The primary determinants for achieving smart city status are intelligent electricity networks and the utilization of smart energy sources. The purpose of the study conducted by *Sami et al.* in (Sami et al., 2021) is to offer a thorough examination of various technologies and techniques, along with their corresponding solutions, that can be utilized to address the challenges associated with renewable energy sources, including their variability and load fluctuations. The study specifically focuses on how these technologies and techniques can be effectively integrated into the microgrids of smart cities to meet the energy demands. *Allan et al.* in (Allan et al., 2021) provide a comprehensive examination of the existing digital technologies and tactics that may be utilized to enhance the rates of bystander detection of sudden cardiac arrest (SCA), facilitate quick activation of energy management system (EMS), commence high-quality cardiopulmonary resuscitation (CPR), and effectively identify, retrieve, and operate automated external defibrillators (AEDs). The potential utilization of wearable devices and household smart speakers in the continuous monitoring of vital signs for those who are susceptible to life-threatening arrhythmias, with the capability to transmit alerts to selected contacts or energy management system (EMS), may be a significant development in the future. *Saeed et al.* put out the concept of a centralized EMS for implementation in intelligentized buildings (Saeed et al., 2023).

This chapter unfolds the details of each EMS (Energy management system), focusing on the various issues that result from the realization of smart city (Fig. 10.1). For example, GEMS (Grid EMS) is the energy management of the power infrastructure, primarily aimed at maintaining the power quality (voltage, frequency, stability, etc.) of the power grid. BEMS (Build EMS) is a system for energy management of

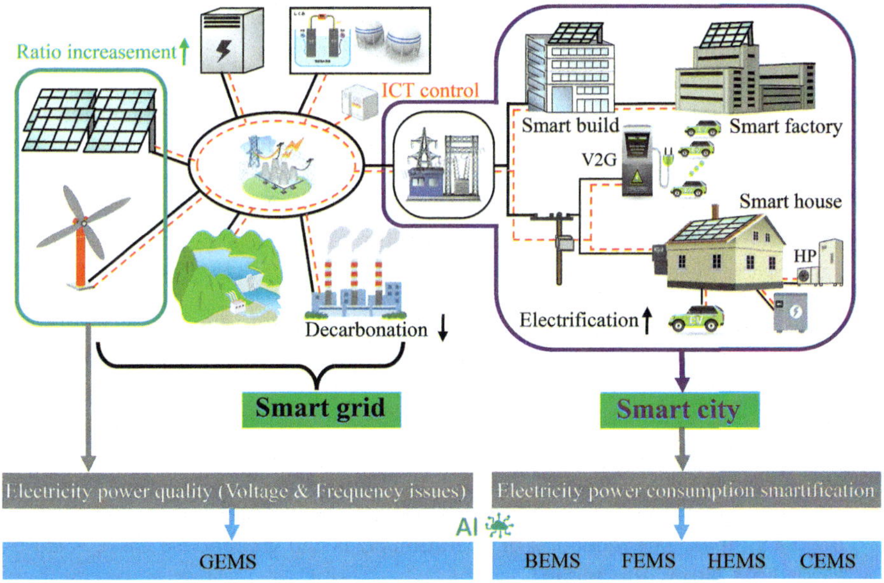

Fig. 10.1 Various energy management systems for smart city realization

air conditioning, lighting, motive power, etc. in offices, buildings, stores, etc. In addition to visualizing the overall energy consumption of a building, BEMS can support optimal energy operation by controlling air conditioning, lighting, and other equipment in the building. FEMS (Factory EMS) is a system that achieves optimization in conjunction with manufacturing schedules by rationalizing energy consumption by monitoring the energy utilization and operating status of production facilities at factories and by improving energy intensity by performing predictive maintenance of production and facilities. HEMS (House EMS) primarily aims at reducing energy consumption and electricity bills. It can streamline the consumption of electricity, helping to reduce CO_2 emissions. And CEMS (Community EMS) is a technology for ecology and residential comfort.

10.2 Functions and Overarching Perspective of EMS in Achieving of Carbon Neutrality

Efforts are now being undertaken to facilitate the establishment of a sustainable society that operates on carbon-neutral principles. These endeavors primarily involve the utilization of PV power, and wind power generation sources and the adoption of energy conservation practices. The adoption of alternative energy solutions is on the rise among customers in many sectors, including factories and buildings. This trend encompasses not only the substitution of traditional energy-efficient equipment but also the integration of privately owned power generation and storage systems, such as rooftop solar panels and energy storage systems. To achieve enhanced energy conservation, it is imperative to build an integrated system for managing and operating the supply and demand of electric energy in utility facilities. Hence, alongside the conventional Energy Management System (EMS) function, which primarily focuses on energy visualization, the optimal operation planning function, which involves the planning and control of utility facilities based on energy generation and demand forecasts, will assume a more prominent role than previously. A classical architecture of EMS is given as Fig. 10.2.

The PDCA (Plan-Do-Check-Act) cycle of an energy management system consists of the following steps. The primary purpose of the EMS is to effectively manage and oversee each individual step, as visually depicted in Fig. 10.3.

- Plan: Understanding the organization's situation, establishing an energy policy, energy management team, and developing an action plan to improve energy performance based on an analysis of energy utilization and consumption, e.g., energy consumption side operating schedule determination, optimal power supplier dispatch.
- Do: Preparing the essential resources, managing the capacity, communication, etc., and operating according to the action plan, including design and procurement activities. The reduction of operation costs and greenhouse gases is almost carried out during this phase.

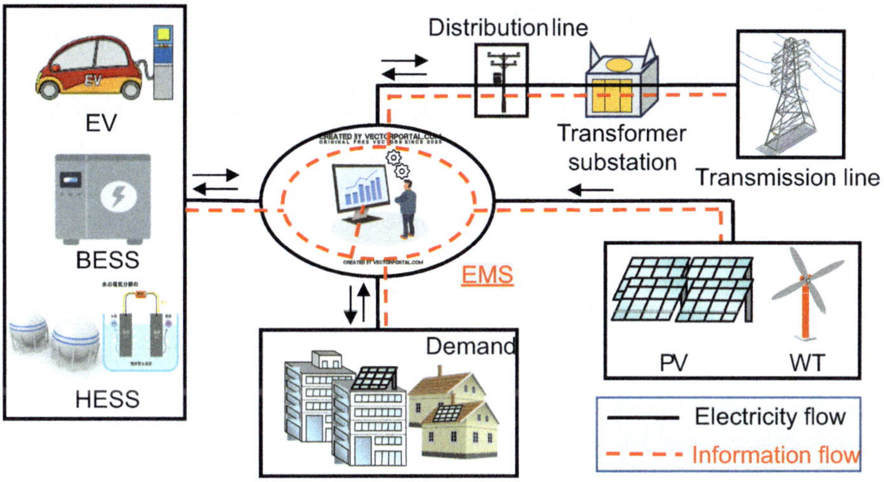

Fig. 10.2 Architecture of a classical energy management system

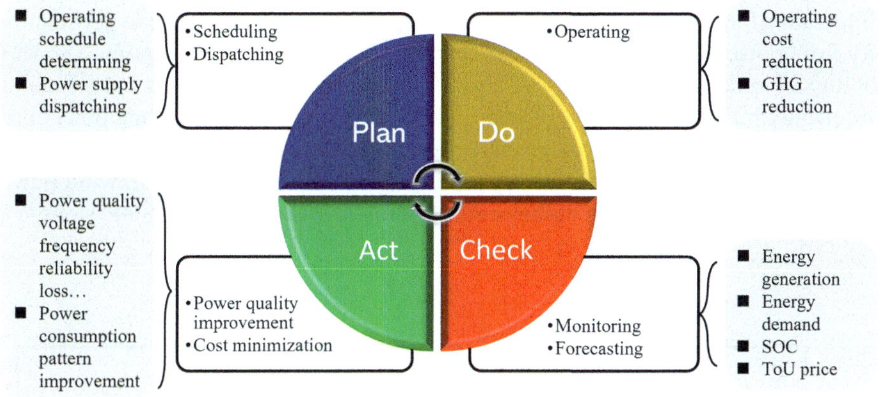

Fig. 10.3 Operation process & functions of EMS in accordance with PDCA cycle

- Check: Monitoring, measuring, analyzing, and evaluating energy performance and operational status, and conducting internal audits and management reviews. Performing forecasting and real-time results analysis, for example, in-house PV generation, in-house wind power generation, energy consumption demand, SOC (state of charge) of energy storage systems, and ToU (time of use) prices in the electricity markets.
- Act: Taking action to address nonconformities and improving energy performance and energy management systems. Based on the examination (C) of the outcomes derived from the execution (D) of the plan (P), it is recommended to

formulate a strategy for enhancing the power quality management system (as referenced in reference (Rathor & Saxena, 2020)). This approach should address aspects such as voltage, frequency, stability, losses, and other relevant factors within the administrative area of the Energy Management System (EMS).

The FEMS, BEMS, and HEMS systems implement energy management strategies that take into account demand response (DR) requests from the CEMS system. These strategies aim to achieve energy conservation control by utilizing different utility facilities, such as renewable energy sources and energy storage systems, at each facility. The CEMS (Community Energy Management System) exercises authority over the entirety of the region in order to achieve a balance between energy supply and demand. This enables the realization of local energy generation for local consumption. In instances of electricity surplus within the region, the CEMS regulates power generation by reducing its output, therefore facilitating the storage of excess power in the energy storage system. In the event of insufficient electricity supply within the area, the CEMS will augment electricity generation and discharge power from the energy storage system. Additionally, demand response (DR) requests will be issued to each of the EMSs as deemed required. Instead of pursuing the objective of achieving total domestic generation for domestic consumption, it is imperative for the CEMS (Community Energy Management System) to get electricity from power producers and electric utilities, even for backup purposes. The variability of renewable energy power generations, such as PV and wind power, is frequently influenced by climate change. The alignment between actual power production and expected output levels poses significant challenges (Figs. 10.4 and 10.5). The significance of technologies such as PV output forecasting, wind power production forecasting, and energy consumption demand forecasting and storage in the EMS is obvious, shown in Fig. 10.2.

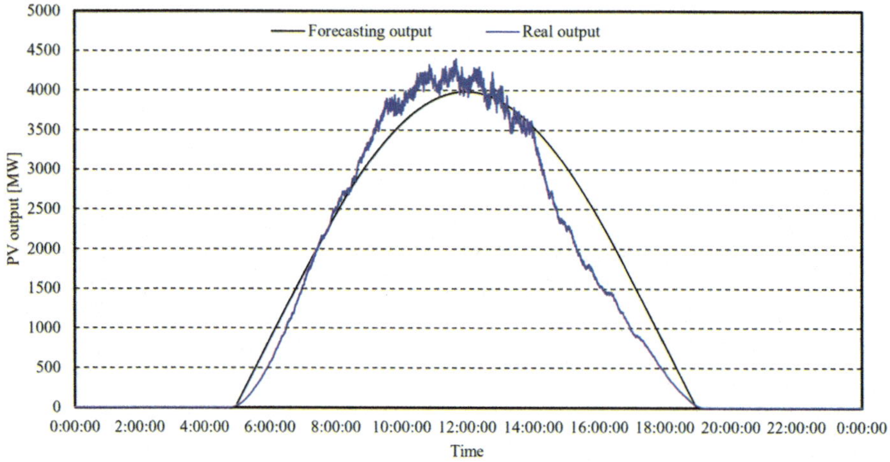

Fig. 10.4 PV forecasting output and real output

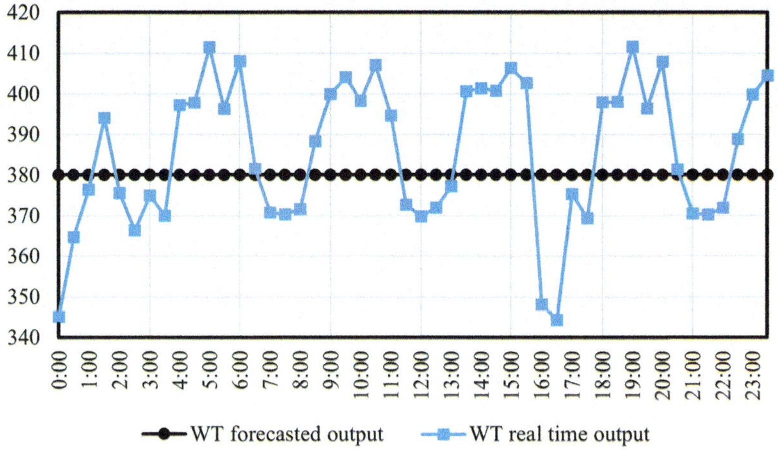

Fig. 10.5 Wind power forecasting output and real output

10.2.1 PV Power Output Forecasting

There are two distinct categories of forecasting methodologies utilized for predicting the PV power systems output. The first category, known as direct forecasting, involves the direct estimation of the quantity of electricity generated by the PV system. Conversely, the second category, referred to as indirect forecasting, involves the prediction of solar radiation levels, which are subsequently converted into estimates of the output of electricity. In the context of direct forecasting, it is important to establish a model that captures the correlation between energy generation levels and many elements specific to each solar module. In contrast, indirect forecasting has gained prominence in recent years as a widely employed approach for predicting. This approach necessitates the establishment of a model that captures the correlation between meteorological data and solar radiation. It may be categorized into three distinct techniques based on the type of weather information employed.

- Weather Forecasting Utilization Methodology: This method aims to forecast the quantity of power produced by considering meteorological factors, namely the presence of sunny or gloomy weather conditions. One limitation of weather forecasting is its inherent inability to provide continuous predictions, resulting in a restricted capacity to attain a high level of forecasting precision.
- Numerical Weather Forecast Data Application Methodology: The Japan Meteorological Agency employs a method of calculating precise numerical weather prediction data by partitioning the Earth's horizontal plane into a grid consisting of many levels above the surface, with each layer spaced several kilometers apart. This methodology is employed for the purpose of predicting electricity generation. One notable feature of this approach is its ability to utilize a wide range of meteorological variables, thereby enhancing the potential for

achieving a high level of prediction accuracy. Conversely, it is important to effectively handle a substantial quantity of climatic factors. Sophisticated statistical processing techniques are also required. This approach utilizes numerical weather forecast data as the beginning condition and performs calculations with a higher resolution grid spacing. One notable benefit of this approach lies in its capability to provide highly precise predictions at minute-level granularity. However, it is important to acknowledge that this advantage is counterbalanced by the inherent drawback of increased computational expenses associated with its implementation.

- Forecasting Results Evaluating Methodology: Typically, predictions are evaluated using several metrics such as the mean error (ME), mean absolute error (MAE), root mean square error (RMSE), and other similar indications. In the context of solar power generation, the error can be standardized by either the installed capacity or the measured values for each specific time interval. It is important to acknowledge that the assessment of error values exhibits significant variation contingent upon the chosen method of normalization. When applying normalization techniques using observed values for each time period, it is possible to calculate the average value of the normalized error for each time period. Alternatively, the relative error for the entire period can be assessed by categorizing the error based on the amount of solar radiation during that period. In addition, it is important to consider the time period preceding dawn and following dusk when doing the assessment. Therefore, while assessing the inaccuracies in solar radiation/photovoltaic forecasts. Therefore, it is crucial to exercise caution when assessing forecast errors in solar radiation/photovoltaic predictions due to the lack of standardized methodology.

10.2.2 Wind Power Output Forecasting

The utilization of wind power for electricity generation has the notable benefit of producing zero carbon dioxide (CO_2) emissions, making it an environmentally friendly energy source. Furthermore, the adoption and implementation of wind power technologies are steadily growing. In forthcoming times, the incorporation of wind power generation into energy management systems will be vital for effective planning. However, because to the significant variability in wind power output, it is imperative to forecast its generation in advance in order to facilitate effective planning.

Forecasting methods for wind power generation can be broadly classified into two categories: methods that use only wind speed and power generation data from wind farms, and methods that forecast from a wide-area perspective of atmospheric motion. Typically, the former is utilized for short-term predictions ranging from 2 to 3 h ahead, whilst the latter is employed for longer-term forecasts spanning several days. Energy management systems often formulate operational plans for the subsequent day or week. This section pertains to the latter. The forecasting technique may be broadly categorized into three distinct parts:

- Forecasting of wind velocity throughout an expansive geographical region.
- Forecasting the velocity of wind in a localized region encompassing wind turbine installations.
- Calculate the power generation based on the wind speed.

10.2.3 Load Demand Forecasting

The issue of energy demand load fluctuation (Fig. 10.6) caused by uncertainties in power consumption activities is a significant concern within the context of Energy Management Systems (EMS). The utilization of load forecasting technologies is crucial for the effective operational planning of energy management systems. Energy management systems in smart communities encompass many components that play a crucial role in ensuring efficient energy use. These components include the Community Energy Management System (CEMS), which is responsible for monitoring and regulating the regional supply-demand equilibrium and addressing demand response needs. Additionally, there are Building Energy Management Systems (BEMS) designed to cater to the energy requirements of buildings, Factory Energy Management Systems (FEMS) tailored for industrial facilities, and Home Energy Management Systems (HEMS) specifically designed for residential users. The CEMS system effectively manages the equilibrium between energy supply and demand by utilizing regional energy load predictions. On the other hand, the FEMS, BEMS, and HEMS systems run suitable energy supply facilities in accordance with their individual load projections. This approach aims to minimize energy wastage and accomplish peak shifting and peak cutting by establishing a connection between

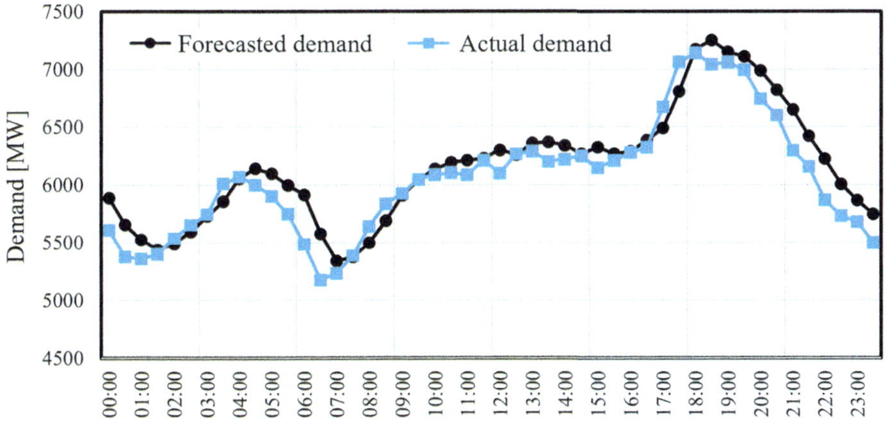

Fig. 10.6 Demand forecasting output and real output

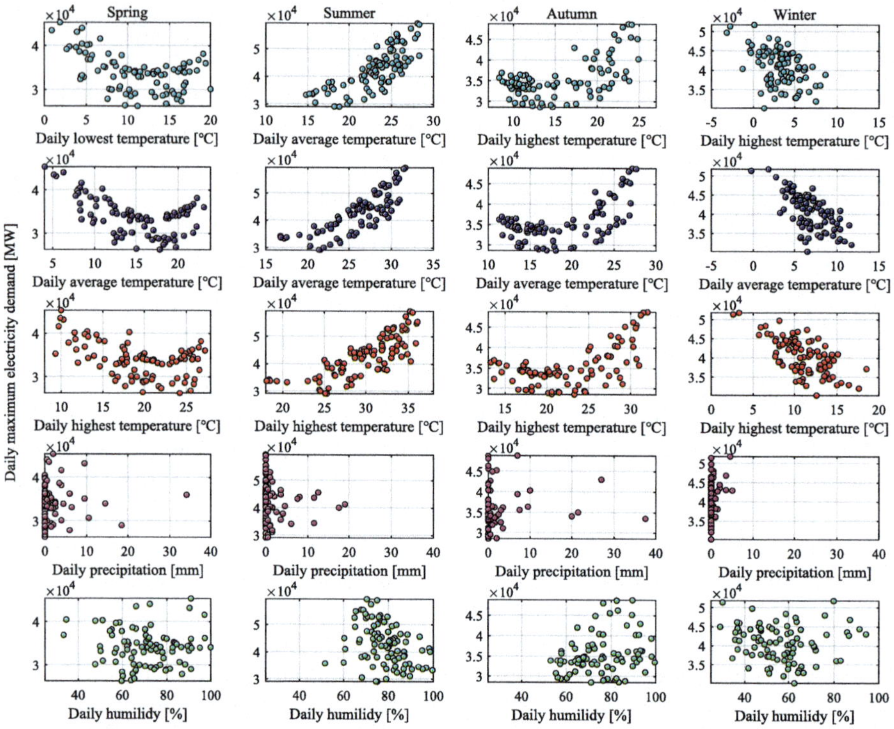

Fig. 10.7 Seasonal forecast of daily electricity demand. (Jie et al., 2023)

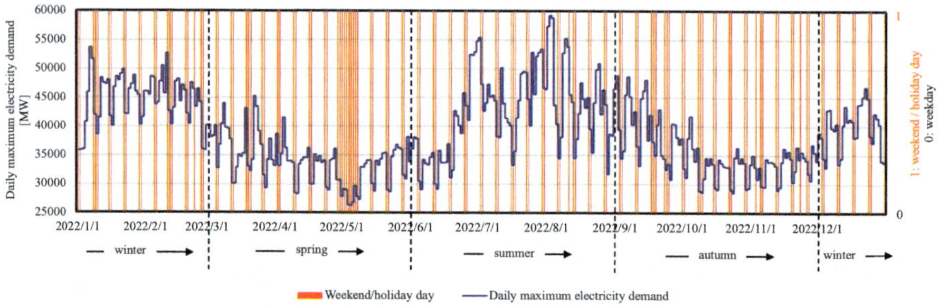

Fig. 10.8 Daily electricity demand forecast. (Jie et al., 2023)

energy demand and supply. Load forecasting technology is utilized in a diverse range of energy management systems (EMS).

Load forecasting technology is considered a crucial technology for effective energy management in different Energy Management Systems (EMSs). Load forecasting approaches may be categorized into three main types: (1) time-series load trend predictions, (2) correlation-based forecasts, such as regression models, and (3) forecasts based on historical data of comparable dates. Method (1) is a modeling approach that captures the temporal patterns of load in a time-series data. It leverages this model to provide predictions for future time points, primarily focusing on short-term forecasting applications. Method (2) refers to a regression modeling approach that utilizes characteristics, such as temperature, which exhibit a significant association with load, as input variables. The purpose of this technique is to develop a model that can accurately estimate future load based on the aforementioned factors. Method (3) employs a strategy that emphasizes the daily resemblance of load data and conducts a search inside the database for historical data from days that bear similarity to the anticipated target day. This approach is utilized to predict future outcomes. And Figs. 10.7 and 10.8 show the load demand forecasting results based on seasonal phase and daily phase separately (Jie et al., 2023).

10.2.4 Energy Storage Technology

Power storage systems are of significant importance in smart communities since they facilitate the regulation of the equilibrium between energy demand and supply. Additionally, several services have been proposed to utilize the stored energy.

10.2.4.1 Load Fluctuation Smoothing

Typically, electrical users utilize energy in accordance with their own outcomes and operational patterns, resulting in periodic fluctuations in power demand. These fluctuations manifest as periodic spikes and drops, such as those observed during day and night cycles. In the context of a smart community, as depicted in Fig. 10.9, the process of leveling power demand can be achieved through the utilization of an electricity storage system, resulting in enhanced facility operating rates and a reduction in capital expenditure.

- *Short-term period load fluctuation smoothing*: By effectively recharging the excess electricity produced by PV power generation using a battery energy system, it is possible to optimally utilize PV power output and achieve reductions in peak power consumption and load leveling (Fig. 10.9). The effectiveness of photovoltaic (PV) power generation is constrained by unfavorable weather conditions, including precipitation in the form of rain or snow, as well as the absence of sunshine during night-time hours. Therefore, it is imperative to obtain electricity from the power company in the conventional method under adverse weather conditions, regardless of the time of day. However, the storage of excess electricity in a battery enables its application in unfavorable weather circumstances or during periods of reduced photovoltaic (PV) output, such as during nighttime.
- *Long-term period fluctuation smoothing*: In the context of long-term cycle period energy storage measurement, it is possible to utilize surplus power for the pur-

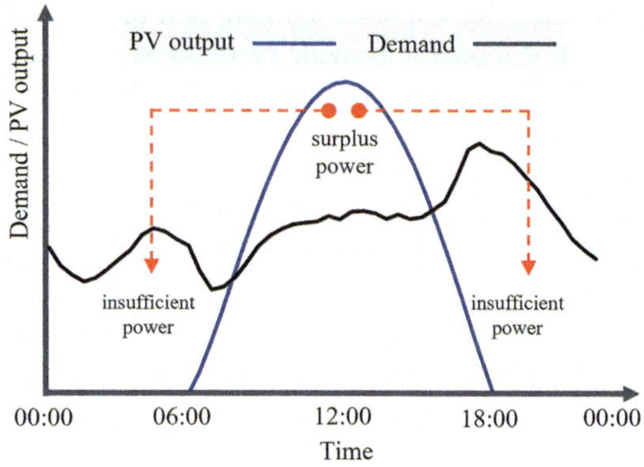

Fig. 10.9 Daily electricity demand peak cut by short-term energy storage system

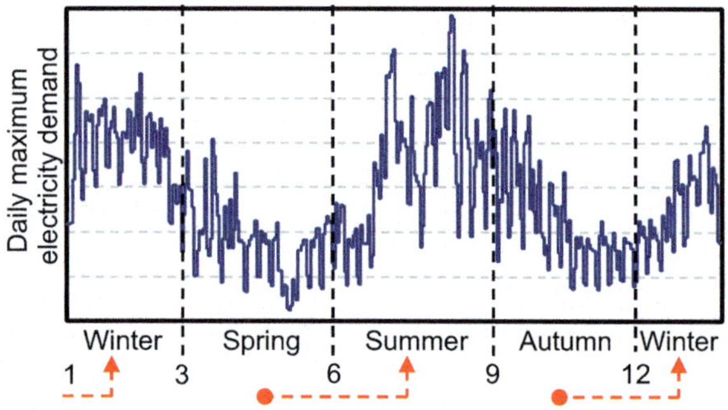

Fig. 10.10 Seasonal electricity demand peak shift by long-term energy storage system

pose of manufacturing hydrogen. This may be achieved through the process of water electrolysis within a designated unit referred to as the hydrogen production unit. The hydrogen that is generated as a result is then stored in a specialized storage container known as the hydrogen tank. In situations where power is of utmost importance, the stored hydrogen is utilized as a fuel source to generate energy within a fuel cell. The advantageous attributes of hydrogen storage for the extended preservation of electrical energy stem from its capacity to prevent self-discharge, a quality that sets it apart from conventional storage batteries. The utilization of water electrolysis is applied to produce hydrogen as a means of storing extra electricity generated from photovoltaic (PV) output for subsequent usage in fuel cells. During emergency situations or periods of high demand, these fuel cells have the capability to produce power while just emitting water, without the concurrent emission of carbon dioxide. Hydrogen storage devices have demonstrated efficacy in utilizing excess electricity over extended periods, hence mitigating energy wastage, as seen in Fig. 10.10.

10.2.4.2 Renewable Energy Output Fluctuation Smoothing

The power output of renewable energy sources, such as solar power and wind power, can be subject to changes due to natural factors, such as variations in solar radiation and wind conditions. These fluctuations have the potential to upset the balance between power supply and demand within the power system, leading to impacts on the voltage and frequency of the system. The installation of power storage systems enables the mitigation of output variations through the process of recharging and discharging power, so achieving a more consistent and stable output.

10.3 xEMS

In the practical implementation of energy management, distinct levels of management align with various types of Energy Management Systems (EMS) (*TAIGA*). Various energy management systems (EMS) should prioritize distinct aspects and include diverse technology. A comprehensive comprehension of the underlying principles and a broad familiarity with diverse EMS (Energy Management Systems) are crucial for the establishment of a modernization intelligent urban environment. This section provides a comprehensive introduction to the ideas and facts pertaining to xEMS that are essential for the development of a smart city.

10.3.1 CEMS

This section elucidates the operational mechanism of Community Energy Management Systems (CEMS) as a pivotal component of energy management inside a smart community (Fig. 10.11) (*Fuji Electric*). There exists a necessity to advance technological solutions that may effectively maximize energy use, namely in addressing power deficits and harnessing the potential of intermittent renewable energy sources, all while ensuring the preservation of power integrity. Furthermore, in terms of mitigating risks associated with large-scale centralized power sources, there is a growing interest in exploring alternative approaches such as decentralized systems that combine small and medium-sized power sources for local production and consumption. Additionally, autonomous power infrastructures are being investigated as a means to ensure a consistent power supply to communities, even during emergency situations. Smart communities are a form of community energy management systems that strive to uphold the standards and convenience of electric power while simultaneously working toward the establishment of a low-carbon society. These systems integrate smart grid technology to ensure a consistent and reliable supply of electric power from the production side, while also incorporating various energy management systems (EMS) on the consumer side to promote efficient

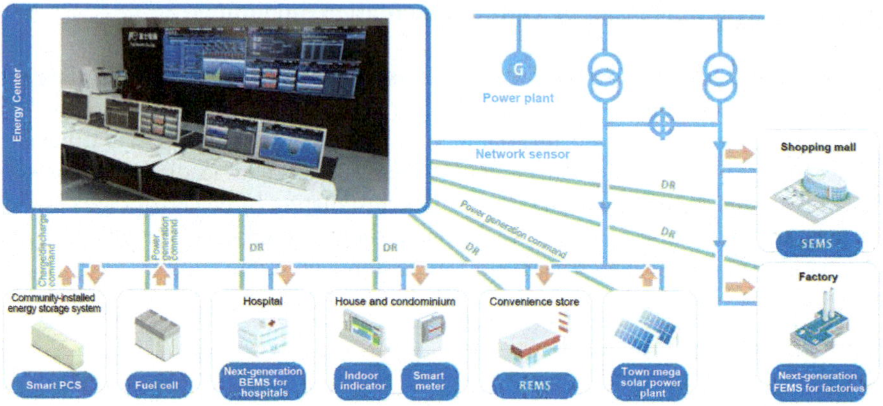

Fig. 10.11 CEMS. (Source: (*Fuji Electric*))

energy consumption. In order to implement a comprehensive community energy management system (CEMS), the utilization of information and communication technology (ICT) is imperative. This technology enables the coordination of diverse CEMS components, including buildings, industries, and residences, as well as the effective control of energy sources such as electricity and heat. The utilization of ICT (Information and Communication Technology) is crucial in facilitating the coordination of many components within the CEMS framework, including buildings, factories, and residential areas.

Figure 10.1 (Sect. 10.1) illustrates the correlation between the electricity system, which is governed by the smart grid, and the smart community, which is regulated by the Community Energy Management System (CEMS)

As depicted in Fig. 10.1, the Community Energy Management System (CEMS) is a sophisticated framework that enhances the efficiency of the overall energy system. It achieves this by establishing connections between various entities such as buildings, factories, general consumers, and distributed power sources within a specific geographic area. These connections are facilitated through power distribution lines and information and communication technology (ICT). Additionally, CEMS operates in conjunction with the smart grid, which governs the entire power system. The primary goal is to achieve a simultaneous improvement in power quality and decrease in CO2 emissions by implementing energy efficiency measures and peak shaving techniques.

The utilization of renewable energy sources has significant importance in the establishment of a low-carbon society. Nonetheless, the development of renewable energy, exemplified by solar power generation, exhibits inherent volatility and poses challenges in terms of predictability. Consequently, the widespread integration of renewable energy generations would unavoidably impact the quality of electricity power system by causing variations in frequency, excess power generation, voltage elevation, and similar effects. The discussion around the enhancement of energy supply resilience has emphasized the importance of decentralizing power generation and improving energy efficiency at the consumer level. Additionally, there is a growing recognition of the imperative to encourage local electricity production for local consumption. A crucial requirement entails the establishment of a framework that facilitates the coordination of decentralized power sources, such as solar power generation, with power consumption management at the consumer level. This mechanism aims to maximize energy use through collaborative efforts among consumers during periods of power scarcity.

CEMS operates on the foundation of the power grid and establishes connections between decentralized power sources within the area, encompassing various customer segments such as residential households, commercial buildings, and industrial factories, as well as electric vehicles (EVs) that can serve as energy storage units. The primary objective of CEMS is to optimize energy consumption within the region by effectively managing the energy supply and demand balance. In order to achieve this objective, ICT (Information and Communication Technology) will be employed for the purpose of exchanging information pertaining to these entities. Additionally, ICT will be integrated with incentives aimed at fostering the efficient energy utilization and facilitating the visualization of energy consumption at individual consumer units. CEMS will function as the fundamental framework for the establishment of intelligent communities, which are local communities that aim to achieve a low-carbon society while simultaneously enhancing their economic worth.

The construction of CEMS aims to achieve comprehensive optimization, targeting the enhancement of energy efficiency in the electric power sector, as well as in the heat energy sector, which constitutes 50% of the total energy demand, and in the transportation sector, which has experienced significant growth in demand in recent times. The anticipated social infrastructure to be implemented by CEMS is projected to rely on a diverse array of technologies and goods. One illustration of this is the necessity for novel technologies to effectively manage many components of the power system, including consumers, distributed power sources associated with renewable energy system and energy storage system. This may be achieved through the utilization of power electronics technology and information and communication technology (ICT).

10.3.2 BEMS

Building Energy Management System (BEMS) is designed to uphold a secure and pleasant indoor atmosphere within buildings, while concurrently promoting energy conservation and enhancing operational and managerial efficiency. The term "building" in this context pertains to non-residential structures, including but not limited to offices, department shops, supermarkets, hotels, hospitals, schools, and exhibition facilities, as defined by the Building Standards. Within the realm of energy management, non-residential buildings encompass various establishments such as factories and offices. However, it is important to note that the word "building" often excludes factories.

Building Energy Management Systems (BEMS) are responsible for monitoring and regulating a wide range of equipment and devices within a building. These systems aim to efficiently manage and regulate all aspects of the building's facilities in a holistic and integrated manner. In essence, Building Energy Management Systems (BEMS) are designed to efficiently regulate various building facilities, including but not limited to air conditioning, lighting, electricity, sanitation, elevators, escalators, disaster prevention, and crime prevention (Fig. 10.12). BEMS achieves this by individually optimizing the control of each facility and also by overseeing and managing the entirety of building facilities through the coordinated control of associated equipment and devices.

The air conditioning system is responsible for overseeing and regulating various components of the air conditioning equipment, including chillers, hot/cold water transfer pumps, air conditioners, blowers, flow control valves, temperature sensors, humidity sensors, and others. Its primary objective is to create a comfortable air-conditioned environment within the building. Additionally, the system collaborates with access control equipment to implement energy-saving measures, such as automatically deactivating the air conditioning in unoccupied rooms. The system also regulates energy saving by the automated deactivation of air conditioning in unoccupied rooms, in coordination with the access control equipment.

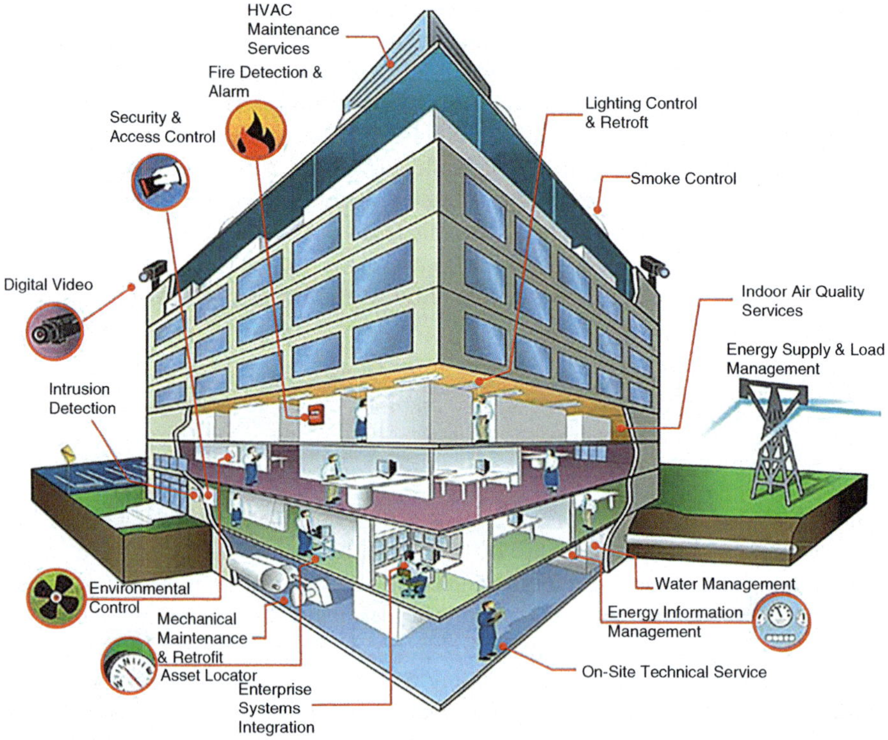

Fig. 10.12 BEMS. (Source: (*Laplace BEMS*))

Building Energy Management Systems (BEMS) offer several advantages to various stakeholders, including building occupants, building owners, and building managers. Building Energy Management Systems (BEMS) provide building occupants with a multitude of benefits. These include the provision of a living environment that ensures safety, security, and comfort. Moreover, BEMS foster an atmosphere that boosts work efficiency, promote creativity, and accommodate the evolving work styles of a varied workforce. In order to achieve this objective, the system regulates the air conditioning and lighting within the rooms, ensures a consistent power supply to information equipment, facilitates network connectivity, maintains 24-h air conditioning, and optimizes the functioning of building utilities. The implementation of this system for building owners yields enhanced value for their properties through the enhancement of convenience, comfort, safety, and energy efficiency. Additionally, it offers adaptable assistance for building operations, ensuring the fulfillment of tenants' requirements while attracting high-quality renters. Energy conservation plans that focus on minimizing wasted energy and optimizing the operation of facilities and equipment, utilizing data on building operational performance, along with building maintenance plans that rely on information regarding equipment failures and deterioration, are instrumental in mitigating building maintenance and management expenses. Furthermore, in the realm of building management, which entails overseeing the functioning of building facilities, the integration of monitoring and management systems for these facilities can greatly enhance operational efficiency and reduce labor requirements. This, in turn, leads to a reduction in overall management costs. Additionally, such integrated systems enable swift and appropriate responses to any abnormalities that may arise in the facilities or equipment, thereby ensuring optimal performance and functionality. Moreover, the BEMS (Building Energy Management System) has a noteworthy capability to mitigate building energy usage and the emission of greenhouse gases, including CO_2, in light of pressing worldwide environmental concerns. These concerns are widely recognized as significant societal issues among the many stakeholders involved.

10.3.3 FEMS

The functions expected of Facility Energy Management Systems (FEMS) extend beyond the basic tasks of visualizing energy data, implementing high-efficiency equipment in isolation, and supporting energy conservation efforts within individual departments. FEMS also aim to enhance overall energy utilization, decrease energy expenses, and promote collaborative and sustained energy conservation initiatives by analyzing real-world energy consumption data from diverse angles (Fig. 10.13). Moreover, by the precise anticipation of energy demand in the immediate future, taking into account factors such as weather conditions, manufacturing schedules, and other relevant variables, and by implementing a system that allows for dynamic optimization of energy supply to align with operational needs, it

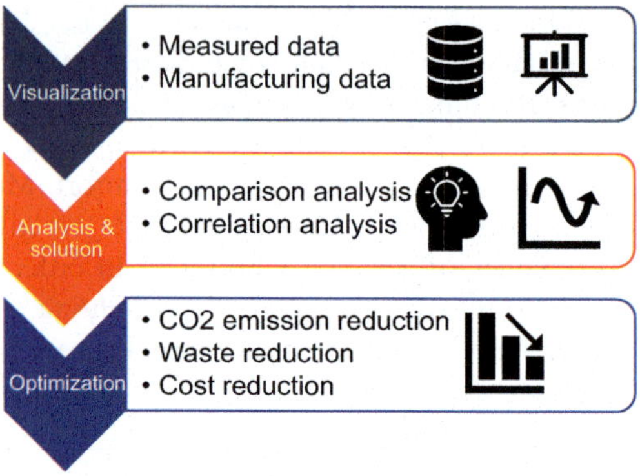

Fig. 10.13 FEMS steps

becomes feasible to achieve a state of continuous and exceptionally efficient energy management.

Step 1: Visualization The foundation of FEMS is rooted in the process of data visualization. The aforementioned statement serves as the fundamental basis for the quantification and aggregation of crucial data pertaining to energy management, with the aim of motivating operators to undertake suitable measures. The Field Energy Management System (FEMS) encompasses a wide range of aspects including energy, heat (both cold and hot), air conditioning, and other related components. Additionally, it incorporates factory production statistics and weather data, among other factors. This comprehensive approach aims to facilitate the development of a more efficient and optimized system. The primary objective in implementing this system is to initially refine it by identifying the minimal essential measurement sites, with the eventual goal of optimizing the energy consumption of the whole production. Specifically, the construction of a measurement system for electricity systems is rather straightforward; however, measuring thermal systems (such as cold water, hot water, steam, etc.) is a laborious and expensive process. Therefore, it is crucial to strategically choose the most efficient locations for measurements in order to optimize the design. In order to effectively communicate measurement data to operators, it is essential to provide the information in a clear and concise manner. Additionally, managing energy consumption per unit of output may be achieved by not only including time-series energy consumption statistics but also by analyzing breakdowns based on several categories such as plant, building, process, and so on.

Step 2: Analysis and Solution After the collection and accumulation of measurement data necessary for daily management and control, and the achievement of energy data visualization through the process of "visualization," the subsequent phase involves the analysis of the data from various angles. This analysis aims to identify instances of inefficient energy utilization and areas within the factory that can be improved. The methods employed for analysis encompass several approaches, such as comparing current data with historical data from the corresponding time period, doing seasonal comparisons, comparing data across different departments, facilities, and floors, evaluating data in relation to facility operating conditions, performing correlation analysis with influencing factors, and conducting correlation analysis of energy metrics.

Step 3: Optimization The function facilitates the optimization of energy supply and demand inside a factory by considering the unique features and efficiency of each facility. This optimization is achieved via a comprehensive examination of many factors. The objective function for optimization in this context may involve minimizing energy expenditures, minimizing CO_2 emissions, minimizing maintenance costs, and other related factors.

In the recent past, a discernible surge in the utilization of sustainable energy resources has occurred frequently, storage batteries, thermal storage, and cogeneration, driven by the goals of energy conservation and business continuity planning (BCP). Consequently, the system design of utilities in factories has become increasingly intricate. The accurate prediction of energy demand in the near future (1–2 days) is crucial, considering historical performance data. This prediction relies on identifying influential factors such as weather data and production plans that significantly impact energy consumption. Subsequently, leveraging these predictions, optimal planning support functions can be developed for energy supply facilities. During the planning phase of chiller facilities that utilize outdoor heat from renewable energy generation and free cooling, it is imperative to take into account various factors. These factors include not only the demand but also the efficiency and characteristics of each supply facility. Additionally, operational constraints such as time interval, frequency, priority operation, and power supply constraints must be considered in order to optimize the forecast.

10.3.4 HEMS

The Home Energy Management System (HEMS) is a technology that facilitates the interconnection of intelligent household appliances over an information network. Its primary objective is to enhance the efficiency of energy consumption within residential settings by employing controller-based operations. Smart home appliances refer to household appliances that can be connected to a network and controlled by a Home Energy Management System (HEMS) controller. This category encompasses white goods like air conditioners and washing machines, as well as energy generation and storage equipment such as solar power generation devices and household storage batteries. Figure 10.14 presents a figuration of HEMS. The primary objective of Home Energy Management Systems (HEMS) is typically to optimize energy usage by conserving electricity or, in situations when supply is restricted, to effectively reduce consumption (thus proactively mitigating future

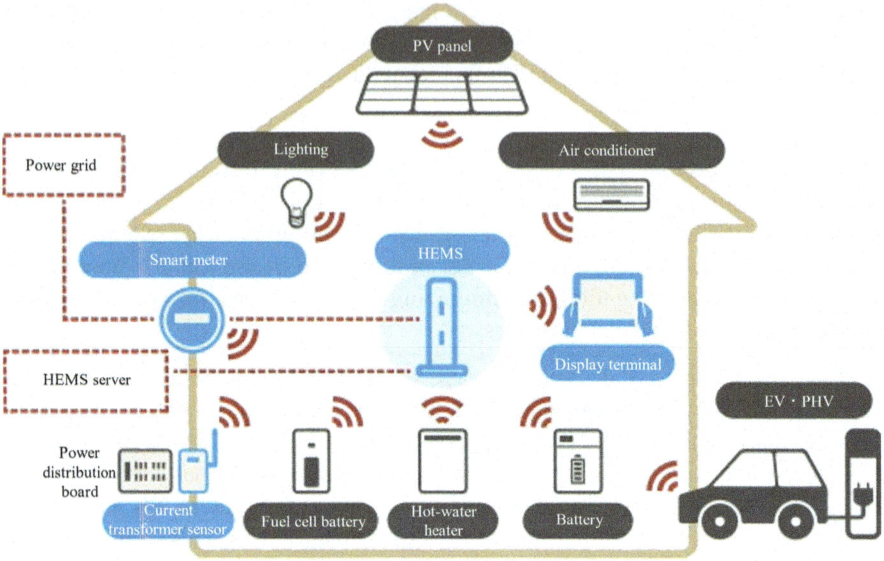

Fig. 10.14 HEMS. (Source: (*NIES HEMS*))

demand). Additionally, there is an anticipation among many stakeholders for the development of novel services in this domain.

The pursuit of carbon neutrality has prompted increased interest in residential energy management, primarily driven by the instability in power supply resulting from the decline in thermal power output. Furthermore, there has been a progressive transition in residential housing from the traditional focus on energy consumption to a more contemporary emphasis on energy generation, storage, and distribution. This movement is shown by the integration of solar power generation systems and storage batteries into residential properties. Furthermore, the widespread implementation of solar power generation is anticipated to foster the emergence of novel energy resources.

Furthermore, the proliferation of solar power generation has resulted in an excess supply of energy within the power system as a whole. Consequently, there is a growing recognition of the imperative to regulate the reverse power flow in small-scale residential PV systems. This scenario necessitates the implementation of many processes, including the modification of electricity consumption timing and the scheduling of charging and discharging periods. Consequently, a power management and control mechanism is required. In the event that a residential property is equipped with an electric vehicle (EV), the stored electrical energy can be utilized either for internal consumption within the house or for external distribution to the broader power grid.

The application of HEMS as a management and control device has historically been seen as a specialized tool. However, there is currently a shift in focus toward the utilization of easily accessible and convenient cellphones and tablet PCs for this purpose. These interfaces have the potential to function as client/user interfaces in conjunction with a central server, or alternatively, as management and control devices. The functions and processes of HEMS are undergoing substantial modifications in response to shifts in the surrounding environment. Consequently, there is a growing belief that the practical implementation of HEMS is progressively expanding.

The concept of household Energy Management Systems (HEMS) encompasses several meanings across different contexts. In the context of managing and regulating household appliances, HEMS may be seen as an extensive automated system that enables consumers to operate equipment within allowed parameters. HEMS, or Home Energy Management System, can be characterized as an enhanced manifestation of a home appliance manufacturer's capacity to monitor and regulate the condition of individual appliances through the integration of the controller and appliances. This integrated unit serves as a comprehensive resource for managing and controlling diverse home appliances and facilities. Currently, it is imperative to establish the capability of seamless integration with residential infrastructure, irrespective of the diverse array of manufacturers involved. Manufacturers throughout diverse sectors, encompassing home appliances, autos, and housing, have been actively engaged in the advancement, pilot initiatives, and commercial implementation of Home Energy Management Systems (HEMS). However, it is noteworthy that, in the majority of instances, the integration of HEMS is predominantly confined to their respective product portfolios. Nevertheless, within a typical residential dwelling, there exists a conglomeration of various equipment and amenities sourced from many entities. Consequently, the ability to effectively monitor and control devices and equipment from many manufacturers is crucial for the successful implementation of HEMS (Home Energy Management Systems) within a smart house.

Interconnectivity of equipment and facilities from many manufacturers is crucial for effective management and control. In order to achieve this objective, efforts are being made to standardize communication interfaces and adopt procedures aimed at ensuring the security of information.

Furthermore, the implementation of diverse management and control methods must be carried out in a manner that ensures both safety and effectiveness. The term "safety" refers to the condition in which devices and equipment, operating in an interconnected environment and manufactured by multiple entities, are utilized within a defined range that ensures their safe operation. This entails limited exchange of mutual information among interconnected devices and equipment, while also ensuring that their usage does not have any detrimental effects on the surrounding environment, including users and residential areas. Ensuring the safe usage of equipment and mitigating any potential adverse effects on the surrounding environment, including users and nearby residences, is of utmost importance.

Furthermore, the term "effectively" denotes the crucial need to actively enhance the surrounding environment of individuals and dwellings. Consequently, in the practical implementation of Home Energy Management Systems (HEMS), it is imperative to ensure the safe and efficient execution of tasks such as activating the heating system and activating the cooling system within the residential environment, inhabited by individuals of the general population. Ensuring this in the actual world, beyond the confines of typical ICT technologies, presents a distinct challenge. In light of the prevailing conditions, a pressing future endeavor is expanding the accessibility of management and control techniques for Home Energy Management Systems (HEMS) by extending their availability as application services to other entities outside the original equipment and facility makers.

10.4 Conclusion

The worldwide concern for addressing global warming has led to a widespread focus on energy and environmental concerns. The EMS market aligns with the prevailing worldwide trend, exhibiting a consistent growth in size on an annual basis. The importance of smart communities is increasing, particularly in the realm of constructing resilient towns and cities that can withstand natural catastrophes. These initiatives aim to rebuild places that have been severely impacted by such calamities.

Due to this rationale, we contend that this chapter has significance since it surpasses conventional comprehension by providing a comprehensive overview of the diverse technological aspects pertaining to smart communities, as opposed to just textual descriptions. This chapter provides a summary of the enabling technologies for FEMS, BEMS, and HEMS. These systems aim to achieve energy efficiency for a range of customers, including factories, buildings, and homes. Additionally, it should be noted that CEMS, which is responsible for coordinating the local energy supply, is also discussed in this chapter. Furthermore, this article presents an overview of the frequently employed technologies in Energy Management Systems (EMSs), including generation forecasting, demand forecasting, and power storage systems for renewable energy.

References

Allan, K. S., O'Neil, E., Currie, M. M., Lin, S., Sapp, J. L., & Dorian, P. (2021). Responding to cardiac arrest in the community in the digital age. *The Canadian Journal of Cardiology, 38*(4), 491–501.

Choi, J. S. (2019). A hierarchical distributed energy management agent framework for smart homes, grids, and cities. *IEEE Communications Magazine*.

Fuji Electric Co., Ltd. *Community Energy Management System (CEMS)*. JASE-W Japanese Smart Energy Products & Technologies, O-35. https://www.jase-w.eccj.or.jp/technologies/pdf/office/O-35.pdf

Ismagilova, E., Hughes, L., Dwivedi, Y. K., & Ravi Raman, K. (2019). Smart cities: Advances in research—An information systems perspective. *International Journal of Information Management, 47*, 88–100.

Jie, B., Baba, J., & Kumada, A. (2023). *Mathematical analysis of time-series meteorological factors in electricity demand forecasting based on carbon neutrality*. In Proceeding of the 2023 IEEE IAS industrial and commercial power system Asia (IEEE I&CPS ASIA 2023).

Laplace. *Building Energy Management Systems*. https://www.laplacesolutions.com/building-energy-management/

National Institute for Environmental Studies (NIES). *HEMS explanation*. https://tenbou.nies.go.jp/science/description/detail.php?id=17

Nguyen, T. P., Yeh, C.-T., Cho, M.-Y., & Huang, Y.-T. (2020). Complementary grid power prediction using artificial neural network in the energy management system of a disaster prevention smart solar microgrid. *International Journal of Smart Grid and Clean Energy*.

Rathor, S. K., & Saxena, D. (2020). Energy management system for smart grid: An overview and key issues. *International Journal of Energy Research, 44*, 4067–4109. https://doi.org/10.1002/er.4883

Saeed, M. A., Eladl, A. A., Alhasnawi, B. N., Motahhir, S., Nayyar, A., Shah, M. A., & Sedhom, B. E. (2023). Energy management system in smart buildings based coalition game theory with fog platform and smart meter infrastructure. *Scientific Reports, 13*(1), 2023.

Sami, M. S., Abrar, M., Akram, R., Hussain, M. M., Nazir, M. H., Khan, M. S., & Raza, S. (2021). Energy management of microgrids for smart cities: A review. *Energies, 14*(18), 5976.

TAIGA Publishing Co., Ltd. Editorial board of the technical *journal on Smart Grid: Energy management system for smart community*.

Yigitcanlar, T., Kamruzzaman, M., Foth, M., Sabatini-Marques, J., da Costa, E. M., & Ioppolo, G. (2019). Can cities become smart without being sustainable? A systematic review of the literature. *Sustainable Cities and Society, 45*, 348–365.

Bo Jie received his B.E., M.E. and PhD degrees in electrical engineering from Shaanxi University of Science and Technology, China, in 2009, Yokohama National University, Japan, in 2017, and 2022, respectively. His current research interests cover power system operation and control with market mechanism, including power system economics, evolutionary computation, and optimization. He is currently an Assistant Professor with the Department of Electrical Engineering and Information Systems Graduate School of Engineering, The University of Tokyo, Japan. Dr. Jie is a member of IEEJ, J-PVS, and IEEE.

Jumpei Baba received a Ph.D. in electrical engineering from the University of Tokyo in 2001. He joined Tokyo University of Science as an assistant in 2001 and joined the University of Tokyo as an assistant in 2003. He was appointed as a lecturer in 2005, an associate professor in 2006, and has been a professor from 2021. He is a member of IEEE, IEEJ, IEIEJ, and CSSJ. His main interest is power electronics, power systems, energy storage systems, and renewable energy sources.

Akiko Kumada received the B.Eng., M.Eng., and Dr. degrees in electrical engineering from the University of Tokyo in 1994, 1996, and 1999, respectively. After working as a research associate in the same university from 1999 to 2001, she joined the Tokyo Electric Power Company as a researcher from 2001 to 2003. Since April 2003, she has been with the Department of electrical engineering of the University of Tokyo and is now a professor of electrical engineering.

Chapter 11
Innovating Urbanity: The Synergy of Smart Cities and Circular Economy in India

Swati Chauhan, Ojaswini Bansal, and Anshul Puriya

Abstract This chapter explores the dynamic landscape of smart and circular economies within the urban context of India. As India grapples with the challenges of rapid urbanization and resource scarcity, novel approaches are essential for achieving sustainability. This study introduces a pioneering circular economy framework, synthesized from existing literature, to assess the concept's applicability and effectiveness in ten diverse Indian cities. The research provides a comprehensive analysis of the distinctive Circular Economy (CE) framework, offering valuable insights into its integration with smart city principles. The study adopted mixed method in writing this chapter and prepared a circular economy framework to evaluate the concept of CE in 10 cities of India. To select the cities, India Smart City Award (ISAC 2022) under Smart Cities Mission (SCM) and Swachh Survekshan ranking 2022 under Swachh Bharat Mission (SBM)-Urban has been considered.

The findings of this study demonstrate the innovation and adaptability of the proposed circular economy framework, highlighting the strategies and initiatives that each city has employed to foster sustainability, innovation, including waste reduction, resource optimization, and sustainable urban planning. The study also examines the challenges faced during the transition, such as regulatory hurdles, infrastructure limitations, and behavioral change. Furthermore, the research underscores the broader implications of smart and circular economies, encompassing economic growth, environmental conservation, and enhanced quality of life. It emphasizes the importance of multi-stakeholder collaboration and policy interventions to facilitate a smoother transition toward circular and smart urban ecosystems.

This study offers a valuable resource for policymakers, urban planners, and researchers seeking to comprehend and implement sustainable development practices in Indian cities. The unique circular economy framework, based on a thorough literature review, provides a novel perspective on the multifaceted challenges and

S. Chauhan (✉) · O. Bansal · A. Puriya
Atal Bihari Vajpayee Institute of Good Governance and Policy Analysis,
Bhopal, Madhya Pradesh, India
e-mail: swati.chauhan@mp.gov.in; anshul.puriya@mp.gov.in

S. Majumdar et al. (eds.), *Smart Cities*, S.M.A.R.T. Environments,
https://doi.org/10.1007/978-3-031-59846-3_11

opportunities faced by each city in their pursuit of smart and circular sustainability. It contributes significantly to the ongoing dialogue on sustainable urbanization in emerging economies.

Keywords Smart city · Circular economy · Sustainable waste management · Urbanism · Sustainability

11.1 Introduction

Globally the resource pressure is increasing exponentially every year due to rapid population growth, urbanization, industrialization, and the consequent increase in consumption patterns. This is further exacerbated by unsustainable practices and inefficient resource management. The United Nation's Sustainable Development Goals (SDG) from 2016 anticipate that by 2030 and 2050, respectively, there is expected to be a demand for natural resources equivalent to 2–3 planets. Realizing the challenges of resources in the future, the current model of linear economy is giving way to the closed-loop Circular Economy model. The concept of Circular Economy (CE) falls in line with three of the proposed 17 SDGs where SDG 7 encourages efforts to ensure access to affordable, reliable, sustainable, and modern energy for all, SDG 12 ensures sustainable consumption and production patterns, and SDG 11 promotes inclusive, safe, resilient, and sustainable cities and human settlements. CE is a tool to attain the targets set in these SDG's (Siddiqui and Pandit, 2021; Roleders et al., 2022; Dinca et al., 2022). The concept of CE is being adopted by businesses in developing economies like India where traditional economic model is unsustainable due to high resource demands and CE model is emerging that focuses on managing resources and minimizing waste (Goyal et al., 2018).

As simple as this terminology sounds, CE has been interpreted and implemented in different ways across the globe and requires a long-term systemic change making this a rather complicated concept. Countries like China link it to pollution and have mitigation of pollution as the driving force for their CE policies whereas European Union strongly emphasizes on business models and industries to take the center stage in achieving circularity. CE oriented research, distribution, and application are still in the initial stages in developing economies. The various stages of linear economy model and circular economy model are shown in Fig 11.1.

Most of the urban and rural places in India have the concept of linear model taken by the concept of "take-make-dispose" philosophy toward consumption of resources. This model is driven by single use or cradle to grave lifestyle (Ellen MacArthur Foundation, 2013; World Economic Forum, 2014). The presence of large amount of waste in cities and villages is a sign of an overabundance of consumption, irresponsible material use, and inadequate waste disposal systems. Recently the concept of CE has entered in the ecosystem of the country. Currently, the four main stages of managing the Municipal Solid Waste (MSW) which are collection, transportation, processing, and disposal can all be characterized as being in

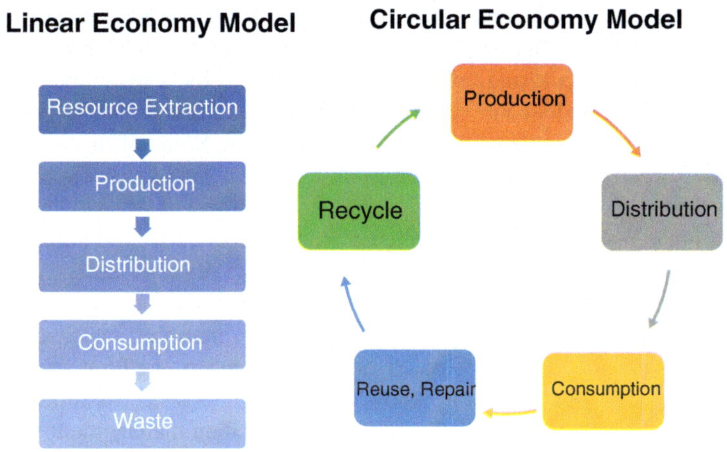

Fig. 11.1 Linear and circular economy model. (Source: Insights taken from Ahmed et al., 2022)

an unorganized state with non-scientific disposal techniques. In India, large cities collect 70–90% of the MSW produced, compared to less than 50% for smaller cities and towns (Mani & Singh, 2016). Upwards of 91% of the MSW that is collected is discarded and left in unauthorized landfills, which is a concerning Waste Management (WM) practice (Sharma, 2018). Despite this situation there are a few exceptionally innovative Solid Waste Management (SWM) practices demonstrated in many municipalities such as Pune, Indore, Surat, etc.

CE emphasizes on less dependence on primary raw material or energy and the first step of waste management is to reduce linking these two concepts at the base level. The foundation of CE is built by establishing that waste is a resource. CE from the perspective of waste can be seen as an innovative way for extending product lifespan through enhancements in design and service and creating its value, such that the product lifecycle loop is a closed one and the trash produced is reused and recycled, and more prudently and efficiently using resources. Every kind of waste generated by us from wastewater, organic, plastic, Construction & Demolition (C&D) waste, etc. can be seen as a raw material to fulfil resource requirement for some other cycle.

Leaders across the world are being forced to rethink the current urban systems and same applies for value creation systems like CE. Since the scale has shifted from product to economies, user from individual to a connected web of people and the world moves toward intelligent systems, we need to integrate circularity with smart cities for long-term sustainability, resilience, and economic viability. The end objective of smart city is to reduce the consumption of resources, cost that goes in the process and improve the performance and quality of city services by leveraging the usefulness of Information and communication technology (ICTs) (Li et al., 2021). If we use the power of data and technology in smart city to achieve the goals of CE at various stages in waste management, these two concepts can work in conjunction resulting in an effective and efficient waste management system. Smart cities employ a range of technologies to improve urban efficiency and sustainability. These include Integrated Command and Control Centres (ICCCs) for data collection and system monitoring, optimization methods for gap analysis, Artificial Intelligence (AI) for decision support, and Internet of Things (IoT) networks for enhanced asset management and inventory evaluation. Some of these technologies contribute directly or indirectly at various stages of circular economy in waste management.

Cities are central to accelerating the CE transition as economic hubs. Smart city strategies that prioritize regenerative systems enhance resource utilization and promote urban economic resilience. Sustainable development strategies centered on inclusive and accountable governance, along with overcoming inefficient resource management and addressing outdated infrastructure, are essential drivers of this shift. Governments must facilitate knowledge-sharing and access to technological advancements like sensors, digitalization, and cloud data management for successful CE implementation. Within this framework, smart cities employ Integrated Command and Control Centres (ICCCs) for data collection and system monitoring, optimization methods for gap analysis, Artificial Intelligence (AI) for improved decision-making, and Internet of Things (IoT) networks for enhanced asset management and inventory evaluation. Together, these technologies contribute to the transition towards a circular economy. technical assistance, alter policies effectively, enable innovative growth, find financial solutions to leverage investments, and have more capacity to foster innovation in order to meet these issues (Siddiqui & Pandit, 2021). To take a more holistic approach aimed at country wide development the Indian government set off the Smart Cities Mission (SCM) in 2015 and the Swachh Bharat Mission (SBM) in 2014. Swachh Survekshan under SBM includes 4354 Urban Local Bodies (ULBs), 94,856 wards having 94.6% door-to-door waste collection, and 88.29% waste segregation at source. 4339 ULBs are Open Defecation Free (ODF) certified, out of which 3533 are ODF+, 1191 are ODF++, and 14 ULBs are water+. Total 8472 solid waste treatment plants are established in these ULBs under which maximum (5621) are for solid waste treatment plant (Ministry of Housing and Urban Affairs, GoI, 2023a, b).

With the introduction of SCM, Digital India, and SBM which considered CE as a solution to inefficient management of waste and have the potential to integrate the CE principles, a space for systematic approach combining these two terms (CE and SWM) was created which was very local in nature. To fulfil the objectives of these mission, cities have started adopting smart practices to gain circularity.

In India a strong local tradition regarding repair and refurbish is pre-existing (like kabadiwalas, repairmen, kitchen gardens, etc.), nullifying the notion of these ideas being western. The SCM and SBM have just created a fostering environment for innovation in adopting the same culture we have been following in a structural manner and at a larger scale. As a result, many smart cities like Indore, Coimbatore, and Surat have successfully integrated smart practices such as ICCC, e-governance, smart bins, route optimization, etc. with the objectives of SBM to achieve circularity. This chapter studies the best practices of SBM and SCM in Madhya Pradesh as the state received “Best State Award” at ISAC award 2022 (Ministry of Housing and

Urban Affairs, GoI, 2023a, b). Seven cities of Madhya Pradesh are developing 779 multisectoral projects worth Rs 15,696 Cr for welfare of the citizens. Indore is continuously grabbing the first position in SBM and SCM; further this city is setting standard in carbon credit, waste management, and public space management in the country. Indore is the commercial capital of Madhya Pradesh and setting a benchmark of sustainable development in the world.

11.2 Madhya Pradesh's Journey in Cleanliness

Madhya Pradesh is one among the few states that made significant efforts to develop and execute the SBM concept. Apart from the ULB's it gave specific attention to the cleanliness of rural areas and took the necessary steps to bring in change with its particular demography and setting. Under the Swachh Bharat Mission-Gramin (SBM-G), the state of Madhya Pradesh has provided safe sanitation service levels and behavioral change communication to more than 6.5 million rural households during the past 5 years. In order to guarantee complete toilet coverage, the state has also identified the families that were left out and plans to provide them with safe sanitation technology. The state has created a policy to preserve Open Defecation Free (ODF) and general cleanliness in all 50,228 of its villages through consistent toilet use, better hygiene practices and well thought out management of solid and liquid. The state as a whole has recognized that in CE each element influences the functionality of the whole system and complete circularity cannot be achieved even if one step of this process is inefficient. Therefore, making itself the top performing state in the country.

Breaking through habits and adopting new methods are naturally challenging and when it is done at mass scale in a municipality it becomes even more complicated in nature. Within the many good performing urban areas in the state, there are cities such as Indore and Bhopal that cracked this by emphasizing on the human society first and then the stages of CE in waste management. Indore has achieved the award of cleanest city since 2016, and India's first ever seven-star Garbage Free City in 2022. It has successfully implemented Intelligent Solid Waste Management (ISWM) which compliments smart systems with the SBM creating circularity such that the waste for which no one wants to be responsible is now generating a revenue of crores of rupees. In addition, because the city managed to maintain the quality of wet waste due to 100% segregation, it generates huge quantity of clean fuel which is used in many sectors like city bus, hotels, and colleges like IIM, etc. Every city in the country is attempting to succeed in SBM in its own unique way, but Indore has set a standard that proves it is doable (Gurlhosur et al., 2020). Many of these cities have achieved success in some stages of CE and provided best practice cases. However, we need to analyze these successful cities in a stage-by-stage manner for gap analysis to understand in which stage they are performing well and which stage requires attention from poor performing cities to achieve complete circularity in SWM.

Further, structure of this paper is as follows: next section is discusses the literature from developing and developed countries point of view followed with objectives and research methodology. Afterwards next section discusses the results and analysis followed with challenges, conclusion, and way forward.

11.3 Literature Review

A smart city is a forward-looking urban development concept that harnesses technology and data-driven solutions to enhance the quality of life for its residents while promoting sustainability and efficiency. These cities leverage advanced technologies such as IoT, AI, and data analytics to optimize various aspects of urban living, including transportation, energy consumption, waste management, and public services. The aim is to create a well-connected, digitally enabled urban environment that enhances the well-being of its inhabitants, reduces environmental impacts, and fosters economic growth. Circular economy (CE) (McDowell et al., 2017), on the other hand, is an innovative economic model that seeks to minimize resource waste and environmental degradation by promoting the continual use, recycling, and repurposing of materials and products. This concept encourages the design of products with longevity in mind, encourages the recycling and refurbishment of goods, and reduces the production of waste and emissions. In the context of waste management, a circular economy can play a pivotal role in promoting sustainability, resource efficiency, and resilience by minimizing waste generation and optimizing resource utilization within the urban ecosystem. Together, the smart cities and circular economies can work hand in hand to create more sustainable, resilient, and livable urban environments for the future.

Smart cities and the circular economy share a symbiotic relationship that holds great potential for sustainable urban development (Dinca et al., 2022; Chauhan et al., 2021). The integration of smart technologies in urban planning and management can significantly enhance the principles of the circular economy. Some of the smart cities tools align perfectly with the circular economy's goal of maximizing resource use and minimizing waste generation. For example, sensor networks in smart cities can monitor energy, water, and waste flows in real-time, enabling better resource management and informed decisions to reduce resource wastage. Smart transportation systems in these cities promote shared mobility and efficient transportation, reducing the need for private car ownership and lowering emissions. This not only improves urban mobility but also aligns with circular economy principles by extending the lifespan of vehicles through shared usage. Additionally, smart waste management systems can enhance recycling and reusing practices by monitoring and optimizing waste collection routes and promoting recycling behaviors among residents. This reduces the volume of waste sent to landfills and incinerators, contributing to a more circular approach to managing resources. Furthermore, digital platforms in smart cities can facilitate the exchange and reuse of goods among residents, creating urban ecosystems where products are shared, traded, or repurposed, rather than discarded after single use. Literature shows many studies focused on smart cities and circular economies concept worldwide. Here authors refereed various studies in developed and developing countries. For better understanding the studies are separately discussed for the developed countries and developing countries in the next section.

11.3.1 Circular Economy in Waste Management in the Developed Countries

A comprehensive global literature review on the convergence of circular economy (CE) and smart cities reveals a rich combination of research exploring the multifaceted relationship between these two transformative concepts (Aceleanu et al., 2019). Studies from around the world emphasize the role of technology as the linchpin in realizing the potential of CE within smart urban contexts. Researchers globally highlight how these technologies enable cities to transition from linear to circular models by tracking resource flows, encouraging sustainable consumption, and fostering resource-efficient supply chains (Chen, 2022; Caputo et al., 2023).

Circular economy research and implementation vary significantly between developing and developed countries. In developed countries, circular economy initiatives are often more advanced and well-funded, driven by a combination of economic incentives, environmental consciousness, and established regulatory frameworks (Liu & Ramakrishna, 2021; Silvestri et al. 2020; McDowell et al., 2017; Van Fan et al., 2019; Khan & Ali, 2022; Ahmed et al., 2022; Smol et al., 2020; Roleders et al., 2022; Hemidat et al., 2022; Bellezoni et al., 2022; Aceleanu et al., 2019; Petit-Boix & Leipold, 2018). Research in these countries tends to focus on optimizing waste management systems, climate change, designing sustainable product lifecycles, and fostering collaboration among industries to reduce resource consumption and minimize waste generation. Developed nations also benefit from greater access to data, technology, and financial resources, enabling them to experiment with innovative circular models like product-as-a-service and closed-loop supply chains. McDowell et al. (2017) examined the CE policies in Europe and China and found that Chinese viewpoint regarding the circular economy encompasses a wide range of aspects, including pollution and various environmental issues in addition to waste and resource management. This perspective is shaped as a reaction to the environmental problems arising from rapid growth and industrialization. In contrast, Europe's understanding of the circular economy has a more limited environmental focus, concentrating primarily on waste, resources, and business prospects. Additionally, we can explore similarities and disparities in policy priorities between the two regions and examine the metrics employed to gauge their respective progress.

In another study Silvestri et al. (2020) examined the regional development of circular economy in European nation through two indices, viz. Circular Economy Static Index (CESI) and Circular Economy Dynamic Index (CEDI). Both the indices considered 11 indicators in three dimensions, viz. Socio-health, Economic, and Environment dimension. The ranking framework classifies all the regions into four groups, viz. Never Give Up, Satiated and Sleepy, the best is yet to come and we do not mind group. The two indices illustrate distinct approaches to assessing Circular Economy (CE) performance, emphasizing the idea that despite varying initial conditions (as indicated by the CESI) in terms of absolute CE performance among regions, some can still make quicker strides toward enhancement. This dual perspective on CE performance implies the need, at the European level, to establish mechanisms that accommodate a more comprehensive and multifaceted understanding of how CE develops in order to accurately categorize its performance. Similarly, Van Fan et al. (2019) examined the cross-disciplinary approaches toward smart, resilient, and sustainable circular economy. This is a review study and mainly focuses on energy, water, waste management, green policy, and pollution minimization strategies. Study suggested that there is a need for additional research that relies on real-life case studies and prototypes, which involve collaboration and agreement among stakeholders and are supported by economically viable feasibility studies. Smol et al. (2020) proposed a CE roadmap to support transformation toward CE in municipal waste management using ReSOLVE framework. Study suggested that awareness about waste management, prevention of food waste, and change in legislation is required for successfully implementation of CE. Similarly, another study carried out in 28 European states by Dinca et al. (2022) examines the role of circular economy in smart cities sustainable development especially on smart environment (reduction of air pollution).

Smart waste management is one of the key areas in achieving circular economy. Hemidat et al. (2022) carried out a review study in 10 Middle East and North Africa (MENA) regions to understand the various practices of SWM and found that waste management is critical issue in these countries and facing many challenges in manpower, funding, insufficient collection system, lack of know-how, and lack of trained and skilled manpower.

Furthermore, the literature delves into the significance of policy frameworks and governance structures, revealing diverse approaches and regulatory landscapes worldwide. Comparative studies of countries like Europe and China highlight the role of governments, local authorities, and international organizations in shaping the trajectory of CE adoption in smart cities (McDowell et al., 2017). Research from different regions elucidates the need for context-specific strategies that align with local socio-economic and environmental conditions. Table 11.1 shows the best CE practices around the world.

To sum up, the global literature on CE and smart cities underscores the universality of the challenges and opportunities presented by these concepts. It emphasizes the pivotal role of technology, economics, environmental considerations, and governance mechanisms in driving the evolution of cities into innovative, resource-efficient, and sustainable hubs worldwide. This rich tapestry of research paints a vivid picture of the transformative potential when CE principles converge with the smart city paradigm on a global scale. The next section discusses similar points in the context of developing countries.

Table 11.1 CE practices around the world

Country	Major CE practices
USA	Creating carpets from recycled plastic, repurposing old shipping containers for housing, manufacturing jeans using discarded plastic bottles, recycling and repurposing second-hand clothing, and Walmart's investment in promoting sustainable practices among its suppliers (Ahmed et al., 2022; Ghosh et al., 2022)
Canada	The production of plastic packaging that is reusable, recyclable, and compostable, as well as the recycling of cartons (Xavier et al., 2021; Cocker & Graham, 2020)
UK	Construction of Plastic Roads (Sasidharan et al., 2019)
Poland	Bio-degradable cutlery (Boro et al., 2020)
France	In agriculture production (Gonçalves et al., 2022)
Portugal	Using Coffee waste for growing mushrooms (Pereira, 2022)
Brazil	Utilizing recycled construction materials in the composition of mortar and concrete, along with paper and plastic recycling (da Conceição Leite et al., 2011; Beja et al., 2020)
Argentina	Solid waste Recycling (Morero et al., 2017, 2020)
Chile	Recycling and reuse of wastewater (Villamar et al., 2018; Livia et al., 2020)
Australia	Using fish skin for making leather bags (Fadeeva & Van Berkel, 2023; Ahmed et al., 2022)
Japan	Recycling of used clothes (Semba et al., 2020)
Singapore	Converting food waste into fertilizer, the collection and recycling of waste, and transporting waste to energy plants (Khoo et al., 2010; Grandhi & Appaiah Singh, 2016; Tong et al., 2018)
South Korea	Recycling paper cups (Kim et al., 2005)

Source: Author's compilation

11.3.2 Circular Economy in Waste Management in the Developing Countries

In the developing country context, the literature review on the integration of circular economy (CE) principles within smart cities reveals a growing body of research that recognizes the significance of these concepts in addressing the unique challenges faced by urban areas in the country (Mandpe et al., 2023; Ahmed et al., 2022; Srivastav et al., 2023; Goyal et al., 2018; Yduvanshi et al., 2016; Siddiqui & Pandit, 2021; Singh, 2021; Mondak, 2021; Jaya Surya & Kranti Kumar, 2021; Patwa et al., 2021). Additionally, these studies also focused on challenges of these countries in terms of basic infrastructure deficiencies, limited access to capital, and a higher dependence on resource-intensive industries (Anand et al., 2018; Hoelscher, 2016). Research in such contexts may emphasize the adaptation of circular principles to local needs, the development of low-cost recycling technologies, and strategies to integrate informal waste management sectors into formal economies. It is essential to address socio-economic disparities and ensure that circular economy initiatives in developing countries do not exacerbate inequalities but contribute to sustainable development, job creation, and poverty reduction. Ahmed et al. (2022) explored the circular economy concept in developing countries mainly focused on Bangladesh and found that the concept is at initial stage and it will require a concentrated effort for the transition from linear to circular economy. Further many studies have been conducted in Bangladesh (Dulia et al., 2021; Repp et al. (2021); Azizuddin et al., 2021) and found that CE model's applicability is very limited in Bangladesh, being exercised mostly through recycling processes in some industries like textiles and clothing (Saha et al., 2021). Another study by Patwa et al. (2021) examined the implementation of Circular Economy (CE) principles within emerging economies. These economies encounter distinct challenges compared to developed ones, including differences in resource availability, diverse government policies, and consumer behavior. Further, Khan and Ali (2022) proposed a framework for the adoption of smart waste management technologies in the context of CE in developing countries. Their research recommends a phased approach for implementation of smart waste technologies. Concentrated efforts are required to establish regulations that effectively tackle the prevailing waste mismanagement issues within the country. The concept of circular economy in low- and middle-income countries is at nascent stage and require a lot of ground level efforts to evolve. Ferranto et al. (2019) in their study talked about the opportunities and devised a model of circular economy in these countries by taking insights from developing and developed countries like Poland, Romania, and Europe.

India has made notable strides in embracing the circular economy concept, recognizing its potential to address both environmental challenges and economic sustainability. The country's circular economy journey has seen significant progress in recent years. Initiatives have been implemented across various sectors, including manufacturing, agriculture, and waste management. One area where India has made significant headway is in the recycling and waste management sector (Modak, 2021; Yaduvanshi et al., 2016). Table 11.2 is showing the best practices of CE in Indian context.

Table 11.2 Best CE practices in India

Project	Major CE practice
Banyan Nation	Produce plastic granules, better plastic from post consumption and Industrial goods. Created a tracking platform in Hyderabad to monitor waste (Modak, 2021)
UNDP India, Hindustan Coca-Cola Beverages (HCCB) and Hindustan Unilever Ltd. In partnership	Created a CE model of plastic waste and running Integrated Recycling Program in 50 cities. They are managing 85,000 tons of plastic waste every year and improving the lives of 37,500 Safai Mitras (Modak, 2021)
UNDP-HCCB-HUL with Indian Tobacco Company (ITC)	Launched a program "Well-being out of waste (WOW)" which is operational in 10 states and 562 municipal wards. The program generated livelihood to 14,500 rag pickers and impacted life of 77 lakhs citizens by collecting a dry waste of 50,196 tons
Bizango Company	Producing bio-degradable, spill proof meal tray to replace disposable plastic meal tray in 550 restaurants across India
GO-Bardhan Bio CNG Plant, Indore	Innovative sustainable business model of CE (MOHUA, Smart Cities Mission, 2023)

Source: Author's compilation

As evident from Table 11.2, the country has a robust informal recycling industry that plays a vital role in collecting and processing materials like plastics, paper, and metals. Efforts are underway to formalize and modernize this sector, improving the working conditions of waste pickers and enhancing recycling efficiency. Additionally, India has introduced extended producer responsibility (EPR) regulations for certain products, shifting the onus of waste management back to manufacturers, thereby promoting circularity in product design and lifecycle management (Siddiqui & Pandit, 2021). Further, Ministry of Housing and Urban Affairs (MOHUA), 2021 released a comprehensive forward-looking action plan report on solid and liquid waste management and a roadmap to transition from linear economy to circular economy; further report talked about the concept of SBM including 3 major Rs (Reduce, Reuse and Recycle) and discussed the concept of waste management. However, there is a need of conducive ecosystem including regulatory framework to achieve the goals of SDG 2030.

Mandpe et al. (2023) examined the present waste management methods in India and evaluates their hurdles in implementing CE as a viable solution for effective waste management. Study found that developing economy like India lack good SWM practices and recycling of waste. This study offers insights into the current and emerging waste management strategies aimed at creating a CE by efficiently processing waste and driving a Sustainable Development (SD) initiative. Additionally, there is a need to assess the willingness of Urban Local Bodies (ULBs) to overhaul their MSW management systems within the framework of the CE concept. While this approach appears promising and relevant for addressing MSW management challenges in developing economies, it hinges on the involvement of stakeholders, the establishment of regulations, and ensuring financial viability.

Economic considerations are also a central focus, with studies underscoring the potential for CE practices to unlock economic value in the Indian context. CE-driven circular business models, such as refurbishing and recycling, cannot only reduce costs but also create employment opportunities in a country with a large and diverse workforce. Researchers frequently highlight the need for tailored strategies that align with India's socio-economic diversity, recognizing that circular practices must be inclusive and accessible to diverse urban populations. Environmental sustainability is a key concern in the Indian literature, given the country's pressing environmental challenges. CE is viewed as a promising approach to mitigate pollution, resource depletion, and environmental degradation. Studies frequently emphasize the potential for CE practices to reduce greenhouse gas emissions, alleviate pressure on natural resources, and improve air and water quality in densely populated urban areas. Moreover, governance and policy frameworks emerge as critical in the Indian context. Researchers highlight the importance of regulatory measures and incentives to encourage CE adoption within smart cities. The role of local governments and public-private partnerships is emphasized, given the decentralized nature of urban governance in India.

In summary, the literature on circular economy and smart cities emphasizes the relevance and potential of these concepts in addressing the country's unique urban challenges. It emphasizes the role of technology, economics, environmental considerations, and governance mechanisms in shaping the future of Indian cities as sustainable, resource-efficient, and technologically advanced hubs that cater to the needs of diverse urban populations. This study is trying to evaluate the status of smart and circular economy in Indian cities. Based on the literature and authors understanding the study derived following objectives in this study 1. Evaluate the existing best practices, projects, and successful case studies related to smart cities and Circular economy (CE) principles in different cities of India; 2. Examine how cities are optimizing resource utilization, reducing waste generation, and promoting sustainable consumption patterns; 3. Explore how different cities are leveraging technological advancements and innovations to enhance solid waste management while minimizing environmental impact; 4. Identify challenges and barriers hindering the adoption of smart and Circular economy (CE) practices in other cities.

11.3.3 City Selection Matrix

The study conducted an extensive literature survey to gather existing knowledge, theories, and frameworks related to smart and Circular economy (CE) concepts. Literature study has been done on basic concept of Circular economy (CE) around the world, Waste Management, and Smart cities followed by detail study of these concepts in developed and developing countries. To finalize the cities for analysis, the IṢAC awards 2022 under SCM and Swachh Survekshan ranking 2022 under SBM-Urban has been considered. Under ISAC all the three cities (Indore, Surat, and Agra) getting awards have been included. Then projects getting award under "ICCC- Sustainable Business Model" and "Sanitation" were included. Under "ICCC- Sustainable Business Model" Ahmedabad, Surat, and Agra were included and Gwalior was excluded because the project was not related to SWM. All the cities under Sanitation category were included. This adds Kakinada and Chandigarh to the list making a total of six cities from ISAC. Now four unique cities not included already in the six ISAC cities were chosen from Swachh Survekshan 2022 national ranking under "more than 10 lac population" category. This makes the total city count to ten covering a diverse group in population, location, size, and awards in SWM (refer Fig. 11.2).

Since Indore is a consistent top performer in both the rankings, it has been described in detail as a role model case study in adopting the CE framework and bringing in sustainable changes through smart city tools. Along with Indore, other nine cities (refer Fig. 11.2) from these two ranking have been considered to evaluate the CE framework and how the cities are managing it.

From the understanding gained through existing studies, cases (theory, interview, experience), and literature the study has proposed a framework of CE merged with smart city concept (refer Fig. 11.3). This framework of "Smart Circularity" can be

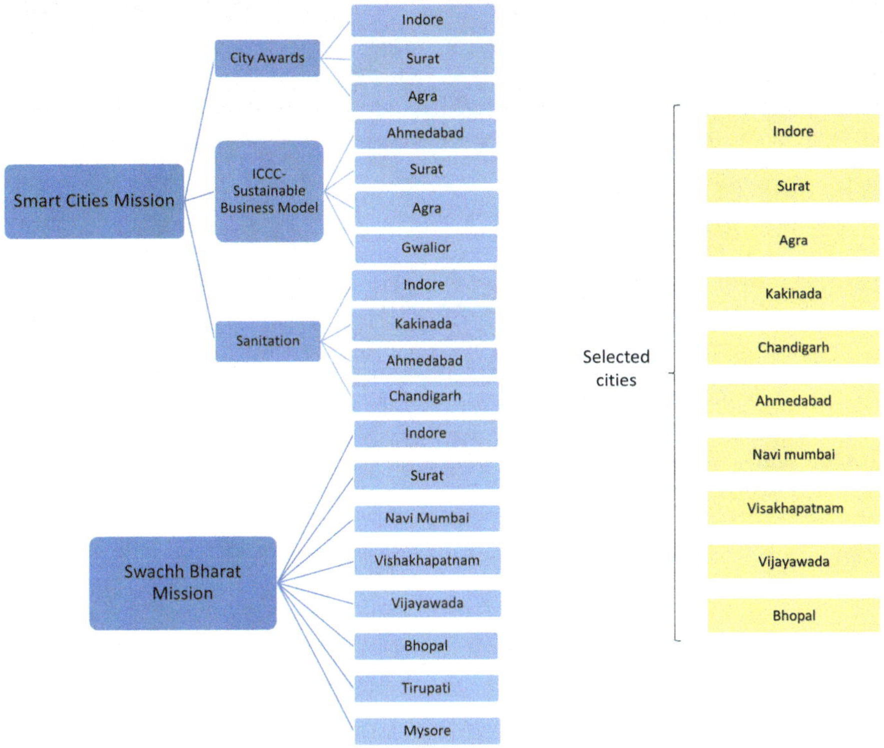

Fig. 11.2 City selection matrix

used in a matrix format to compare and evaluate performance of cities in achieving circularity through intelligent systems and do stage wise gap assessment. It has been realized that these are the major steps where even if one stage is lacking in performance, the city fails to achieve complete circularity.

The steps in circular economy in the framework developed as shown in Fig. 11.3 are described below:

Reduction of Waste Generated
The practice of eliminating waste before it is created or essentially using less material to get the job done. This is the most environmentally preferred strategy. This stage works on the 3R concept apart from design of the products and manufacture processes based on sustainability. It can be analyzed by the recycling policies adopted by the city and reduction methods used at product sources. An additional R (Repair) has been including at this stage.

Waste Segregation at Source
This stage refers to identifying and separating the waste into wet, dry, hazardous, e-waste, etc. as soon as it is generated. The importance of this stage cannot be emphasized enough as it determines the quality of waste which will be put into the

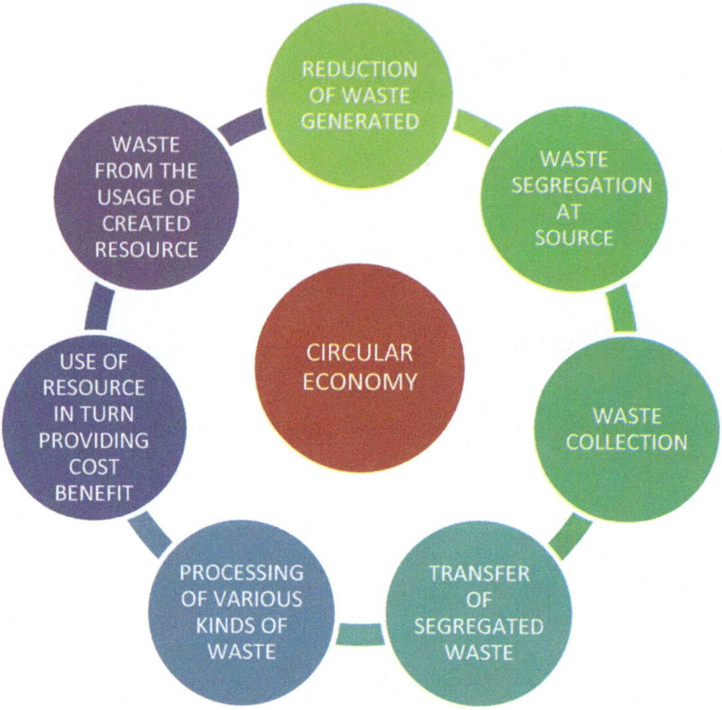

Fig. 11.3 Framework of smart circular economy in Indian cities

circular process. The better the quality of waste, the more feasible it will be to process and to convert it into energy or resource like wet waste into biogas, plastic for cement, road industry, etc. This stage can be analyzed by the percentage of segregated waste at source and also the number of types it is being segregated into.

Waste Collection

Even after the waste is segregated, if the collection is not done efficiently, it has high chance of getting mixed again. This will in turn affect the circularity as the process will become less feasible and give lesser cost benefit. One other reason for the importance of the step is that it necessitates lifecycle treatment of wet waste to convert it into energy. If not implemented properly there are also chances of this stage leading to creation of dumping spots in the city. This stage can be assessed by monitoring of waste transferred in transport and measures to reduce GVPs.

Transfer of Collected Waste

A decentralized system for transfer of the collected waste is very crucial as this stage determines the weight and amount of waste collected which can help in understanding the requirement of processing system. In this stage due to inefficiency, GVP (garbage vulnerable points) can be formed in the city by dumping and burning which are a hindrance to the CE framework. It can be seen by percentage of GVP's reduced or eliminated in the city and percentage of waste successfully transferred to the stations after collection.

Processing of Various Types of Waste
This is the main step of converting waste to resource where the good quality waste of various types is processed and the operation changes the chemical, physical, or biological properties of the material as required. It is either send to processing units to recover a resource or sent directly into industries where it can be used as a raw material. The different types include converting wet waste into biogas in bio methanation plants, plastic and paper to cement, road industries, etc. This can be seen by the percentage of waste that goes into some or the other process as compared to landfills.

Use of Resource in Turn Providing Cost Benefit
The monetary benefit in terms of selling the material as a raw material or by converting it to energy which can either be sold or used by the city services in turn again reducing the cost of that resource determines how much circularity has the city achieved by its SWM system. It can be determined by the capital cost that has been put into the services and facilities for the above steps as compared to the earning from the resources produced.

Management of Waste from the Usage of Created Resource
This is where the cycle of nature is formed as the converted resources will generate waste as a result of the process, they will be used in. Putting this waste back into the cycle will ensure maximum efficiency of the established system and lead to no gaps.

The analysis based on this CE framework will lead to a better understanding of status of Indian cities in these aspects and give conclusion on the challenges and requirements that need to be fulfilled to achieve circularity by all cities. Further to compete at a global scale, through learnings form the international case example, a way forward has been proposed suggesting the methods and smart tools that will be necessary to be adopted by Indian cities and states in order to create a position in the world ranking in the future.

11.4 Case Study: Indore

Like any other city in India, before 2016 Indore also struggled with waste management, leading to unhygienic conditions, increased pollution, and negative impacts on public health. The city generates nearly 1900 MT of solid waste every day and was nearly choking by haze brought on by the burning of enormous quantities of plastic garbage. Since then, Indore has come a long way in successfully converting trash to treasure by using smart solutions for its problems related to SWM. After securing 25th position in Swachh Survekshan 2016, Indore was declared the cleanest city in 2017. It has managed to do the impossible and has maintained the position ever since. Indore has not only distinguished itself by excelling in all Swachhata metrics, but it has also set new benchmarks of cleanliness for other cities to imitate. With the use of smart technologies, such as ICCC, GIS, route optimization, and many other tools, the Indore Municipal Corporation (IMC) has successfully integrated the various steps of waste management like collection, transportation, processing, and disposal of solid waste generated in the city.

The process of Indore changing its identity by becoming the cleanest city in India not just once but 7 times has been a breakthrough for any city in India in the sector of SWM. But this was just the first feather in the cap as the city of Indore continued to become the first in the country to start generating money out of waste and take step toward Circular Economy and then again cracked the carbon credit system.

Indore could manage to do this by taking each and every step of CE and perfecting its function and efficiency by using technology and smart tools. The combination of these smart tools with rigorous citywide behavior change campaign is the two pillars of success for Indore in SWM. As the city officials understood the complication in inducing change of habit they started with increasing awareness and knowledge amongst citizens through Information, Education and Communication (IEC) leveraging on festival gatherings, wall paintings, songs in collection vehicles, Nukkad Natak, etc. Along with this they trained the staff to guide the citizens on the importance of segregation when they were not willing to do it at home, if needed multiple times. This was supported with the backbone of citizen application where service delivery could be tracked and monitored centrally. With the nudges turning into stricter warnings and eventually penalties, and the collection reaching every household and commercial user, the concept of segregation started becoming a normal concept. The city became the first with population greater than 10 Lakh to achieve segregation at source in 100% of its households majorly segregating in three categories, i.e., wet, dry, and domestic hazardous. It used ISWM module for tracking collection vehicles for (Door to Door) D2D service. Some basic information about Indore is shown in Fig. 11.4.

As citizen participation increased, a dedicated team of Safai Mitras was established to manage waste collection, street cleaning, remove Garbage Vulnerable Point (GVPs), and ensure a waste-free environment throughout the city. The city implemented several technological solutions to optimize these operations. These include RFID tags and GPS tracking for real-time waste monitoring, biometric systems for personnel management, a GPS-integrated Vehicle Tracking Monitoring System (VTMS) to address waste collection vehicle complaints, and GIS-based tools for route optimization and asset management. These initiatives collectively strengthened the city's waste management system, ensuring efficient collection and maximizing resource recovery.

The behavior change nudge was kept in place by making the citizens used to a clean city and imposing spot fines and citywide promotion. With active monitoring on citizens, monitoring of tippers was also imposed. The city was supported in this by smart tools such as geo-fencing of waste bins and monitoring of weigh bridge at multiple decentralized garbage transfer stations. This resulted in excessive reduction in the Garbage vulnerable points and dumping spaces. From these transfer

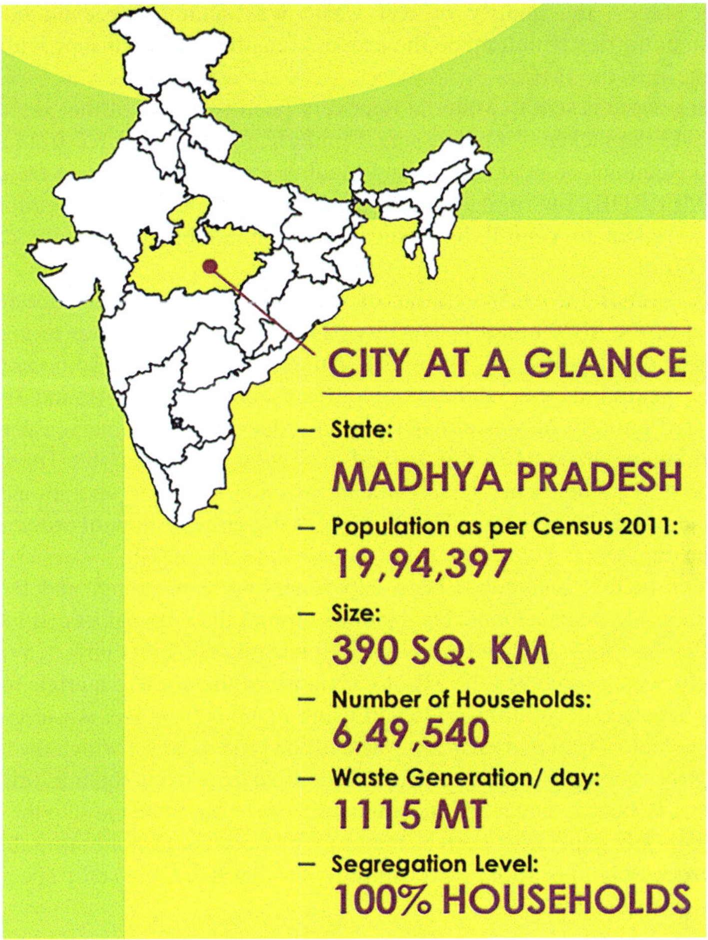

Fig. 11.4 Indore's demography, area, and location map

stations the processing of segregated waste was planned made more efficient by tailor made reports for SWM.

The wet waste is transferred to Garbage Transfer Station (GTS) and then directly taken to the bio methanation plant to be processed. The Bio CNG plant, a unique and award winning project, has a capacity to generate 17,000–18,000 kg of Bio CNG and 10 tons of organic manure which contains 95% pure methane gas with its slurry being turned into compost is running profitably. This fuel is used to run 150 city buses saving a large cost to the city. The Indore Municipal Corporation (IMC) earned Rs 14.45 crore from waste disposal last fiscal, including Rs 8.5 crore from carbon credits in the international market and Rs 2.52 crore as annual premium from a private company for providing waste to the bio CNG plant. This sets a pilot example that by increasing the efficiency of the previous stages, the quality of wet waste was seen to be even better than European countries which made the process feasible and in turn providing monetary benefit to the IMC.

The dry waste is sent to Material Recovery Facility where further segregation is done by 343 rag pickers for processing. Similarly, plastic recovered from dry waste is sold to Neemuch cement factory and Madhya Pradesh Rural Road Development Board (MPRRDB) for rural road construction. The Domestic Hazardous Waste (DHW) is taken to central biomedical waste treatment facility in sealed and weighed drums.

Indore exemplifies the integration of smart city principles for effective implementation of the SBM mission. The city has streamlined processes to map carbon footprints and sell earned carbon credits on the trading market. Additionally, Indore promotes "neki ki deewar," the use of cloth bags, cloth bag ATMs, and 3R gardens (constructed entirely of waste materials) to reduce, recycle, and reuse materials. These initiatives encourage a shift towards a circular economy (CE). The success of Indore's SWM model generates significant revenue, ensuring operational sustainability and preventing closure. This reinforces the circular flow of waste materials within the city.

Apart from this Indore has been experimenting, innovating, and leading the country in waste management. The landfills around the city had turned into mountains which had foul odor and were a health hazard. The city converted this into a picnic spot which promotes the 3R of CE by using the waste materials in creating planters, benches, decoration, and even using manure from wet waste established system to create a fruit garden. As e-waste is one type of waste which the city is not able to process completely, innovative solutions such as using them as raw material in making bill boards, street art, etc. are in process to put this kind into the CE cycle as well. For bulk generators such as major food streets and Choithram sabzi mandi, on-site processing units monitored centrally through ICCC are setup specific to the type of waste generated.

11.5 Analysis and Discussion

After a comprehensive description of the Indore model of waste management table 3 compares nine other cities on various parameters of CE in waste management as developed in the smart circularity framework. The study prepared a CE framework for evaluating the smart and circular economy in various cities of India. This framework has been derived by evaluating the developed and developing countries CE model. This framework has seven stages of waste reduction, waste collection, segregation, processing, revenue generation, and circularity, under which ten cities have been evaluated. Table 11.3 presents the city wise comparison under seven stages. All the seven stages are further classified into measurable indicators for better data capturing and comparison. The first stage which is talking about reduction of waste generation in 4 R, viz. Reduce, Reuse, Recycle, and Repair. The analysis

Table. 11.3 Comparison of Indian cities on circular economy Framework

S. No.	Stage of CE cycle	Indicator	Indore	Surat	Agra	Kakinada	Chandigarh	Ahemdabad	Navi Mumbai	GVMC Vishakhapatnam	Vijaywada	Bhopal
Census Population (2011)			0.19 Cr.	0.45 Cr.	0.16 Cr.	0.03 Cr.	0.1 Cr.	0.56 Cr.	0.11 Cr.	0.17 Cr.	0.1 Cr.	0.18 Cr.
Stage 1	Reduction of waste generated	Reduce	Yes	Yes	Yes	Yes	Yes	Yes	Yes	Yes	Yes	Yes
		Reuse	Yes	Yes	Yes	Yes	Yes	Yes	Yes	Yes	Yes	Yes
		Recycle	Yes	Yes	Yes	Yes	Yes	Yes	Yes	Yes	Yes	Yes
		Repair	No	No	No	No	No	Yes	No	Yes	Yes	Yes
Stage 2	Waste segregation at source	% of waste segregated at source	100%	100%	88.05%	100%	100%	100%	100%	100%	100%	100%
		Number of segregations	6	5	4	3	2	4	3	3	16	3
Stage 3	Waste collection	% of waste collected Door to door	100%	100%	100%	100%	100%	100%	100%	100%	100%	100%
Stage 4	Transfer of segregated waste	Monitoring of waste in transport	Yes	Yes	Yes	No	Yes	Yes	No	Yes	Yes	Yes
		Measures to reduce GVPs	Yes	Yes	Yes	No	Yes	Yes	No	Yes	Yes	Yes

(continued)

Table 11.3 (continued)

S. No.	Stage of CE cycle	Indicator	Indore	Surat	Agra	Kakinada	Chandigarh	Ahemdabad	Navi Mumbai	GVMC Vishakhapatnam	Vijaywada	Bhopal
Stage 5	Processing of various kinds of waste	% of waste processed	100%	100%	100%	81.12%	99.99%	100%	100%	100%	100%	100%
		Wet waste process in place	Yes	Yes	Yes	Yes	Yes	Yes	Yes	Yes	Yes	Yes
		Dry waste process in place	Yes	Yes	Yes	Yes	Yes	Yes	Yes	Yes	Yes	Yes
		hazardous waste process in place	Yes	No	No	Yes	No	Yes	Yes	Yes	Yes	Yes
		Construction & Demolition Waste	Yes	No	No	No	Yes	Yes	No	Yes	No	Yes
Stage 6	Use of resource in turn providing cost benefit	Cost benefit vs the capital & service maintenance expenditure	Yes	Yes	No	Yes	No	No	No	Yes	Yes	No
Stage 7	Waste from the usage of created resource	System in place for waste to go back into the cycle (yes/no)	Yes	Yes	No	No	No	Yes	Yes	Yes	Yes	Yes

Source: Authors Compilation by reviewing the following documents: Compendium of Best practices of Smart Cities Mission (2023), ISAC (2022), Swachh Survekshan (2022), Tummala et al. (2019)

portrays that all cities are implementing measures to reduce, reuse, and recycle waste. However, only a few cities (Ahmedabad, GVMC Vishakhapatnam, Vijayawada, and Bhopal) have measures in place to repair waste. Second stage focusses on segregation of waste under which two indicators, viz. percentage of waste segregated and number of segregations, are considered. All cities have achieved significant waste segregation at source, with most cities achieving 100%. Vijayawada (16) has the highest number of segregations at source and Chandigarh follows the minimum 2 segregation. Various studies have shown that waste segregation is a crucial step in achieving CE (Flores & Tan, 2019; Pereira et al., 2019). Waste segregation allows for the efficient separation of different types of materials, such as plastics, paper, glass, organic waste, and electronic waste. By categorizing waste streams, materials can be more effectively recycled or repurposed, reducing the amount of waste that ends up in landfills or incinerators. This, in turn, conserves valuable resources, reduces the strain on raw material extraction, and mitigates the environmental harm caused by waste disposal. Third stage focusses on waste collection where all cities are achieving 100% door-to-door collection. In the fourth stage the segregated waste is transported to the processing facilities and the waste is measured at source and destination and various tools are employed to reduce the Garbage Vulnerable Points (GVPs). Two cities Kakinada and Navi Mumbai mentioned that they do not have a monitoring system for transportation of waste and they have not employed any technological solution or tool to reduce the GVPs.

Stage five focusses on processing of waste which further classified into percentage of waste processed and kind of waste (wet, dry, hazardous, and C&D waste). The majority of cities effectively manage 100% of their waste, with the notable exception of Kakinada, which processes approximately 81.12% of its waste. All cities have established waste management procedures for both wet and dry waste. Nonetheless, it is worth noting that not every city has developed protocols for handling hazardous waste and Construction & Demolition (C&D) Waste. Processing of waste is a critical and foundational element of the circular economy concept. It plays a pivotal role in achieving the overarching goals of sustainability. In a CE, the emphasis is on extending the lifespan and utility of products and materials, and waste processing is the linchpin that allows for this. Moreover, waste processing promotes the closure of resource loops. By recovering and recycling materials, waste processing ensures that resources stay within a closed system, being reused or repurposed instead of ending up in landfills or incineration.

Next two stage are very crucial for the CE where waste is used as a revenue generation model and further utilization of created waste into the close loop of CE. Indore is the best example of revenue generation where Bio CNG plant has been established and generated fuel is utilizing to run the city buses. Amongst the sample cities only Surat, GVMC Vishakhapatnam, and Vijaywada are able to generate the revenue; however, other cities are consistently focusing to achieve this stage. Further, last stage focusses on utilization of waste further in the cycle; Surat, Ahmedabad, Navi Mumbai, GVMC Vishakhapatnam, Vijaywada, and Bhopal have systems in place for waste to go back into the cycle however Agra, Kakinada, and Chandigarh have no system in place to further utilize the waste in cycle. The above analysis shows that most cities are taking measures to achieve the goal of Circular economy (CE); however, the role of smart toolsSmart tools is crucial in achieving this. Many studies in the literature discussed the role of innovation and technologies in achieving this.

To further understand the role of smart tools in achieving CE in SWM in its implementation in various Indian cities has been discussed. Table 11.4 compiles the smart tools used by various Indian cities. Smart tools are proving to be indispensable in aiding cities on their journey toward achieving the circular economy. These innovative technologies are instrumental in reshaping urban landscapes by promoting sustainability, resource efficiency, and waste reduction. Smart tools facilitate data-driven decision-making, allowing cities to monitor resource consumption, waste generation, and environmental impact in real-time. With the wealth of data at their disposal, urban planners and policymakers can make informed choices that optimize resource management, reduce waste, and promote the 4R principles—reduce, reuse, recycle, and repair. Across these Indian cities, a diverse array of smart tools and technologies is being harnessed to revolutionize solid waste management and bolster their efforts toward achieving a more circular and sustainable urban environment.

In Indore, Surat, and Ahmedabad, the Integrated Command and Control Center (ICCC) is being utilized to issue challans for littering and spitting, creating a cleaner urban space. Moreover, these have introduced a robust monitoring system that oversees 600 door-to-door waste collection vehicles, optimizing routes, and reducing missed collection points. The route optimization has not only enhanced efficiency but also resulted in substantial cost savings, underlining the financial benefits of these smart tools. Agra has implemented a vehicle management system for real-time tracking of Solid Waste Management (SWM) vehicles, enhancing operational control and efficiency. Kakinada has taken an integrated approach to solid waste management, combining ICT components and features such as Intelligent bins, RFID, and GIS solutions. This comprehensive system optimizes the entire value chain of solid waste management and improves operational effectiveness by transmitting real-time data to the command and control center.

Chandigarh has embraced Information and Communication Technology (ICT) and analytical tools, introducing automated decision-making for SWM operations. The integration of ICT with door-to-door collection, material recovery facilities, and optimal route planning has led to improved waste collection and resource management. Further, Navi Mumbai's use of Remote Sensing and GIS technologies aids in selecting efficient landfill sites and transportation paths, a crucial step in efficient waste management. GVMC Vishakhapatnam's adoption of battery-operated e-rickshaws and QR code-embedded RFID tags on households significantly enhances waste collection, tracking, and management. Further, Vijayawada's online waste management system, combined with QR code scanners, allows real-time monitoring through a dashboard. Additionally, the installation of smart bins sends alerts when they are full, ensuring timely collection. Bhopal incorporated a GPS-based Automated Vehicle Location System and smart sensors to track and manage

Table 11.4 Usage of smart tools in Indian cities

City	Smart tools used
Indore	ISWM module for tracking collection vehicles for D2D service till processing unit SCADA—Integration with field sensors to increase efficiency in processing stage Safai Mitra and 311 Application for citizens to crowdsource data and compile management information on collection Asset management utilizing GIS (Geographic Information System) for geo-fencing waste containers, vehicles, and personnel Optimization of routes and implementation of a Vehicle Tracking Management System (VTMS) Overseeing and tracking weighbridge and processing facilities via the Integrated Command and Control Centre (ICCC) A system for generating alerts based on predefined rules SMART city Indore streamlined the procedure to map the Carbon footprints and sell the carbon credits earned in the carbon trading market
Surat	ICCC is being used for challans for littering, spitting on roads, etc Monitoring of 600 Door-to-Door vehicles 2800 trips and over 10,000 points of interest to collect garbage from 16 lakh households Use of Route Optimization for Door-to-Door Waste Collection Missed locations for waste collection reduced from 1000+ to less than 10/ day Routes optimization has been done leading to saving of 3900 km/ day which is ₹ 4.9 Cr./ year
Agra	Vehicle management system to track real-time location of SWM vehicles
Kakinada	Integrated solid waste management based on the collection and transportation of municipal solid waste using ICT components and integrating them into one system Intelligent bins, RFID, and GIS solutions to optimize processes along the entire value chain of solid waste management Finally, data on all operations are transmitted to the Command and Control center to take necessary action and keep the SWM functioning effectively and efficiently
Chandigarh	ICT and analytical tools for enabling automated decision-making have been introduced to help effectively monitor solid waste management operations Optimal routes of waste collection vehicles are created to reduce the operation cost of door-to-door waste collection
Ahmedabad	Spotting spitting through ICCC Geo-tagging of all garbage collection vehicles Establishment of a robust mechanism wherein the garbage collection vehicles have to collect garbage from a designated location at a pre-specified time Since the complaint redressal portal and the garbage collection system work in tandem, any grievances filed by the citizens are resolved quickly and therefore yield satisfaction to the citizens

(continued)

Table 11.4 (continued)

City	Smart tools used
Navi Mumbai	Using Remote Sensing and GIS technologies to select efficient landfill sites and transportation paths
GVMC Vishakhapatnam	The Greater Visakhapatnam Municipal Corporation (GVMC) has launched 60 electric rickshaws powered by batteries for managing solid waste in specific city regions. These e-rickshaws are employed for waste collection and transportation, and they are equipped with facilities for battery replacement and maintenance GVMC has begun the process of affixing QR code-integrated Radio Frequency Identification (RFID) tags to the walls of each residence. This initiative aids in the monitoring and management of waste collection
Vijayawada	The Vadodara Municipal Corporation (VMC) has set up 45 intelligent bins at different sites, which notify the authorities immediately when they are filled to capacity An online waste management system has been implemented that enables sanitation workers to utilize QR code scanners during the collection of garbage bins from residences. This process can be overseen in real-time via the commissioner's dashboard
Bhopal	Automated Vehicle Tracking System: This system, based on GPS technology, guarantees the collection, separation, and processing of waste at each stage Intelligent Waste Collection Monitoring System and Smart Sensors: These sensors, installed in public bins and operating in real-time, provide authorities with the precise location of the bin as soon as it becomes full

Source: Author Compilation using Compendium of Best practices of Smart Cities Mission, 2023 and other documents

waste collection. These sensors provide real-time location information, enabling efficient waste segregation and processing.

The above discussion shows that innovative technologies are playing crucial role in transforming urban landscapes, promoting sustainability, and facilitating resource optimization and waste reduction. Through their multifaceted contributions, smart tools are fostering the principles of the circular economy in Indian cities. Smart tools enable efficient resource management by providing real-time data on resource consumption, waste generation, and environmental impact. This data-driven approach empowers cities to make informed decisions, optimize resource utilization, and identify opportunities for waste reduction, reuse, and recycling, thus aligning with the circular economy's fundamental principles. Furthermore, sharing economy models optimize asset utilization and contribute to the circular economy's objectives. By engaging citizens with real-time data, smart tools promote eco-conscious choices, encouraging individuals to actively participate in sustainable practices. This engagement aligns with the circular economy's emphasis on involving all stakeholders in the transition to a more sustainable model. These technologies represent a significant leap toward realizing the circular economy model in Indian urban areas.

11.6 Challenges

As evident from the case studies, after roll-out of SBM the Indian cities have improved on waste collection and disposal, and after SCM they have improved upon the transportation and monitoring of waste. But these are the best performing cities which are ranked higher and have been awarded at national level in waste management and smartness and there are only 100 cities being made smart under SCM and they receive grants from the central governments under SCM projects. So, the rest cities which are not in 100 smart cities have a challenge to fund the innovation and experiments in SWM on their own.

Many cities are yet to initiate certain steps of the circular economy, such as processing construction and demolition wastes. The need for circular business models in construction and demolition waste management has also been emphasized by Surya & Kumar, 2021 to address the increasing demand and limited resources in the construction sector. The waste management regulations for municipal, plastic, and e-waste are not aligned with circular economy principles, posing a challenge to their integration within administration which has also been highlighted by Priyadarshini and Abhilash (2020). Centre for Science and Environment (2021) has also highlighted that the informal sector in waste management is a significant contributor to the circular economy and enabling use of smart tools in their working seems to difficult. Also, the lack of clear laws and policies to integrate them into the system and protect the rights of waste pickers continues to be a hurdle. Therefore, overcoming these challenges is crucial for achieving circularity and sustainable development within the Indian economy.

11.7 Conclusion and Future Directions

This chapter has proposed a CE framework to evaluate the status of smart circular economy in ten selected Indian cities. The study prepared a seven-stage framework of CE and examined the status of 10 cities under these stages. Along with this the study discussed the case of Indore city which is consistently six-time winners in Swachh Survekshan ranking. Indore city is the commercial capital of Madhya Pradesh consistently setting a standard of sustainable development for other cities. Achieving a self-sustainable model is the need of hour and Indore is proving this by generating "Wealth out of waste" model. The findings show the increasing importance of CE principles and the pivotal role that smart tools play in transforming the urban landscape. Indian cities are demonstrating a strong commitment to resource efficiency, waste reduction, and sustainability, as evidenced by their adoption of innovative technologies and strategies. As we move forward, it is clear that the fusion of circular economy principles and smart tools will continue to be a driving force for sustainable urbanization in India and beyond. By embracing these concepts, cities are not only working toward reducing their environmental footprint but also redefining the future of urban living, where resources are conserved, waste is minimized, and the quality of life for all residents is enhanced. This research provides a robust foundation for further exploration and implementation of circular economy strategies in the Indian urban context and serves as a testament to the potential for a more sustainable and prosperous future for cities across the nation.

As seen through the case examples in this chapter, smart urbanism has become an integral part of the process of circular economy in waste management in the country. India has one of the best practices in every stage of the process, but it still has a long way to have the best practice in these areas at global level. This is just a starting point where the advantages of sensors, ICCC's, RFID tags, etc. have been used to effectively and efficiently convert waste into wealth. As the world moves toward even smarter concepts of Artificial Intelligence, Machine Learning, and Internet of Things, it paves way for circular economy to also become smarter. In line with recent technological advances, intelligent monitoring based on machine learning for recurring and probable GVP's, prompt service delivery and data-driven perspective on monitoring and investigating of waste management is the next milestone. If the technological advances allow us few years down the line it can be further seen to identify areas of improvement, provide context-based strategies and solutions to the ULB's depending on type of waste, locality, existing system in the city, etc. which can then prove to make our cities completely smart having maximum possible economy generated out of waste. This would help in achieving the most desired zero waste concept.

References

Aceleanu, M. I., Şerban, A. C., Suciu, M. C., & Biţoiu, T. I. (2019). The management of municipal waste through circular economy in the context of smart cities development. *IEEE Access, 7*, 133602–133614.

Ahmed, Z., Mahmud, S., & Acet, H. (2022). Circular economy model for developing countries: Evidence from Bangladesh. *Heliyon, 8*(5).

Anand, A., Sreevatsan, A., & Taraporevala, P. (2018). *An overview of the smart cities mission in India* (pp. 1–17). Centre for Policy Research.

Azizuddin, M., Shamsuzzoha, A., & Piya, S. (2021). Influence of circular economy phenomenon to fulfil global sustainable development goal: Perspective from Bangladesh. *Sustainability, 13*(20), 11455.

Beja, I. A., Motta, R., & Bernucci, L. B. (2020). Application of recycled aggregates from construction and demolition waste with Portland cement and hydrated lime as pavement subbase in Brazil. *Construction and Building Materials, 258*, 119520.

Bellezoni, R. A., Adeogun, A. P., Paes, M. X., & de Oliveira, J. A. P. (2022). Tackling climate change through circular economy in cities. *Journal of Cleaner Production, 381*, 135126.

Boro, M., Devi, R. J., & Sharma, L. S. (2020). Biodegradable cluteries and tablewares as substitute for plastic: An exploratory study on green solutions. *International Journal of Research and Scientific Innovation (IJRSI), 7*, 27–29.

Caputo, F., Magliocca, P., Canestrino, R., & Rescigno, E. (2023). Rethinking the role of technology for citizens' engagement and sustainable development in smart cities. *Sustainability, 15*(13), 10400.

Centre for Science and Environment. (2021). *Integration of informal sector in solid waste management*. Retrieved from https://www.cseindia.org/integration-of-informal-sector-in-solid-waste-management-10886

Chauhan, A., Jakhar, S. K., & Chauhan, C. (2021). The interplay of circular economy with industry 4.0 enabled smart city drivers of healthcare waste disposal. *Journal of Cleaner Production, 279*, 123854.

Chen, X. (2022). Machine learning approach for a circular economy with waste recycling in smart cities. *Energy Reports, 8*, 3127–3140.

Cocker, J., & Graham, K. (2020). Circular economy in Canada. In *Circular economy: Global perspective* (pp. 87–122).

da Conceição Leite, F., dos Santos Motta, R., Vasconcelos, K. L., & Bernucci, L. (2011). Laboratory evaluation of recycled construction and demolition waste for pavements. *Construction and Building Materials, 25*(6), 2972–2979.

Dincă, G., Milan, A. A., Andronic, M. L., Pasztori, A. M., & Dincă, D. (2022). Does circular economy contribute to smart cities' sustainable development? *International Journal of Environmental Research and Public Health, 19*(13), 7627.

Gurlhosur, V. S., Qadri, W. M., Dohare, D. (2020). *Municipal solid waste management of Indore:A Review. International Journal of Ecology and Environmental Science, 2*(1), 218–223.

Dulia, E. F., Ali, S. M., Garshasbi, M., & Kabir, G. (2021). Admitting risks towards circular economy practices and strategies: An empirical test from supply chain perspective. *Journal of Cleaner Production, 317*, 128420.

Fadeeva, Z., & Van Berkel, R. (2023). Towards circular economy of food systems: An explorative appraisal of opportunities in fish, seafood value chains. In *Sustainable food value chain development: Perspectives from developing and emerging economies* (pp. 61–86). Singapore.

Ferronato, N., Rada, E. C., Portillo, M. A. G., Cioca, L. I., Ragazzi, M., & Torretta, V. (2019). Introduction of the circular economy within developing regions: A comparative analysis of advantages and opportunities for waste valorization. *Journal of Environmental Management, 230*, 366–378.

Flores, M. G., & Tan, J. (2019). Literature review of automated waste segregation system using machine learning: A comprehensive analysis. *International Journal of Simulation: Systems, Science and Technology, 11*.

Ghosh, T., Hanes, R., Key, A., Walzberg, J., & Eberle, A. (2022). The circular economy life cycle assessment and visualization framework: A multistate case study of wind blade circularity in United States. *Resources, Conservation and Recycling, 185*, 106531.

Gonçalves, A., Galliano, D., & Triboulet, P. (2022). Eco-innovations towards circular economy: Evidence from case studies of collective methanization in France. *European Planning Studies, 30*(7), 1230–1250.

Goyal, S., Esposito, M., & Kapoor, A. (2018). Circular economy business models in developing economies: Lessons from India on reduce, recycle, and reuse paradigms. *Thunderbird International Business Review, 60*(5), 729–740.

Grandhi, B., & Appaiah Singh, J. (2016). What a waste! A study of food wastage behavior in Singapore. *Journal of Food Products Marketing, 22*(4), 471–485.

Hemidat, S., Achouri, O., El Fels, L., Elagroudy, S., Hafidi, M., Chaouki, B., et al. (2022). Solid waste management in the context of a circular economy in the MENA region. *Sustainability, 14*(1), 480.

Hoelscher, K. (2016). The evolution of the smart cities agenda in India. *International Area Studies Review, 19*(1), 28–44.

India's Cleanest City Indore Earns Huge Bucks from Its Waste. (2022, October 1). https://www.outlookindia.com/. https://www.outlookindia.com/national/india-s-cleanest-city-indore-earns-huge-bucks-from-its-waste-news-227209

Jaya Surya, R., & Kranti Kumar, M. (2021). Implementation of circular business models in construction and demolition waste management of India. *International Journal of Engineering and Management Research, 11*(2), 87–93. https://doi.org/10.31033/ijemr.11.2.12

Khan, F., & Ali, Y. (2022). A facilitating framework for a developing country to adopt smart waste management in the context of circular economy. *Environmental Science and Pollution Research, 29*(18), 26336–26351.
Khoo, H. H., Lim, T. Z., & Tan, R. B. (2010). Food waste conversion options in Singapore: environmental impacts based on an LCA perspective. *Science of the Total Environment, 408*(6), 1367–1373.
Kim, S., Oah, S., & Dickinson, A. M. (2005). The impact of public feedback on three recycling-related behaviors in South Korea. *Environment and Behavior, 37*(2), 258–274.
Li, X., Bao, J., Sun, J., & Wang, J. (2021). Development of circular economy in smart cities based on FPGA and wireless sensors. *Microprocessors and Microsystems, 80*, 103600.
Liu, L., & Ramakrishna, S. (Eds.). (2021). *An introduction to circular economy*. Springer Singapore.
Livia, S., María, M. S., Marco, B., & Marco, R. (2020). Assessment of wastewater reuse potential for irrigation in rural semi-arid areas: The case study of Punitaqui, Chile. *Clean Technologies and Environmental Policy, 22*, 1325–1338.
Mandpe, A., Paliya, S., Gedam, V. V., Patel, S., Tyagi, L., & Kumar, S. (2023). Circular economy approach for sustainable solid waste management: A developing economy perspective. *Waste Management & Research, 41*(3), 499–511.
Mani, S., & Singh, S. (2016). Sustainable municipal solid waste management in India: A policy agenda. *Procedia Environmental Sciences, 35*, 150–157.
McDowall, W., Geng, Y., Huang, B., Barteková, E., Bleischwitz, R., Türkeli, S., et al. (2017). Circular economy policies in China and Europe. *Journal of Industrial Ecology, 21*(3), 651–661.
Ministry of Housing and Urban Affairs, GoI. (2023a). *Smart Cities Mission*. Retrieved October 6, 2023, from https://smartcities.gov.in/
Ministry of Housing and Urban Affairs, GoI. (2023b). *Swachh Survekshan 2022*. Retrieved October 5, 2023, from https://ss2022.sbmurban.org/#/dashboard
Modak, P. (2021). Circular economy practices in India. In *An introduction to circular economy* (pp. 555–575).
MOHUA. (2021). *Circular economy—In municipal solid and liquid waste*. Available at: https://mohua.gov.in/pdf/627b8318adf18Circular-Economy-in-waste-management-FINAL.pdf
MOHUA (2023). *Smart Cities Mission- compendium of best practices*. Available at https://smartcities.gov.in/sites/default/files/2023-09/Compendium%20of%20Best%20Practices_book_web_version_1.pdf
Morero, B., Vicentin, R., & Campanella, E. A. (2017). Assessment of biogas production in Argentina from co-digestion of sludge and municipal solid waste. *Waste Management, 61*, 195–205.
Morero, B., Montagna, A. F., Campanella, E. A., & Cafaro, D. C. (2020). Optimal process design for integrated municipal waste management with energy recovery in Argentina. *Renewable Energy, 146*, 2626–2636.
Patwa, N., Sivarajah, U., Seetharaman, A., Sarkar, S., Maiti, K., & Hingorani, K. (2021). Towards a circular economy: An emerging economies context. *Journal of Business Research, 122*, 725–735.
Pereira, G. R. D. S.M (2022). *Circular economy business models in the coffee industry: The case of delta cafés* (Doctoral dissertation).
Pereira, W., Parulekar, S., Phaltankar, S., & Kamble, V. (2019, February). Smart bin (waste segregation and optimisation). In *2019 Amity international conference on artificial intelligence (AICAI)* (pp. 274–279). IEEE.
Petit-Boix, A., & Leipold, S. (2018). Circular economy in cities: Reviewing how environmental research aligns with local practices. *Journal of Cleaner Production, 195*, 1270–1281.
Priyadarshini, P., & Abhilash, P. C. (2020). Circular economy practices within energy and waste management sectors of India: A meta-analysis. *Bioresource Technology, 304*, 123018. https://doi.org/10.1016/j.biortech.2020.123018

Repp, L., Hekkert, M., & Kirchherr, J. (2021). Circular economy-induced global employment shifts in apparel value chains: Job reduction in apparel production activities, job growth in reuse and recycling activities. *Resources, Conservation and Recycling, 171*, 105621.
Roleders, V., Oriekhova, T., & Zaharieva, G. (2022). Circular economy as a model of achieving sustainable development. *Problemy Ekorozwoju, 17*(2).
Saha, K., Dey, P. K., & Papagiannaki, E. (2021). Implementing circular economy in the textile and clothing industry. *Business Strategy and the Environment, 30*(4), 1497–1530.
Sasidharan, M., Torbaghan, M. E., & Burrow, M. (2019). *Using waste plastics in road construction*. Institute of Development Studies.
Semba, T., Sakai, Y., Ishikawa, M., & Inaba, A. (2020). Greenhouse gas emission reductions by reusing and recycling used clothing in Japan. *Sustainability, 12*(19), 8214.
Sharma, K. K. (2018). Solid waste management in India. *International Journal of Research in Social Sciences, 8*(7), 672–690.
Siddiqui, A., & Pandit, R. K. (2021). Smart Cities in India: Linkages with circular economy. In *Sustainability in the built environment in the 21st century: Lessons learned from India and the region* (pp. 185–200). Springer International Publishing.
Silvestri, F., Spigarelli, F., & Tassinari, M. (2020). Regional development of circular economy in the European union: A multidimensional analysis. *Journal of Cleaner Production, 255*, 120218.
Singh, R. (2021). Municipal solid waste management in the City of Indore- a case study. *Journal of Civil Engineering and Environmental Sciences, 7*, 8–17.
Smart City Indore. (n.d.). *Solid waste management*. Smartcityindore.org. https://www.smartcity-indore.org/solid-waste/
Smol, M., Duda, J., Czaplicka-Kotas, A., & Szołdrowska, D. (2020). Transformation towards circular economy (CE) in municipal waste management system: Model solutions for Poland. *Sustainability, 12*(11), 4561.
Srivastav, A. L., Markandeya, P. N., Pandey, M., Pandey, A. K., Dubey, A. K., Kumar, A., Bhardwaj, A. K., & Chaudhary, V. K. (2023). Concepts of circular economy for sustainable management of electronic wastes: Challenges and management options. *Environmental Science and Pollution Research, 30*(17), 48654–48675.
Tong, H., Shen, Y., Zhang, J., Wang, C. H., Ge, T. S., & Tong, Y. W. (2018). A comparative life cycle assessment on four waste-to-energy scenarios for food waste generated in eateries. *Applied Energy, 225*, 1143–1157.
Tummala, R., Asadi, S. S., & Chandra, D. S. (2019). An integrated approach for municipal solid waste management: A model study from Vijayawada. *International Journal of Recent Technology and Engineering (IJRTE), 7*(6C2).
Van Fan, Y., Lee, C. T., Lim, J. S., Klemeš, J. J., & Le, P. T. K. (2019). Cross-disciplinary approaches towards smart, resilient and sustainable circular economy. *Journal of Cleaner Production, 232*, 1482–1491.
Villamar, C. A., Vera-Puerto, I., Rivera, D., & De la Hoz, F. (2018). Reuse and recycling of livestock and municipal wastewater in Chilean agriculture: A preliminary assessment. *Water, 10*(6), 817.
Xavier, L. H., Ottoni, M., & Lepawsky, J. (2021). Circular economy and e-waste management in the Americas: Brazilian and Canadian frameworks. *Journal of Cleaner Production, 297*, 126570.
Yaduvanshi, N. R., Myana, R., & Krishnamurthy, S. (2016). Circular economy for sustainable development in India. *Indian Journal of Science and Technology, 9*(46), 1–9.

Swati Chauhan is an alumni of Indian Institute of Forest Management, Bhopal; she completed her doctoral research on Capital Structure, Efficiency and Sustainability of Indian Microfinance Institutions. She has over 14 years of experience in academics and research. She has worked as an academician and researcher in various professional institutes. She has published 20+ research papers in different academic journals of national and international repute. She holds an MBA majoring in Finance, from UP Technical University, Lucknow along with UGC- NET- JRF. As an active member of professional bodies like AIMS International, NHRDN and Indian Accounting Association, she has participated and presented research papers in various National and International conferences. Her areas of interest include Microfinance, Financial Econometrics, Financial Inclusion, Financial literacy and Circular Economy.

Ojaswini Bansal is an Architect and Urban Designer currently working as Research Associate at Atal Bihari Vajpayee Institute of Good Governance and Policy Analysis, Bhopal, India. She has previously worked as Smart Cities Fellow at Ministry of Housing and Urban Affairs, New Delhi, India. She is an alumnus of School of Planning and Architecture, Bhopal.

Anshul Puriya is an early career researcher currently working as Advisor in the Centre for Urban Governance, Atal Bihari Vajpayee Institute of Good Governance and Policy Analysis, Bhopal, India. He is an Architect and Urban Planner by education and an alumnus of MANIT-Bhopal and IIT-Roorkee, India. He works on issues related to Urban Governance, Urban Planning and Urban Management including housing, inclusive design and planning, sustainability. He has published and presented his works in National and International conferences and journals and government reports focussing his current research on Urbanization, Resilience, Housing, and Well-being.

Chapter 12
Smart City Initiatives Entangled in Political Priorities: Case-Based Analysis on Visakhapatnam, Andhra Pradesh

Shilpa Sarvani Ravi, Shikha Bhagat, and Rashmi Rai

Abstract Visakhapatnam, often referred to as Vizag, is a port city located in the Indian state of Andhra Pradesh. It is a major industrial and commercial hub known for its natural beauty, beaches, and historical significance. This city with a GDP of $43.5 billion and the ninth richest city in India has been identified as one of the potential Indian smart cities in 2015. The Indian government's Smart Cities Mission is a program for urban renewal and retrofitting that aims to transform a limited number of cities throughout the nation into smart cities by implementing various infrastructure and technical upgrades. This chapter uses a case-based approach and describes Visakhapatnam's experience since being selected and declared as a smart city in the first round of the Smart Cities Mission, which was announced in January 2016 to present. The ruling political party's state capital shift to Visakhapatnam proposals evoked a massive political controversy. This study provides a timeline of growth in Visakhapatnam smart city construction hampered by intergovernmental coordination, political priorities and budget allocation, local politics and stakeholder engagement, election cycles, and partisan politics. All potential secondary information including smart city proposals, government website data on smart city missions, and current news articles were gathered, and secondary sources were employed to examine the data. This chapter critically examines the detailed real-time implications and political environmental challenges faced by the city of Visakhapatnam on the threshold of a smart city mission.

Keywords Smart city · Visakhapatnam · Smart Cities Mission · Political controversy · Political priorities · Government initiative

S. S. Ravi (✉) · S. Bhagat · R. Rai
School of Business and Management, Christ University, Bangalore, India
e-mail: Shilpa.sarvani@christuniversity.in; shikha.bhagat@christuniversity.in; rashmi.rai@christuniversity.in

S. Majumdar et al. (eds.), *Smart Cities*, S.M.A.R.T. Environments,
https://doi.org/10.1007/978-3-031-59846-3_12

12.1 Introduction

Let's begin by quoting the immortal words of Gurajada Apparao, a literary giant in Telugu, *"Desam ante matti kadoyi, desam ante manushuloyi,"* which translates to *"A nation is not the land or the soil, but it is the people inhabiting that land."* Even now, more than a century after they were written in 1910, these words still have meaning.

Fast forward to the present era, Andhra Pradesh is the first Indian state created on linguistic grounds following Potti Sriramulu's 1953 hunger strike. The state is now embroiled in political controversies driven by Chief Minister YS Jaganmohan Reddy's decision to establish a decentralized capital system, with Visakhapatnam as the executive capital, Amaravathi as the legislative capital, and Kurnool as the judicial capital which is majorly resulting in decelerating the smart city development of the city Visakhapatnam (Nextias, 2023).

The most populous and major metropolitan area in the Indian state of Andhra Pradesh is Visakhapatnam, often spelled Vizagapatam, Vizag, Visakha, or Waltair. It is located between the Eastern Ghats and the Bay of Bengal shore (Pletcher, 2023). It is the fourth-most populous city in South India and, after Chennai, the second largest city on India's east coast. It is also one of the four smart cities in Andhra Pradesh selected as part of the Smart Cities Mission and serves as the district's administrative hub for the Visakhapatnam district (Atal Innovation Mission, 2018). The city was ranked eighth among the top 20 places shortlisted to be built as a smart city. It is also referred to as "The Jewel City of the East Coast" and "City of Destiny." "It made it to position ten in the smart city ranking for India in 2022 after being chosen as a part of the Smart Cities Mission following a competitive selection procedure that took place between 2015 and 2018. With a $43.5 billion GDP, it is also one of the top 100 fastest-growing cities worldwide (Software Technology Parks of India, 2023).

Additionally, it is India's ninth richest city. It is a vibrant city that blends both business and commercial activities, making it an important center in India's eastern region. Visakhapatnam has been working to enhance its physical infrastructure as a smart city (VMRDA, 2023).

12.1.1 Key Aspect of Visakhapatnam as a Smart City

12.1.1.1 Infrastructure Development

As a part of the Smart Cities Mission, the Greater Visakhapatnam Smart City Corporation Limited (GVSCCL) undertook several significant projects and has been working to enhance its physical infrastructure as a smart city (CW Team, 2023). This involves improving the city's roadways, transportation infrastructure, and public utilities to increase inhabitants' connectivity and mobility (Subhashini,

2023). A special-purpose entity has been created by the Visakhapatnam smart city corporation, GVSCCL, to carry out Vizag smart city projects. To develop the Asilmetta Area commercially, the government has requested expressions of interest to receive 2.7 acres of good land to establish the region (GVSCCL, 2017). The distance between the property and the bus terminal is only 500 m (Khurana, 2023).

12.1.1.2 Smart Transportation

To control traffic congestion, lessen pollution, and increase general transportation efficiency, the city has been attempting to install intelligent transportation systems (ITS). This effort includes initiatives like the adoption of electric buses and smart traffic lights (Elets News Network, 2021). Bus terminal operations, fleet management, re-routing, and frequency assignment are some of the things Visakhapatnam does well. This technology (ITS) is used to monitor and control 100 GPS-equipped public buses that are operating in the municipal corporation region. There are 50 bus stops and 5 bus terminals that are incorporated into this integrated bus transport system (Ministry of Electronics and Information Technology, 2018). It facilitates easy oversight of the bus transportation system, makes real-time information available to the public, and aids in future public transportation planning (NIUA, 2023).

12.1.1.3 Waste Management

To improve waste collection and disposal, automated waste management systems have been developed (Veesa et al., 2016). This includes installing trash cans with sensors that tell authorities when they need to be emptied to minimize the number of times garbage trucks make wasteful journeys (The Hindu, 2022). There are 50 semi-underground bins and 50 Radio Frequency Identification Devices (RFID)-based compactor bins in the ABD area of Visakhapatnam. Sensors determine how much waste is placed in these bins and warn the collecting system before the bins overflow. On-demand and automated bin servicing is also offered (NIUA, 2023).

12.1.1.4 E-Governance

To increase the efficiency and openness of administrative procedures, the city has been actively supporting e-governance initiatives. This involves using technology to improve citizen involvement, online payment choices, and digitization of government services (Greater Visakhapatnam Municipal Corporation, 2023a). E-government uses the enterprise resource planning (ERP) concept so that citizens have the ability to select domain-specific service requests, complaints, or information requests in each domain. These requests are then turned into actions and promptly delivered. The status of citizen's requests can be tracked, and department heads can monitor the effectiveness of their departments' services (NIUA, 2023).

12.1.1.5 Sustainability

Focusing on sustainability and environmental preservation is advised for smart cities (Cheela et al., 2021). The city of Visakhapatnam has been pursuing green projects like solar power generation, energy-efficient street lighting, and the promotion of environmentally friendly modes of transportation (Rao, 2023a). The National Institute for Urban Affairs (NIUA) has conducted an action research project on Visakhapatnam's "transformative climate employing participator data-driven decision-making platforms (T-cap)." It has been chosen as part of the International Climate Initiative (IKI), a significant German government commitment to international climate finance, to design and develop a participatory planning framework for the city's climate-informed decision-making that can serve as an example for other cities around the world (Rao, 2023b).

12.1.1.6 Smart Health Care

The city has been utilizing technology to enhance healthcare access and amenities. To give locals better healthcare services, telemedicine, and electronic health records are being promoted. At Visakhapatnam's King George Hospital, Samsung India announced its Samsung Smart Healthcare program. The initiative sought to give persons belonging to economically and socially disadvantaged backgrounds access to high-quality, inexpensive healthcare. The tech giant has donated free digital ultrasounds and digital X-rays, as well as other cutting-edge medical devices, to several government hospitals around the nation (Yovizag, 2018).

12.1.1.7 Tourism and Culture

Tourism and cultural promotion are other goals of smart cities. Technology has been used by Visakhapatnam to improve the tourist experience, including digital guides, intelligent kiosks, and upgraded infrastructure at tourist hotspots. These Visakhapatnam celebrations, which are distinguished by fervor, liveliness, rituals, and feasts, even aid in advancing the city's cultural scene. In addition, a lot of fairs are conducted there to show off the various arts and crafts that the city and state are known for. Another way to showcase the region's cultural diversity is through various cultural programs. The regional Dhimsa dance performed by the hill tribes is one of the best instances of this. Even musical instruments like the Dappu, Mori, and Thuduma are also played by the various tribal groups. They take pride in donning traditional attire (District Visakhapatnam, 2023a).

12.1.1.8 Safety and Security

A vital component of a smart city is ensuring the security and safety of its citizens. To increase safety, Visakhapatnam has been spending money on surveillance systems, emergency response services, and community policing. Surveillance is one of the most crucial real-time Integrated Command and Control Centre (ICCC) subsystems. This is directly related to law and order, traffic, safety, and security. There are 500 cameras deployed at 83 crossroads throughout the city, with 400 fixed cameras and 100 PTZ cameras with a 500-m zoom range that can swivel in any direction. The police, traffic, and disaster management team's preconceptions about crowd gathering, unattended baggage, counterflow, any illicit activities are the subject of the surveillance video analytics (NIUA, 2023).

12.1.1.9 Education and Skill Development

The city's long-term growth depends on encouraging education and skill development. E-learning platforms, virtual classrooms, and career training programs are examples of smart initiatives (Minz, 2016). On April 19, 2023, digital classrooms were opened in Araku and Ananthagiri mandals, igniting much-needed excitement among kids. The event was in line with Aarohan's year-long effort in the territory of the primitive tribes, which spans over 120 villages in Araku and Ananthagiri mandals in Andhra Pradesh. Aarohan works to improve the lives of children and women (The Hans India, 2023).

12.1.1.10 Data Analytics

To make wise judgments and enhance urban planning, the city is gathering and analyzing data from numerous sources. This data-driven strategy aids in more effective resource allocation and service provision (Babu et al., 2019). ICCC has accumulated, digitized, and integrated data of various types on the common spatial platform of the city. Visakhapatnam can greatly benefit from this overarching collection of numerous data types that have been standardized for cross-sector analysis and decision-making. It is fortunate for Visakhapatnam that this data infrastructure has been developed in an open, geographic, and standardized manner that can be accessed by numerous users (NIUA, 2023).

12.1.2 *Study Area*

The city is located between the Bay of Bengal and the Eastern Ghats. The city is situated at coordinates 17.7041 N and 83.2977 E. The city has 98 wards, 8 zones, and a total size of 682 km^2 (263 sq. mi) shown in Table 12.1. The municipality of Visakhapatnam was founded in 1861, and the administration of Visakhapatnam was formally inaugurated in 1803. It is known as Greater Visakhapatnam Municipal Corporation (GVMC). The city of Visakhapatnam (GVMC) is located between latitudes of 17° 58′ 21″ and 17° 24′ 45″ N and longitudes of 82° 51′ 0″ and 83° 27′ 0″ E. Administratively, the city of Visakhapatnam (GVMC) is divided into six zones, each of which is further divided into 72 wards. There are 6 wards in zone 1, 12 in zone 2, 12 in zone 3, 19 in zone 4, and 5 in zone 5 (Greater Visakhapatnam Municipal Corporation, 2023b).

In 2020, the metropolitan population of Visakhapatnam was 2,175,000, reflecting a 2.35% growth compared to 2019. Subsequently, in 2021, it further increased to 2,226,000, with a 2.34% rise from the previous year. As of 2023, the present metropolitan population of Visakhapatnam stands at 2,331,000, denoting a 2.33% increase from 2022. As per the projections by the United Nations, it is expected to increase by 2,440,000 in 2025 but with a slight decrease in the annual rate of 0.01% due to GDP as shown in Fig. 12.1 (Macrotrends, 2023).

Table 12.1 Visakhapatnam district at a glance

Geographic area	682 km^2
Administrative wings and divisions	98
Total populace	2,331,000
The ratio of women population	49.82
Population density	3300/km^2 (8500/sq. mi)
Sex ratio	978/1000
Literacy rate	81.79%
Working population	1,889,879

Source: www.gvmc.gov.in

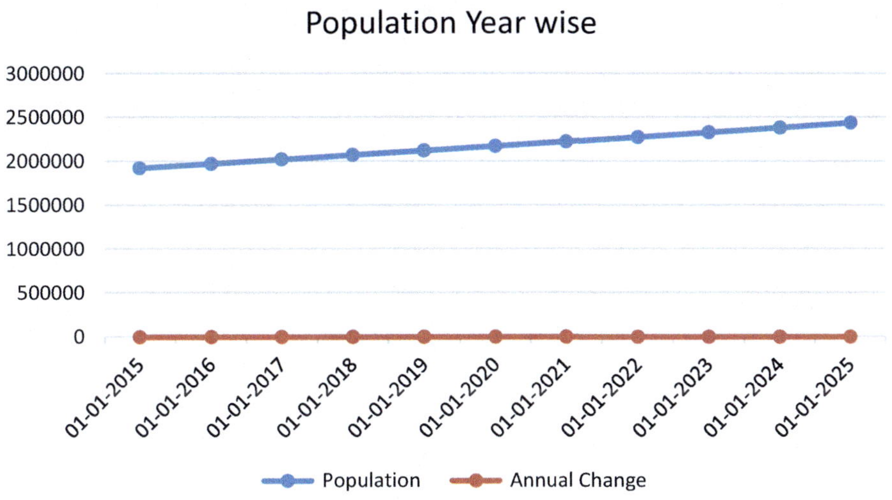

Fig. 12.1 Population and annual growth rate. (Source: www.macrotrends.net)

12.1.3 Achievements Done in Visakhapatnam Till 2023 (Fig. 12.2)

Fig. 12.2 Achievements of Visakhapatnam till 2023. (Source: Compiled by author)

12.1.3.1 Proposals and Plan of Action for the Future (Fig. 12.3)

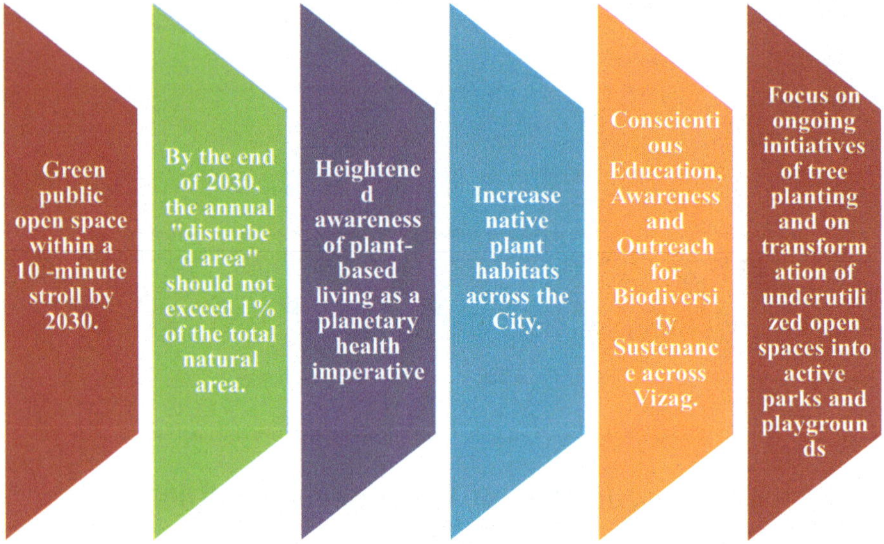

Fig. 12.3 Plan of action for the future till 2030. (Source: Compiled by author)

12.2 Political Framework in Visakhapatnam District, Andhra Pradesh State

In order to guarantee a resounding win for the Yuvajana Sramika Rythu Congress Party (YSRC) in the state general elections of 2024, Chief Minister YS Jagan Mohan Reddy and his YSR Congress are prioritizing the north Andhra-Vizag region as the center point for the party's campaign and strategy. To win the 2024 elections, leaders of the Telugu Desam party, including Nara Chandrababu Naidu and K Pawan Kalyan of the Jana Sena, are focusing on north Andhra-Vizag. Notably, the coastline region, Vizag, and North Andhra played significant roles in each election. Out of the 34 assembly segments in the three districts of north Andhra that were up for election in 2019, the YSRC won 28 seats and the Telugu Desam Party (TDP) won 6. Since the party's founding, Uttarandhra has been a TDP stronghold, but it suffered a devastating loss in the 2019 elections. To draw in the crowds, Chandrababu and Pawan are now making frequent trips to Vizag and north Andhra. The chief minister has been trying to relocate the executive capital to Vizag for the past 3 years due to legal challenges. He planned to advance economic initiatives and take office from Vizag in September 2023 (Ilyas, 2023a).

The Yuvajana Sramika Rythu Congress Party (YSRC) received 49.32% of the vote in the former Srikakulam district, 51.51% in Vizianagaram, 44.26% in Visakhapatnam, 43.48% in East Godavari, and 46.35% in West Godavari district in 2019 elections. In Srikakulam, the TDP received 42.67%, 39.47% in Vizianagaram, 36.74% in Visakhapatnam, 36.76% in East Godavari, and 36.30% in West Godavari. It's interesting to note that in East and West Godavari districts, respectively, the Jana Sena received double-digit voting percentages in 2019 with 14.84% and 11.68%, respectively (Election Commission of India, 2019).

The development of north Andhra and Vizag is the chief minister's main concern, and he has directed ministers and lawmakers to visit the general public to inform them of the YSRC government's development initiatives for the area and solicit their support for the party in the upcoming elections. After taking office in Vizag, the CM believes that the political landscape will change in favor of YSRC. Jagan Reddy gave Vizag first priority and hosted the Global Investors Summit there on March 3 and 4, 2023, with notable businessmen and industrialists in attendance. Additionally, the G-20 summit took place on March 28 and 29, 2023, where CM erected the Mulapet (Bhavanapadu) port in the Srikakulam district on April 19, 2023. He also instructed to begin the construction of the Mahendratanaya offshore reservoir project, Budagatlapalem, an Etcherla mandal fishing port, and the Vamsadhara upliftment program. The Bhogapuram International Airport, Project Works for Taraka Rama Thirtha Sagaram, the Chintapalli Fish Landing Center in Pusapatirega manda on the Chintapalli shore, located in the center of the Vizianagaram, Vizag, Srikakulam districts, and the Bhumi Puja for the Adani Data Center in Vizag were also inaugurated on May 3, 2023, by the chief minister (Ilyas, 2023b).

On May 11, 2023, the chief minister went to Vizag to inaugurate the Sea Harrier Museum. In June 2023, he also opened a 200-bed super specialty hospital and research facility in Srikakulam to address the issue of kidney disease for the local population. He also proposed to launch an exclusive planned freshwater delivery program in Uddhanam for kidney sufferers. In July 2023, the CM also laid the cornerstone for the tribal university in the Vizianagaram district. He also checked on the renovations to the Vizianagaram Medical College facilities. In August 2023, the Chief Minister also laid the cornerstone for the six-lane national highway which connects Bhogapuram and Vizag.

In September 2023, the CM took over the state administration, and he is working tirelessly to unite the Yuvajana Sramika Rythu Congress Party (YSRC) bases in North Andhra, the coastline districts, and Vizag. Nara Chandrababu Naidu, leader of the Telugu Desam Party who is also working hard to restore his former glory through Idemi Kharma gatherings, has made multiple trips to north Andhra, Vizag, and the coastal region (Ilyas, 2023c).

12.3 Research Methodology

Both primary and secondary sources were used to gather both qualitative and quantitative data. Interviews with city citizens and officials yield primary data. The major sources for secondary data included a variety of Indian government information portals, city-specific government websites, department websites, publicly available research papers, publicly available reports and statistics, and surveys carried out by a variety of analytic groups. Government agencies at the federal and state levels, including the Greater Visakhapatnam Municipal Corporation (GVMC), District-Visakhapatnam (DV), Visakhapatnam Metropolitan Region Development Authority (VMRDA), Ministry of Electronics and Information Technology (MEIT), Greater Visakhapatnam Smart City Corporation Limited (GVSCCL), and National Institute of Urban Affairs, (NIUA) have also produced numerous reports.

To achieve the objective, methodology and metrics outlined in this chapter in accordance with the study area across the dominant Political turbulences in smart city development by studying numerous economic, social, political, and urban elements for the transformation of smart cities, the existing scenario of Visakhapatnam city is taken into consideration. A structured method using small case-based illustrations has been used, involving a thorough evaluation of the political and budgetary factors hindering the transformation of a city into a smart city. Consideration of the city's readiness to realize the vision of the smart city in the context of the local political instability that is currently prevailing was evaluated through detailed interview techniques.

12.4 Results and Discussion

12.4.1 Assessment of the Current Political Situation in Visakhapatnam City

Visakhapatnam, which was once a fishing village, has evolved into a smart city. Vizianagaram, Srikakulam, Ganjam, and Koraput, which are now separate districts, were once a part of Vizagapatam, which was once the largest district in the nation (District Visakhapatnam, 2023b). Further analysis is conducted to assess the current situation and circumstances of the Visakhapatnam city.

12.4.1.1 Analysis of Smart City Budgetary Allocations in Smart City Projects

The state government and the federal government each contribute the following amounts to projects for India's smart cities (Fig. 12.4):

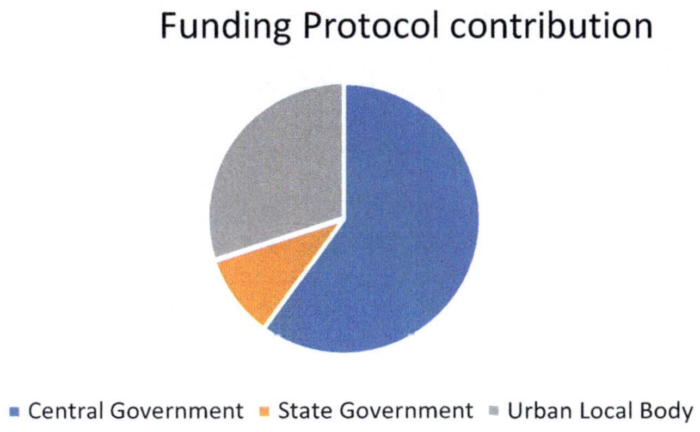

Fig. 12.4 Smart city funding protocol contributions. (Author's compilation)

Central: 60% of the funding for smart city projects comes from the national government.
State government: 10% of the money are contributed by the government of the state.
Urban local body (ULB): The final 30% of the funding comes from the urban local body, which is responsible for carrying out the smart city projects.

The establishment of Vizag Municipality was in 1858. The Vizag Municipality was founded to satisfy the population's basic infrastructure needs. In 1979, it became a Municipal Corporation. The municipality of Greater Visakhapatnam (GVMC) serves as the city of Visakhapatnam's main governing body. Limitations on GVMC's primary configuration range from 111 to 640 square kilometers (District Visakhapatnam, 2023c). It is an arm of the Visakhapatnam Urban Development Authority's planning department as well. The urban local body responsible for governing Visakhapatnam (Vizag) is the "Greater Visakhapatnam Municipal Corporation" (GVMC). The civic body in charge of directing the city's management and growth is the GVMC. It is responsible for various functions including urban planning, infrastructure development, waste management, water supply, sanitation, and more within the municipal limits of Visakhapatnam (Rao, 2021). With the help of the local urban authority and a large financial commitment from the federal and state governments, smart cities can be developed all across the nation.

12.4.2 Challenges and Analysis

The city operations center, solar street lights, smart schools, part retrofitting, and creation of open spaces are among the many of the projects that GVMC has already finished under its umbrella of the smart city. While these are a few examples of the effective execution of a smart city mission, there are also prominent challenges prevailing in the construction of the smart city. A strong political framework and funding system are notably identified to be essential to guarantee that smart cities deliver the promised advantages to their citizens and the environment (Suma et al., 2016). Political cooperation is crucial to address the entangled smart city development mission.

Case #1: CCTV Coverage

Vizag was the first city in Andhra Pradesh to receive full CCTV surveillance, according to records. To maintain a watchful eye on all important areas and traffic intersections in the city, up to 2100 CCTV cameras in four distinct categories were installed (Kumar et al., 2019). With a population of over 20 lakhs, Vizag has experienced a tremendous increase in the number of automobiles on the road. The majority of national and international events are also held there. To speed Vizag's transformation into a smart city, the Greater Visakhapatnam Municipal Corporation (GVMC) erected around 500 CCTV cameras as part of the Smart City Project by March 2018, and the municipal police deployed 1638 CCTV cameras as part of the AP Fibre Grid Project (Deccan Chronicle, 2018).

Although some of the CCTV cameras will be used to identify vehicles by reading their license plates, most of them will be utilized for surveillance. Police claimed that automated number plate recognition (ANPR) cameras help to lessen crimes including auto theft and careless driving. While some cameras will help the police follow offenders using facial recognition technology, others are installed to identify automobiles that run red signals. Commuters will be provided with real-time traffic information so they may take different routes to their destinations, according to sources. A private business by the name of Matrix reportedly installed the CCTV cameras and has plans to install video analysis software as well.

Current Status

The current status reported by the newspaper *The Hindu* highlighted that more than 200 CCTV cameras that were deployed throughout the city are no longer working (Gilai, 2023). The police have also requested that the relevant authorities investigate the problem. The majority of CCTV cameras, according to the authorities, were entirely closed, but some of them were not correctly transmitting the feed. The conclusion of this case analysis deliberates on the interests in priorities of operation and maintenance issues by local bodies resulting in the non-continuation of the smart city initiatives.

Case #2: Cycling Initiative in the City

The local governing body organized a hundred bicycles as part of the smart city plan in July 2018 to encourage riding on the Beach Road. This program intends to safeguard the health of citizens. At four locations, including the YMCA, Pandurangapuram, and Kalimatha Temple, the GVMC has set up 25 bicycles. Young people and kids can borrow bicycles for free and use them to bike down the beach at any of the four places. People must download the Hexi mobile app to ride a bicycle (Ganguly, 2018).

Fig. 12.5 Cycle zone converted into pavement at Waltair junction east gate, Visakhapatnam. (Source: Author)

Current Status

Unfortunately, this program is not continuing in the city and the cycles were also not observed in any of the locations. The discontinuity is observed, and the place allocated for the cycle zone is covered by the pavement as displayed in the attached picture 12.7 at the location Waltair East gate area. This case also evidently emphasizes the continuity and maintenance issues in the construction of the smart city. The below picture illustrates the issue where the smart city initiative was disrupted due to a disparity of interest among different bodies (Fig. 12.5).

Case #3: Visakhapatnam: The Capital Conundrum

As part of its three-capital policy, the southern Indian state of Andhra Pradesh has chosen Visakhapatnam as its next capital. The idea was made in December 2017 while Andhra Pradesh's chief minister, Jagan Mohan Reddy, was following the three-capital plan. This plan was designed to encourage balanced regional expansion and help Amaravati, the current capital, become less congested. Instead of focusing on growth in a single area, the state administration has attempted to distribute its advantages over the entire state. The goal of this action is for the government to raise the level of living for all Andhra Pradesh residents. According to Andhra Pradesh's three-capital plan, the legislative capital will be in Amaravati, the executive capital will be in Visakhapatnam, and the judicial capital will be in Kurnool. The administrative center of the state would be in Visakhapatnam, the executive capital, while policymaking will take place in Amaravati, the legislative capital. Kurnool, the state's capital of justice, will serve as the center of the judicial system. The new executive capital of Andhra Pradesh, Visakhapatnam, is a prominent city with a perfect location on India's east coast. It is a sizable port city that is recognized for both its scenic attraction and its busy harbor. The city also has an excellent transportation infrastructure, which makes it an ideal location for the administration of the state government as stated by the ruling party leaders (Deccan Chronicle,

2023a). The city was also selected as the executive capital due to its proximity to Hyderabad, the state's commercial hub.

The public and the political community have had varying opinions about Andhra Pradesh's three-capital strategy. While some have applauded the decision as a positive step, others have opposed it due to the costs it is anticipated to involve. This decision is leading to insecurity, chaos, and protests in the city against the Government (Deccan Chronicle, 2023b). The state administration has reassured the citizens that it is prepared to implement the three-capital plan and that doing so complies with its objectives for building a better, more just Andhra Pradesh. This was severely protested by opposition parties, political leaders of state, and central in all possible ways which resulted in mixed reactions and high levels of ambiguity in public. The ongoing delay in the shift of administrative capital to Vizag City for the last 5 years after the proposal left the city development and citizens in a paradox contributing to decelerating the growth.

Case #4: The Polavaram Project Delay

Vizag is anticipated to gain a lot from the completion of the Polavaram Project. With the addition of more electricity, water, jobs, and infrastructure, the local economy will experience a significant boost. Additionally, it will have enhanced facilities and better infrastructure, making it a significant tourist attraction. Additionally, Vizag will benefit from increased commerce and investment, better economic opportunities, and improved connectivity with the rest of the state (Polavaram Project Authority, 2023).

On the Godavari River in Andhra Pradesh, about 42 kilometers upstream of the Sir Arthur Cotton Barrage, where the river flows the Eastern Ghats' final peak and reaches the plains, is the Polavaram Project, a national-level multipurpose irrigation project that is currently under development (Ministry of Jal Shakti, 2023). The project site is located at 810–46′E and 170–13′N latitude and longitude, respectively.

The Polavaram project is in dire straits, and the project is expected to delay and raise a huge political breakout in the city of Visakhapatnam among the political parties of YSRCP, TDP, and BJP where leaders are faulting each other for the non-functionality. This also turned out to be the disparity of interests between the state and the central political parties which resulted in the central government freezing funds for the project completion (Raghavendra, 2023).

The Polavaram project is considered to be the major solution to meet the drinking needs of Vizag city and the irrigation water needs of the Visakhapatnam district. Moreover, the proposal highlights the hydroelectricity sustainable project which now, unfortunately, turned out to be a political controversy for obvious reasons. The completion of this project will help the city achieve its smart city goals effortlessly with elegant infrastructural upgradation, self-sufficient hydro requirements, eco-tourism choices, and sustainable tourism options. This will also boost revenues of the local economy and promote cultural and natural heritage through promoting tourism across the city.

The Polavaram multipurpose project on the Godavari River, which is regarded as the lifeline of the state, was originally put out in 1941, and initial research was done between 1942 and 1944. As part of his ambitious "Jala Yagnam," the then-Chief Minister YS Rajasekhara Reddy decided to take up the project in 2004. Arun Jaitley, the former Union Finance Minister, announced that Polavaram would be funded by the Centre and finished by 2021. This project is still under construction due to the political controversies among state government and central-state disparities, and the proposed expected completion is 2025, which is again prone to extensions (Sadam, 2021).

Case #5: The "Special Package" Issue: Politics and Coordination Challenge
Due to the rising complexity of cities' involvement as stakeholders in the value network, with all the agencies and domains they imply, coordination is necessary to ensure that the concept of the smart city works smoothly and that all of its benefits are implemented. The stakeholder relationships are one of the primary aspects that decide whether or not such initiatives succeed or fail. Currently, insufficient political support due to concern over losing power condenses the phase of smart city success. When the integration of all stakeholders is successfully implemented, several benefits can be realized, including holistic integrated planning, a commitment of visionary, committed, and ambitious key actors (politicians, developers, etc.), formation of win-win coalitions, and participation of citizens and other stakeholders in the decision-making process from the beginning to the implementation phase. Cities are intricate organizations, and decisions involving several departments have a tendency to take time and frequently conflict with business sales cycles (Apparasu, 2016).

One of the illustrations of lacking the coordination between governments resulting in unrest in the situation in the city of Visakhapatnam is the "special status" issue. A "central government special package" is referred to as a tailored and personalized financial assistance program or a bundle of policies offered by the country's central government to a particular state, region, or industry (Business Line, 2018). To address particular economic, social, or developmental difficulties and to foster growth and development in the receiving area, this unique status is created to support infrastructure, financial, and overall development. The then newly divided state of Andhra Pradesh in 2014 was offered a "special package" in lieu of "special status" in the year 2017 which led to towering unrest in the city of Visakhapatnam as being the major revenue contributor of the state. This situation also illustrated the issue of lack of understanding between the state and central governments resulting in outbreaks of protest all over the city affecting the revenues and economic situations (The Economic Times, 2018).

12.4.3 Major Results of Public Interviews and Review

This study conducted interviews with 22 citizens of Visakhapatnam and conducted a qualitative analysis to evaluate the results and derive the outcomes of citizen intuitions. The results included the consequences of political entangles in smart city mission execution. The responses mentioned below excluded the repeated opinions.

- *Brain-drain*

- An engineering girl student when asked about the Vizag smart city initiative and status mentioned that *"Not really concerned. As we have very minimal options of good technical opportunities in the city- all my batchmates including me are planning to shift to Chennai or Hyderabad or Bangalore for a better career."*
- *Uncertainty of investors*
- A woman entrepreneur running a third-party logistics company expressed that *"Both Vizag and Amaravathi (capital proposed cities), and eventually the state, is losing in this political fiasco of capital building. People are left in a limbo. Investments are at halt."*
- *Delayed decision-making*
- A Ph.D. student of a renowned university opined that *"Everything looks fantastic and overwhelming on paper but the timelines of executions are always extended and delayed."*
- *Lack of consistency*
- The general manager of an IT company (BPO) located in Vizag stated that *"The plan of action is implementing the smart city is excellent but the caste-based partisan's politics are ruining the show."*
- *Funding challenges*
- An employee of the local municipality added that *"There is a lot of positive feedback on the execution of smart city initiatives in the city but the release of maintenance and operational funds are getting critical to continue the work."*
- *Reduced civic engagement*
- Women running a playschool stated that *"There is always some or the other ongoing issue in the political environment hindering city development and I lost track to identify what the new."*
- When an HSBC employee was approached, he replied that *"I was a school teacher before, joined and enjoyed HSBC BPO for the last eight years and now I'm moving along with my family as the company closed its office in Vizag and moved to Hyderabad. We are losing on career if we are in Vizag."*
- A retired government employee added that *"Real estate prices are always high Vizag but after smart city and capital announcements, I observed that the rates have gone up manifold."*
- A retail manager in a renowned hypermarket expressed that *"As made smart city the city sees development. The Global summit happening in the city is giving a vibe of positivity."*

- A school teacher expressed that *"After declaring as a smart city, Vizag's cleanliness improved."*
- Another central government official who was born and raised in Visakhapatnam claimed that *"The city is now frequently witnessing the movement of VIPs, which is causing more traffic jams."*
- Another retired railway employee felt that *"This smart city initiative should enhance the IT sector in the city to boost so that his children can come back to the city and stay with him, but not happening."*
- A restaurant franchisee owner felt that *"This initiative improved tourism in the city and long way to go."*

12.5 Conclusion

Under the umbrella of the smart city, GVMC already claimed the completion of projects such as the city operations center, solar street lights, smart classrooms, retrofitting of parts, and development of open spaces. City procurement cycles can be up to 3 years, which may dissuade inventive, under-resourced enterprises from pursuing development opportunities in smart cities. Coordination can be challenging due to the operational silos inside the city, and it becomes even more challenging when the private sector is involved. Despite the fact that there is still little research on smart cities that addresses governance-related issues, several cities have already benefited from the adoption of ICTs that enhance governance.

This research examined the considerations of political hitches and analyzed the implications such as uncertainty for investor resource allocation difficulties, stalled infrastructural projects, brain-drain conditions, inefficient regulations, negative image of the city, reduced civic engagement, and waste of public funds and may also lead to the economic downturn. Political decisions are to be taken putting the needs of the city and its citizens ahead of partisan differences in order to promote local growth and development. For political issues to have the least detrimental impact on a city's development, cooperation, sound governance, and transparent decision-making are essential. Additionally, political accountability and an informed electorate can assist in moving the city in a more favorable direction. Civic involvement is one such example. This study also exposed a few situations and political frameworks which are the stumbling blocks to the success of the smart city vision achievement of the Indian Government. While the smart city conceptualization benefits the city, the political considerations and accommodations will ensure the city to accelerate its growth.

References

Apparasu, S. R. (2016, September 08). Centre announces special financial package for Andhra Pradesh. *Hindustan Times*. https://www.hindustantimes.com/india-news/centre-announces-special-financial-package-for-andhra-pradesh/story-Qid1bpFteotZGwvGOlNdhL.html. Accessed 29 Sept 2023.

Atal Innovation Mission. (2018). *The list of 110 cities se-lected under the smart city mission*. https://aimapp2.aim.gov.in/aic/files/List%20of%20selected%20Smart%20Cities.pdf. Accessed 17 Sept 2023.

Babu, N. V., Boyidi, S., & Bendalam, S. (2019). *A geospatial study on solid waste management for the Visakhapatnam smart city*. In Proceedings of the international conference on remote sensing for disaster management: Issues and challenges in disaster management (pp. 625–634). Springer. https://doi.org/10.1007/978-3-319-77276-9_55

Business Line. (2018, March 07). *Centre committed to special package for Andhra Pradesh: Finance Minister*. https://www.thehindubusinessline.com/news/fm-centre-committed-to-special-package-tofor-andhra-pradesh/article22970071.ece. Accessed 25 Sept 2023.

Cheela, V. S., Ranjan, V. P., Goel, S., John, M., & Dubey, B. (2021). Pathways to sustainable waste management in Indian smart cities. *Journal of Urban Management, 10*(4), 419–429. https://doi.org/10.1016/j.jum.2021.05.002

CW Team. (2023). Visakhapatnam Smart City reveals an inovative park project. *Construction World*. https://www.constructionworld.in/urban-infrastructure/smart-cities-projects/visakhapatnam-smart-city-reveals-innovative-park-project/43454. Accessed 18 Sept 2023.

Deccan Chronicle. (2018, March 17). *Visakhapatnam first to have maximum CCTV coverage*. https://www.deccanchronicle.com/nation/current-affairs/170318/visakhapatnam-first-to-have-maximum-cctv-coverage.html

Deccan Chronicle. (2023a, September 21). *CM Jagan likely to shift to Vizag from Dasara*. https://www.deccanchronicle.com/nation/current-affairs/210923/cm-jagan-likely-to-shift-to-vizag-from-dasara.html

Deccan Chronicle. (2023b, September 22). *Jagan's shift to Vizag: Police force being strengthened in the city*. https://www.deccanchronicle.com/nation/current-affairs/220923/jagans-shift-to-vizag-police-force-being-strengthened-in-city.html

District Visakhapatnam. (2023a). *Cultural tourism*. https://visakhapatnam.ap.gov.in/culturaltourism/#:~:text=Utsav%20reflects%20the%20culture%20of,burning%20of%20a%2050%20ft

District Visakhapatnam. (2023b). *History*. https://visakhapatnam.ap.gov.in/history/#:~:text=Coming%20to%20etymology%20of%20the,family%20deity%20called%20Visakeswara%20to. Accessed 28 Sept 2023.

District Visakhapatnam. (2023c). *District at a glance*. https://visakhapatnam.ap.gov.in/. Accessed 12 Oct 2023.

Election Commission of India. (2019) *Statistical report of Andhra Pradesh Legislative assembly Election 2019*. https://eci.gov.in/files/file/11673-andhra-pradesh-legislative-assembly-election-2019/. Accessed 10 Oct 2023.

Elets News Network. (2021, July). Sustainable transport for better tomorrow: visakhapatnam in focus. *egov.eletsonline*. https://egov.eletsonline.com/2021/07/sustainable-transport-for-better-tomorrow-visakhapatnam-in-focus/. Accessed 21 Sept 2023.

Ganguly, N. (2018, July 4). Bicycle diaries. *The Hindu*. Retrieved from: https://www.thehindu.com/society/bicycle-diaries/article24329459.ece. Accessed 21 Sept 2023.

Gilai, H. (2023, May 28). Over 200 CCTV cameras are lying defunct in Visakhapatnam, say police officials. *The Hindu*. Retrieved from: https://www.thehindu.com/news/cities/Visakhapatnam/over-200-cctv-cameras-are-lying-defunct-in-visakhapatnam-say-police-officials/article66901011.ece. Accessed 21 Sept 2023.

Greater Visakhapatnam Municipal Corporation. (2023a). *Smart initiatives*. https://www.gvmc.gov.in/wss/static_content/About%20Vizag.jsp?parentMenuId=200-2&pageName=About%20Vizag.jsp&menutitle=About%20Vizag. Accessed 21 Sept 2023.

Greater Visakhapatnam Municipal Corporation. (2023b). *Zones*. https://www.gvmc.gov.in/wss/static_content/About%20Vizag.jsp?parentMenuId=200-2&pageName=About%20Vizag.jsp&menutitle=About%20Vizag. Accessed 21 Sept 2023.

GVSCCL. (2017). *Vishakhapatnam: Greater Visakhapatnam Smart City Corporation Limited (GVSCCL)*. Retrieved from: https://smartnet.niua.org/5-e2f39306-0b2b-4b90-9510-a8958e8470ac

Ilyas. (2023a, January 2). 2023 Crucial to political parties in AP as stage getting set for big battle of 2024. *Deccan Chronicle*. https://www.deccanchronicle.com/nation/politics/010123/2023-crucial-to-political-parties-in-ap-as-stage-getting-set-for-big-b.html

Ilyas. (2023b, March 15). CM Jagan plans shifting to Vizag in July. *Deccan Chronicle*. Retrieved from: https://www.deccanchronicle.com/nation/in-other-news/140323/cm-jagan-affirms-shift-to-vizag-warns-poorly-performing-ministers.html

Ilyas. (2023c, May 8). CM Jagan to make Vizag as epicentre for big political battle ahead. *Deccan Chronicle*. Retrieved from: https://www.deccanchronicle.com/nation/current-affairs/070523/cm-to-make-vizag-as-epicentre-for-big-political-battle-ahead.html

Khurana, J. (2023). What is a smart city? Features, selection & more. *Magic Bricks*. https://www.magicbricks.com/blog/what-is-smart-city/127301.html. Accessed 20 Sept 2023.

Kumar, B. P., Student, P. G., Parishad, G. V., & Srinivas, M. (2019). *Traffic problems at junctions in Visakhapatnam and their solutions*. https://doi.org/10.33564/IJEAST.2019.v04i06.048

Macrotrends. (2023). *Visakhapatnam, India Metro Area Population 1950–2023*. Retrieved from: https://www.macrotrends.net/cities/21436/visakhapatnam/population. Accessed 21 Sept 2023.

Ministry of Electronics and Information Technology MEIT. (2018). *Smart cities mission ministry of electronics and information technology*. https://www.meity.gov.in/

Ministry of Jal Shakti. (2023). *About project—Polavaram irrigation project*. https://ppa.gov.in/WPSCore/Common/WebPages/Home/AboutProject.aspx. Accessed 21 Sept 2023.

Minz, S. (2016). The cosmopolitan smart city: Visakhapatnam. *Proptiger*. https://www.proptiger.com/guide/post/the-cosmopolitan-smart-city-visakhapatnam. Accessed 21 Sept 2023.

Nextias. (2023, February 2). *Three-capital plan of Andhra Pradesh*. https://www.nextias.com/current-affairs/02-02-2023/three-capital-plan-of-andhra-pradesh/. Accessed 10 Sept 2023.

NIUA. (2023). Smart education campuses in Visakhapanam under the smart city mission—An appraisal. In: *Urban management, smart cities and academia towards action and research* (in press). National Institute of Urban Affairs.

Pletcher, K. (2023). Visakhapatnam geography and travel. In: *Britannica*. https://www.britannica.com/place/Visakhapatnam. Accessed 15 Sept 2023.

Polavaram Project Authority. (2023). *About Polavaram project*. Retrieved from: https://ppa.gov.in/WPSCore/Common/WebPages/Home/ProjectMap.aspx. Accessed 13 Oct 2023.

Raghavendra. (2023, July 31). YSRCP, TDP failed to complete Polavaram project. *The Hindu*. https://www.thehindu.com/news/national/andhra-pradesh/ysrcp-tdp-failed-to-complete-polavaram-project/article67138879.ece. Accessed 21 Sept 2023.

Rao, U. (2021, August 01). *Smart city works in Vizag head towards completion*. Retrieved from: https://timesofindia.indiatimes.com/city/visakhapatnam/smart-city-works-in-vizag-head-towards-comple-tion/articleshow/84931883.cms?utm_source=contentofinterest&utm_medium=text&utm_campaign=cppst

Rao, U. (2023a, July 29). *AP to emerge as a $243 bn econo-my by 2027: The Times of India*. http://timesofindia.indiatimes.com/artcleshow/102221820.cms?from=mdr&utm_source=contentofinterest&utm_medium=text&utm_campaign=cppst

Rao, U. (2023b, May 18). Vizag gets urban living lab for sus-tainability planning. *The Times of India*. https://timesofindia.indiatimes.com/city/visakhapatnam/vizag-gets-urban-living-lab-for-sustainability-planning/articleshow/100316796.cms?from=mdr

Sadam, R. (2021, March 26). Missed deadlines, cost over-runs, tiff with Centre—Why Polavaram isn't yet Andhra's 'lifeline'. *The Print*. https://theprint.in/india/governance/missed-deadlines-cost-overruns-tiff-with-centre-why-polavaram-isnt-yet-andhras-lifeline/622462/
Software Technology Parks of India. (2023). *About STPI Visakhapatnam*. Ministry of Electronics and Information and Technology. https://stpi.in/en/about-hyderabad-vishakhapatnam. Accessed 17 Sept 2023.
Subhashini, B. (2023). From ports to power: Five upcoming infrastructure projects In Andhra Pradesh to spur economic growth and development. *#Swarajya*. https://swarajyamag.com/infrastructure/from-ports-to-power-five-upcoming-infrastructure-projects-in-andhra-pradesh-to-spur-economic-growth-and-development. Accessed 20 Sept 2023.
Suma, S., Balaram, P., Karuna, K., & Ch, S. (2016). *Urban environmental challenges and management in India (A study on Visakhapatnam smart city)*. https://euacademic.org/PastIssueList.aspx?artid=38
The Economic Times. (2018, March 08). *Government promises funds, but TDP dissatisfied with central offer*. https://economictimes.indiatimes.com/news/politics-and-nation/ready-to-give-special-package-to-andhra-pradesh-not-special-status-arun-jaitley/articleshow/63203837.cms?from=mdr. Accessed 24 Sept 2023.
The Hans India. (2023, April 19). *Visakhapatnam: Digital classrooms for tribal students inaugurated*. https://www.thehansindia.com/andhra-pradesh/visakhapatnam-digital-classrooms-for-tribal-students-inaugurated-793533https://www.thehansindia.com/andhra-pradesh/visakhapatnam-digital-classrooms-for-tribal-students-inaugurated-793533
The Hindu. (2022, March 22). Waste-to-energy plant starts functioning in Vizag. *The Hindu*. https://www.thehindu.com/news/cities/Visakhapatnam/waste-to-energy-plant-starts-functioning-in-vizag/article65249551.ece. Accessed 21 Sept 2023.
Veesa, K. S., Dimmala, C. S., Antipas, O. J. H., Mohan, V. R., & John, S. M. (2016). Drinking water practices and sanitation in slums of Visakhapatnam, Andhra Pradesh a developing smart city in India. *International Journal of Community Medicine and Public Health, 3*(11), 3188. https://doi.org/10.18203/2394-6040.ijcmph20163934
VMRDA. (2023). *Metropolitan region development authority Visakhapatnam infrastructure development*. http://vmrda.gov.in/about.aspx. Accessed 18 Sept 2023.
Yovizag. (2018, February 20). *Samsung announces Smart healthcare program at KGH Visakhapatnam*. https://www.yovizag.com/samsung-smart-healthcare-program-visakhapatnam/

Dr. Shilpa Sarvani Ravi, a distinguished Marketing Professor at Christ University, is a dynamic expert in Marketing Management. Armed with a Ph.D. in Retail Marketing from GITAM University, she seamlessly integrates academic rigor with real-world perspectives. Drawing from 4 years as an In-Store Brand Manager at Colgate-Palmolive Limited, Dr. Shilpa imparts practical insights into theoretical concepts in the classroom. She has a teaching tenure in renowned Bengaluru universities and employs innovative methodologies, emphasizing case studies and industry collaborations. Having published 11 research papers in prestigious journals and receiving the best paper award at Jain University's 2023 international conference, Dr. Shilpa Sarvani has showcased her academic and research-oriented contributions at various national and international conferences.

Dr. Shikha Bhagat is an Assistant Professor in the School of Business and Management at Christ (Deemed to be University), Bengaluru. She holds a gold medal from Guru Nanak Dev University, Amritsar, and boasts over 13 years of experience in teaching and research. Her doctoral degree was conferred by Noida International University Greater Noida, Uttar Pradesh, and her research focus encompasses marketing, international marketing, and global business. She has also published several papers and book chapters in preferred International and Indian Journals. She has actively engaged in academia and participated in various forums addressing topics such as consumer behavior and international business marketing techniques. She has also presented her academic and research-based work at numerous national and international conferences.

Dr. Rashmi Rai is currently the Head of the Department at the esteemed School of Business and Management at Bannerghatta Road Campus. She has a passion for academia and administration and brings over two decades of experience.

She holds a double MBA in Marketing from Pune University and Human Resources from All India Institute of Management (AIMA) and a Ph.D. from Mewar University. Her expertise lies in various areas, such as quality of work life, organizational behavior, leadership, and cultural dimensions in business; she has authored and coauthored several articles and book chapters indexed in Scopus. She has also had the privilege of editing a book on quality of life published by Taylor and Francis. However, her journey doesn't stop there. Alongside her teaching commitments, she has conducted numerous workshops that delve into important topics within the management field and has chaired many national and international conferences.

Part IV
The Synergy of Tomorrow: Smart Cities and Sustainable Futures

Chapter 13
The Way Forward to Smart Cities and Sustainability in Hong Kong: Opportunities and Challenges

Kazi Humayun Kabir and Md. Ayatullah Khan

Abstract The transition of Hong Kong into a digital, smart, and sustainable city is discussed in this chapter, along with the positive and negative aspects associated with it, as Hong Kong is moving toward adopting a paradigm that is people-centric to provide better urban services. A comprehensive review of policies and future guidelines has been executed, encompassing a wide range of sources, including governmental, non-governmental, and commercial entities, as well as researchers and academics. Through the adoption of information and technology (I&T) for many years, the city has evolved into a living laboratory, thereby establishing Hong Kong as a mission-driven, globally recognized, and technologically sophisticated city. To ensure a successful outcome, it is necessary to prioritize the smart city pyramid. When it comes to smart cities in Hong Kong, the development and potential of a wide variety of attributes are still in the preliminary phases at present. The administration of the Hong Kong Special Administrative Region (HKSAR) has demonstrated a strong determination to encourage the development of a sustainable, technologically advanced, and globally integrated urban hub for the betterment of future generations. During the process of tracking the efficiency of smart cities in Hong Kong, the government is required to perform periodic updates and premieres of information technology applications, smart mobility, and smart business, respectively.

Keywords People-centric smart cities · Hong Kong SAR · Sustainability · Big data

---

K. H. Kabir
Department of Urban Planning and Design, The University of Hong Kong, Hong Kong, China

Development Studies Discipline, Khulna University, Khulna, Bangladesh
e-mail: kazikabir@connect.hku.hk; khkabir@ds.ku.ac.bd

M. A. Khan (✉)
Department of Geography, Hong Kong Baptist University, Hong Kong, China

Development Studies Discipline, Khulna University, Khulna, Bangladesh
e-mail: ayatullah@life.hkbu.edu.hk

S. Majumdar et al. (eds.), *Smart Cities*, S.M.A.R.T. Environments,
https://doi.org/10.1007/978-3-031-59846-3_13

13.1 Introduction

Rapid urbanization in recent decades has been remarkable in all sociopolitical, environmental, and spatial facets (Angotti, 2018). The share of people living in urban regions worldwide has increased by about 25% since 1950, from 30% in 1950 to 55% in 2018 (according to the World Population) (Abel-Smith, 2018). The phenomenon of global urbanization masks notable regional disparities in the extent of urbanization. Northern America exhibits the highest degree of urbanization among regions, with over 82% of its population residing in urban areas. In contrast, Asia demonstrates a lower level of urbanization, with around 50% of its population living in cities. Africa, on the other hand, remains predominantly rural, with only 43% of its people residing in urban areas as of 2018. The user provided a numerical reference (UN, 2018). Asian cities, along with others, have experienced substantial urban population pressure, leading to the need for advanced technology to effectively handle these challenges. These programs have the objective of advancing the concept of intelligent cities (Komninos, 2013), which acts as a prelude to the evolution of smart cities (Yigitcanlar, 2015). The notion of a smart city is a relatively young concept that continues to undergo active development. Urban residents are increasingly adapting to living in an environment where a significant number of electronic Internet-of-Things (IoT) sensors are utilized to collect data and subsequently leverage acquired information to efficiently manage resources, services, and assets. The involvement of robots, geospatial technologies, artificial intelligence (AI), the IoT, and other technical breakthroughs in the advancement of smart cities is of significant importance (De Jong et al., 2015; Yigitcanlar & Lee, 2014). Numerous urban areas are currently making efforts to become "smart cities," each of which has its own goals, rules, strategies, and aspirations. For instance, Copenhagen is currently in the process of developing a sustainable urban environment and sharing its technological advancements with other countries (State of Green, 2016). Vienna is actively engaged in promoting sustainable development strategies aimed at achieving efficient growth while minimizing resource consumption (Smart City, 2018). Singapore aims to establish a highly connected urban environment that leverages the Information and Communication Technology (ICT) industry (Smart Nation, 2016). Similar efforts are being undertaken in Barcelona, where a comprehensive urban planning approach is being implemented, incorporating the use of the IoT and environmentally friendly technology. These initiatives aim to tackle a diverse range of 22 significant projects and solution domains (Gascó-Hernandez, 2018).

In 2017, the Hong Kong Special Administrative Region (HKSAR) government implemented the Smart City Blueprint, a global initiative with the objective of improving urban resilience via the convergence of urban design, technological innovation, and public policy (Nam & Pardo, 2011). In 2020, the government of the Hong Kong Special Administrative Region (HKSAR) issued a statement concerning the blueprint for the Hong Kong Smart City 2.0. In order to promote economic growth and improve the general welfare of individuals, a comprehensive strategy has been formulated. This strategy comprises a total of 130 measures that are designed to improve urban governance and optimize public services (Government of Hong Kong, 2020). The ambition of creating a digital replica of Hong Kong was highlighted during the Connected Cities Conference held in November 2022, which was organized by KPMG (2022). The process of producing a digital twin for a city includes the replication of both the physical and data components of the urban environment within a digital realm. For instance, Hong Kong International Airport has implemented digital twin technology for a number of years. The use of this technology on a citywide scale in Hong Kong has the potential to yield cost savings and reduce carbon emissions, thereby facilitating the achievement of the city's sustainability objectives (Sharon, 2019). Furthermore, as of 2022, the region of Hong Kong is home to a cumulative count of 3985 startup firms, with approximately 93 of these entities exhibiting a distinct emphasis on endeavors pertaining to the advancement of smart city projects. Several startups located in Hong Kong are presently involved in the execution of smart city initiatives. One notable firm in the field is Archi Fiction, which specializes in mixed reality technology and seeks to fully use the potential of the digital domain. Carbon Wallet, an additional noteworthy firm, offers individuals help in the adoption of a more sustainable lifestyle. Formwork IO is a startup headquartered in Hong Kong that is dedicated to advancing carbon neutrality in the building sector. Varadise is a digital twin-building company that operates inside the specified region (Pandey, 2023).

According to researchers, there is a proposition that the adoption of the smart city concept holds potential as a feasible approach to address the issues arising from urbanization and the need for sustainability (Herrschel, 2013). On the other hand, several studies posit the possibility of smart cities giving rise to a dystopian society characterized by technocratic rule (McLaren & Agyeman, 2015; Vanolo, 2016). Furthermore, numerous scholars highlight the absence of sustainability linked to this groundbreaking urban paradigm. The term "smart" is commonly used by scholars throughout all sectors of society on a global level to describe efforts, ideas, and advancements. The vast majority of these enterprises exhibit a significant dependence on technology. The implementation of smart city initiatives on a global scale has been remarkable, with prominent instances observed in various countries such as India, China, the United Arab Emirates, South Korea, and Mauritius (Kitchin, 2015). Nevertheless, there is a dearth of scholarly investigations regarding the potential advantages associated with adopting a people-centric approach to smart cities in Hong Kong, as well as the challenges that need to be addressed in order to actualize these benefits. This chapter examines the potential benefits of adopting a people-centric smart city approach in Hong Kong, as well as the challenges that must be addressed in order to ensure its efficacy. To secure the long-term economic stability of Hong Kong and the well-being of its citizens, it is essential to thoroughly analyze the potential benefits associated with implementing a human-driven advanced city model. By prioritizing the needs and interests of its citizens, Hong Kong has the potential to enhance its efficiency, connectivity, and livability. This chapter provides valuable evidence for policymakers, urban planners, researchers, and stakeholders involved in the development of the next phase of human-centric, technologically sophisticated, and sustainable cities in Hong Kong. Moreover, this chapter provides valuable insights for governments and other stakeholders in Hong Kong about the allocation of funding and planning of infrastructure development, taking into account the potential hazards associated with smart cities and sustainability.

13.2 Materials and Methods

The current investigation was undertaken through the implementation of a thorough review of policies obtained from several scholarly databases, including Google Scholar and ScienceDirect repositories. Furthermore, an extensive assessment was conducted on reports provided by various governmental, nongovernmental, and commercial organizations. The present chapter utilized the following criteria in order to meet the selection and inclusion requirements: The primary materials utilized in this study's framework were deliberately selected to encompass scholarly articles featuring empirical information as well as a diverse range of reports sourced from governmental, nongovernmental, and commercial entities. However, the inclusion criteria did not encompass review articles, novels, book chapters, book series, and conference proceedings. The second criterion focused on studies that utilized quantitative, qualitative, or mixed-method research methodologies. The justification for selecting only quantitative, qualitative, or mixed-method studies was grounded in the understanding that these particular study designs are less prone to bias and have a higher likelihood of producing replicable and generalizable findings in future research (Carr, 1994). Due to the strengthened basis offered by quantitative data, the incorporation of qualitative findings has made mixed-method research the most appropriate selection. In addition, the incorporation of both quantitative and qualitative research approaches provides a strong foundation for making informed judgments (Bryman, 2016; Rasinger, 2013). To minimize potential misinterpretation and difficulties in the translation process, only materials originally published in the English language were considered for the calculation. The present study selected a research duration of 14 years (2010–2023) to comprehensively investigate the benefits and drawbacks related to the transformation of Hong Kong into a digitally advanced, intelligent, and environmentally sustainable urban setting. The primary emphasis of the selection process was placed on scholarly articles that mostly revolved around Hong Kong. This was done in order to align with the study's objective, which is to critically analyze the potential advantages and disadvantages associated with Hong Kong's transition toward becoming a digital, smart, and sustainable city. Table 13.1 provides a concise summary of the essential criteria.

This chapter utilizes a three-stage review methodology. The evaluation process spanned from October 2022 to May 2023, commencing with the initial stage of identifying keywords for the search method. The identification of relevant keywords related to smart cities, city sustainability, opportunities, and challenges within the

Table 13.1 Inclusion and exclusion criteria

Criterion	Inclusion	Exclusion
Literature/ research type	Journal (research article including quantitative or qualitative or mixed-method studies) Reports from governmental, nongovernmental, and commercial organizations	Journals (systematic review and future modeling) Book series Book Chapter in book Conference proceeding
Language	English	Non-English
Timeline	Between 2010 and 2023	<2010
Country and territory	Within Hong Kong	Outside of Hong Kong

Table 13.2 Keywords and search terms

Database	Keywords and search terms
Science direct	{(Smart cities OR digital towns OR technologically advanced cities) AND (sustainable city OR viable city) AND (opportunities OR possibilities) AND (challenges OR threats) AND (way forward OR action OR step ahead) AND (Hong Kong)}
Google scholar and Google	First search: Opportunities of smart cities and sustainability in Hong Kong Second search: Challenges for smart cities and sustainability in Hong Kong Third search: The way forward for challenges for smart cities and sustainability in Hong Kong

context of smart cities and sustainability in Hong Kong, as well as potential approaches for promoting smart cities and sustainability in Hong Kong, were established by referring to existing research and a thesaurus (refer to Table 13.2). The procedure for discovering pertinent articles entailed the use of many keywords linked to the research subject. After doing a comprehensive search using different search engines, appropriate keywords were utilized to locate pertinent scholarly papers. Following that, articles that were duplicates were excluded from consideration. In the second part of the study, the selection of pertinent publications was conducted based on precise criteria for inclusion and exclusion, as detailed in Table 13.1. In the end, a meticulous analysis was undertaken of the selected papers in order to carry out a qualitative synthesis. The synthesis findings underwent a process of theme analysis in order to identify repeating patterns, emerging trends, and interrelationships. The results have been methodologically arranged, successfully addressing the research objectives, and offering a comprehensive review of the opportunities and challenges linked to sustainability and smart city initiatives in Hong Kong.

13.3 Results and Discussion

13.3.1 Defining Smart Cities

The principle of a "smart city" is gaining more and more interest from governmental organizations and urban planners. Moreover, a thorough examination of the academic literature concerning the idea of "smart city" indicates that there is a continuing discussion about this idea, which may be divided into three primary areas of implementation: government, academia, and industry. As a result, the discussion about smart cities lacks accuracy due to two main factors. To begin with, the term "smart" is used in several situations globally. Furthermore, it is employed by diverse interest groups (Mosannenzadeh & Vettorato, 2014). The phrase "smart cities" is commonly employed in both public and academic publications, but a comprehensive and precise definition is still lacking (Albino et al., 2015). Within scholarly discourse, the term "smart" encompasses a broad spectrum of technological attributes, such as self-configuration, self-treatment, safeguarding oneself, and self-optimizing, among others. Within the realm of industrial writing, which mostly focuses on business and industrial tools, the term "smart" is commonly employed to characterize products and services that exhibit intelligent behavior, including artificial intelligence and cognitive machinery (Nam & Pardo, 2011). Government publications seek to govern urban expansion by providing a definition of the term "smart" within the context of a theoretical urban planning method known as "smart growth." The concept, originating in the United States during the beginning of the 1990s, was formulated as a solution to the issue of urban sprawl. Smart cities frequently need to resolve several inquiries related to individuals, entities, time, locations, rationales, and procedures that are relevant to effectively handling new urban complications. Various pertinent research inquiries have been investigated concerning smart cities, encompassing (i) the imperative of developing smart cities (the reason?); (ii) the fundamental constituents of smart cities (what actually?); (iii) the principal participants engaged in smart cities (who in their right mind?); (iv) the methodologies for constructing smart cities (the way?); and (v) the most appropriate site and timeframe for the advancement of smart cities (where and while?). The user provided a numerical reference without any accompanying textual details. Currently, scholars are concentrating on defining clear boundaries for cross- and multi-disciplinary methods in order to conceptualize the idea of a smart city. This investigation has determined various crucial elements that are commonly associated with smart cities, including smarter administration, intelligent personnel, advanced economic affairs, sophisticated living/livability, efficient transportation, efficient construction, effective schooling, efficient health services, decent ecology, public security, and cultural identity (Cohen, 2011; Neirotti et al., 2014).

Fig. 13.1 Hierarchy of smart city needs

13.3.2 Priority of Smart City Needs

The priority setting of smart city requirements is expected to be a vital factor for cities worldwide, spanning various situations. The diagram presented in Fig. 13.1 illustrates the hierarchical arrangement in which public security, housing, food, and sanitation considerations form the foundational elements for the progress of smart cities. In Hong Kong, the main emphasis has been on several socioeconomic aspects, including employment prospects, standard of living, urban resilience, and sustainability. Furthermore, the latest pandemic has exacerbated these worries (Hartley, 2023). The government of the HKSAR is facing a deficit in its finances. As a result, numerous corporations and businesses are facing challenges in achieving their financial objectives (Leung & Lee, 2023). Therefore, it is recommended to adhere to the hierarchical structure of smart city requirements in Hong Kong (see Fig. 13.1).

13.3.3 Smart Cities and Sustainable Development

The notion of "sustainable development" can have several connotations, depending on the specific context in which it is employed (Brown et al., 1987). Sustainable cities encompass a variety of characteristics that not only tackle many issues but also acknowledge the intricate nature of urban operations, integrating cultural, social, and political elements. These factors include the establishment of a clean and pollution-free environment, the promotion of sustainable urban transportation, and the availability of accessible and inexpensive urban transit (Bătăgan, 2011). The core emphasis of the connection between sustainability and sustainable urban development is on the vital resources needed for different services and processes inside urban areas, including housing, transportation, and recreation (Lytras & Visvizi,

2018). To effectively accomplish the Sustainable Development Goals (SDGs) set by the United Nations by the specified deadline of 2030 (Sachs, 2012), it is crucial to construct a distinct framework for urban development. Since the mid-twentieth century, global societies have seen substantial impacts resulting from a range of environmental, social, and economic crises (Gascó-Hernandez, 2018). The concept of sustainability requires us to reevaluate our lifestyles in order to enhance the environmental and social factors that contribute to human security, well-being, and health (McMichael et al., 2003). The notion of the smart city can be perceived as a relatively recent development that has come about after the rise of the data-driven city, the digitized city, and cities that are sustainable (Fig. 13.2). Nevertheless, the usage of this term has significantly increased, particularly after 2013, increasing the rate of mentions of other phrases such as "sustainable city." Although the principle of smart cities has been implemented, there is still a lack of consensus over how it should be described (Angelidou, 2015). Smart cities can be categorized into three distinct dimensions: (a) technological aspects, which include both hardware and software components; (b) demographic characteristics, which involve innovation, diversity, and schooling; and (c) administrative attributes, which comprise administration and legislation.

To achieve an improved comprehension of sustainability, it is crucial to incorporate the idea of viable urban growth when evaluating cities (Dizdaroglu & Yigitcanlar, 2016). The preceding phrase refers to a procedure of transition where the utilization of supplies, the distribution of funding, the improvements to technological devices, and the execution of institutional changes are linked with both present and future needs. During the 1990s, there was a substantial surge in public curiosity and endorsement of the notion of viable cities. This notion aimed to offer a thorough comprehension of the interrelationship between economic, social, and environmental sustainability. This was achieved via the utilization of a set of indicators for each of these dimensions (Ahvenniemi et al., 2017). Sustainable urban development encompasses the achievement of a balanced and mutually beneficial relationship between the progress of urban areas and the conservation of the environment. Furthermore, it highlights the significance of guaranteeing equity in relation to earnings, job prospects, accommodation, required amenities, social infrastructure,

Fig. 13.2 Sustainable smart city

and transportation in metropolitan areas (Hiremath et al., 2013). The increasing focus on smart cities and associated concepts can be attributed to a range of factors, including urbanization, climate change, limited availability of natural resources, globalization, and intensified competitiveness. Therefore, it is crucial for metropolitan regions to offer improved and flexible services to meet the demands of individuals (Angelidou, 2015). A smart and sustainable city endeavors to attain its objectives in a manner that is characterized by adaptability, reliability, scalability, accessibility, and resilience. The objectives encompassed in this framework involve improving the overall standard of living for the inhabitants, stimulating economic expansion through the creation of job prospects, enhancing the welfare of individuals by facilitating better access to social and community services, advocating for a sustainable approach to development that takes into account climate and environmental considerations, and guaranteeing the availability of vital services and infrastructure such as public transport (Dhingra & Chattopadhyay, 2016).

13.3.4 Smart City Development in Hong Kong: Strengths, Weaknesses, Opportunities, and Threats (SWOT) Analysis

Similar to numerous global cities, Hong Kong is actively endeavoring to transition into a smart city that prioritizes sustainability and leverages technology to enhance the well-being of its residents. The trajectory of Hong Kong in achieving sustainability and developing smart cities presents a range of advantages and challenges. This section explores the benefits of implementing a people-centric approach, with an emphasis on sustainable development, the well-being of citizens, and the utilization of technology to enhance urban environments. Furthermore, it acknowledges the obstacles that need to be efficiently addressed to accomplish the successful execution of such endeavors. Our discussion aims to clarify the potential benefits and problems of creating people-focused, environmentally friendly, and technologically advanced cities in Hong Kong (Fig. 13.3). We have used specific examples and supporting evidence to support our points.

13.3.4.1 Strengths

Hong Kong exhibits superior performance when compared to other international commercial centers in various aspects, including its infrastructure, affordability of technology for connection and information solutions, and overall business environment. The existing infrastructure in Hong Kong provides a strong foundation for the city's advancement regarding a "smart city" paradigm. Hong Kong ranks ninth globally in terms of ICT capabilities, as indicated by the 2017 ICT Development Index (ITU, 2017). Furthermore, Hong Kong's strong and reliable

Fig. 13.3 SWOT analysis on smart city development in Hong Kong

information and communication technology infrastructure has established a solid foundation for the further advancement of the city as a sophisticated place to live. Based on the information provided in the 2016 HK Fact Sheet (Victorian Government, 2016), Hong Kong employs a collective sum of nine submarine cable networks and terrestrial cable systems, which are interconnected with three telecommunications carriers located on the mainland. In addition, Hong Kong utilizes a total of ten satellites for the purpose of facilitating worldwide communication. Hong Kong is recognized for its notable achievement in the realm of Internet connectivity, as it holds the distinction of having the second highest average peak Internet speed on a global scale. Additionally, Hong Kong's average Internet speed is positioned at the fifth rank internationally. The setting up of the foundation, together with Hong Kong's widespread Internet coverage and high-speed connections, has been crucial in advancing the city's smart city initiatives. Additionally, it is noteworthy to mention that the expense associated with Internet connectivity in Hong Kong is quite inexpensive in comparison to other highly developed economies. The presence of a well-established, fixed broadband infrastructure in the city has been identified as a contributing factor (Office of the Government Chief Information Officer, 2019).

13.3.4.2 Weaknesses

Hong Kong exhibits significant advancements in its technological infrastructure, yet it has not yet surpassed other cities in terms of the software required to provide smart city capabilities. Hong Kong lacks a clear strategy for the establishment of advanced societies and does not have significant regulations to support the use of technology for communication and information (ICT) (Leung & Lee, 2023). Furthermore, it can be argued that the governance of Hong Kong exhibits a deficiency in terms of a reliable and robust high-level structure, hence diminishing the effectiveness of smart city endeavors (Hartley, 2023). Moreover, Hong Kong exhibits deficiencies in terms of data integrity and accessibility, legislative adaptability, mechanisms for managing privately owned enterprises, enough financial support for research and development endeavors, and public readiness (Office of the Government Chief Information Officer, 2019).

13.3.4.3 Opportunities

The metropolitan cluster of Hong Kong serves as a crucial entry point to China, highlighting the significance of effective communication and collaboration with the mainland to advance the implementation of the smart city idea within its jurisdiction. Moreover, it is crucial to recognize that the ICT sector in Hong Kong has the potential to exploit the current situation and promote technological progress in mainland China. By using Hong Kong's unique position as a dual-function platform, which facilitates the attraction of foreign investment and the expansion of global operations for mainland enterprises, the aforementioned objective can be accomplished (Office of the Government Chief Information Officer, 2019). In addition, Hong Kong possesses the capacity to augment the efficacy of its scientific research through the cultivation of more intimate partnerships with the Pearl River Delta region, particularly in Shenzhen. An exemplary case may be found in the joint initiatives between educational institutions in Hong Kong and technological enterprises in Shenzhen with the aim of conducting research and development (Mok & Jiang, 2017; Tang, 2022). Furthermore, the swift rise of Shenzhen as a significant center of innovation in China has enabled new opportunities for interdisciplinary collaboration in the field of technology. Currently, global corporations, such as Huawei, have strategically located their corporate headquarters in Shenzhen, a major city located in the southern region of the People's Republic of China. The metropolitan hub of Shenzhen in the People's Republic of China exhibits a robust and enduring economic performance. Nevertheless, it is crucial to acknowledge that there exists a visible disparity between the urban area's economic expansion and the advancements achieved in the establishment of its tertiary education establishments (Wang et al., 2010). Despite the rigorous efforts made by Shenzhen, it might be argued that Hong Kong has a greater capacity for innovation compared to Shenzhen, particularly in its ability to efficiently incorporate foreign concepts within the mainland. When scholars from Hong Kong engage in partnerships with companies in Shenzhen, they are afforded the chance to participate in translational or applied research and have access to the ample funding resources allocated for research and development inside the city (Chandra et al., 2022; Office of the Government Chief Information Officer, 2019; Tang, 2022).

13.3.4.4 Threats

The government must recognize the potential threats that may arise from societal shifts when engaging in collaboration with the mainland and exchanging information, as these activities might contribute to the establishment of a smart city in Hong Kong and assist in the optimization of associated advantages. The concerns mentioned in the text include several obstacles, such as the increasing age of the population, the widening gap in digital access, a decrease in job prospects, and other similar variables (Ip & Cheng, 2022; Ma et al., 2018; Office of the Government Chief Information Officer, 2019; SCMP, 2019). One potential risk that exists pertains to Internet transactions carried out in Hong Kong. The 2019 publication by the South China Monitoring Post (SCMP) reveals that there exists a deficiency in the level of preparedness among enterprises and government organizations in Hong Kong to adequately mitigate cybersecurity risks. There may be potential resistance to the construction of smart cities due to concerns among individuals about potential infringements on their privacy and security stemming from their use of information and communication technologies (SCMP, 2019). Moreover, it is anticipated that by 2036, the percentage of elderly individuals in Hong Kong's population will reach 26%, predominantly due to the diminishing birth rate and extended life expectancy. Moreover, there has been a decrease in the mean size of households in Hong Kong. The issue presented by a progressively aging population is further intensified by a lackluster rate of population expansion. One of the main challenges faced by the smart city effort involves efficiently tackling the growing demand for elder care and services by using technological advancements (Ma et al., 2018). In a similar vein, the growing integration of technology in the operational processes of smart cities has given rise to analogous apprehensions regarding the possible replacement of jobs requiring minimal skills as a result of widespread automation. The development of smart cities has the capacity to bring about changes or even the whole eradication of job prospects (Ma et al., 2018; Office of the Government Chief Information Officer, 2019).

13.3.5 *Way Forward to People-Centric Smart City in Hong Kong*

In recent years, numerous governments have allocated substantial resources to the advancement and growth of smart cities. The concept of a smart city in Hong Kong is widely understood as a plan focused on enhancing municipal services with a primary emphasis on human welfare. To guarantee that the benefits are visible and easily accessible to both residents and tourists, the development should be designed according to the requirements and desires of the local community. The 2016 Policy Address in Hong Kong proposed that the Innovation and Technology Bureau (ITB) conduct a comprehensive evaluation to determine the practicality of constructing a "smart city." This undertaking would necessitate substantial collaboration with various academic institutions as well as public and commercial companies. Following the policy opinion, ITB plans to establish a digital framework and a set of standards to offer guidance for the advancement of Hong Kong as a smart city. The primary emphasis of this study appears to be on the significance of ICT infrastructure. A significant amount of research has been conducted to examine the effects of many factors on urban growth, such as human resources and schooling, interpersonal and social capital, and environmental concerns (Hiremath et al., 2013). The Hong Kong government demonstrates a proactive approach by offering guidelines in six key sectors: "Smart Mobility," "Smart Living," "Smart Environment," "Smart People," "Smart Government," and "Smart Economy" (as shown in Fig. 13.4) (Government of Hong Kong, 2017).

The Hong Kong Special Administrative Region (HKSAR) government has implemented various initiatives to establish smart cities. According to the ASEAN+3

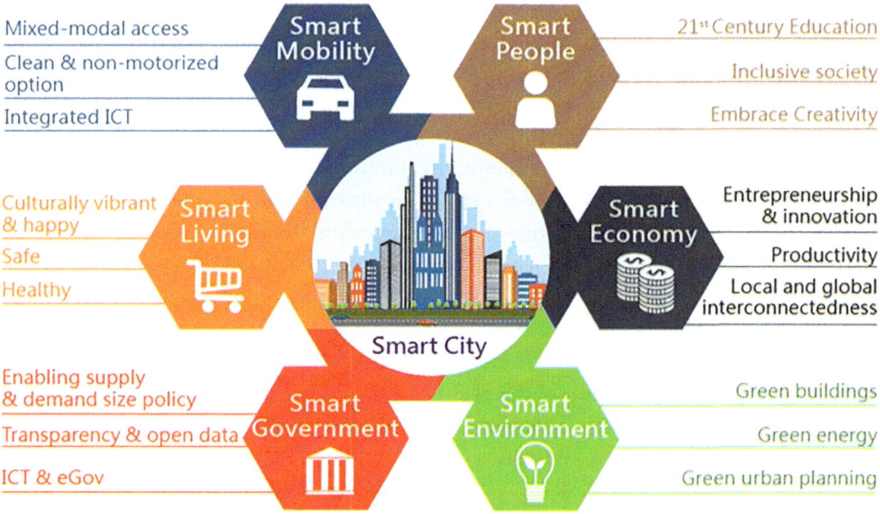

Fig. 13.4 Smart city concept adopted in Hong Kong

Macroeconomic Research Office (AMRO), an organization responsible for monitoring regional macroeconomics, the greater Bay Area of South China is expected to experience higher economic growth in comparison to other regions. Cities have a vital role in promoting economic advancement, thereby necessitating the implementation of advanced urban planning and development activities over extended periods. Nevertheless, the ongoing worldwide pandemic has impeded the advancement of the suggested development initiatives (Hartley, 2023; Leung & Lee, 2023). Mainland China and Singapore are both actively engaged in the promotion of smart city services through collaborative efforts with high-ranking government officials. Shenzhen and Singapore have together produced a comprehensive collection of 29 projects. These programs aim to enhance digital connectivity and market accessibility in both the Southeast Asian region and the Guangdong-Hong Kong-Macao Greater Bay Area. These efforts will utilize advanced technology and AI to efficiently tackle carbon emissions and encourage the widespread use of electric vehicles. Hence, the integration of big data platforms in the cross-border region shows potential for improving the effective handling of financial and logistical operations (Li et al., 2021; Yeh et al., 2021).

The leadership efforts of the HKSAR government have been somewhat ineffective in harnessing the economic prospects presented by smart cities, particularly when compared to neighboring cities like Shenzhen and Singapore in recent years. In the near future, coordinated research among aspiring smart cities may improve the implementation of regional cooperation projects, such as the Common Operational Picture (COP). The COP serves as a specialized digital data platform designed to effectively manage instances of typhoons and flooding (Giaoutzi & Scholten, 2017).

Similarly, Hong Kong, a vibrant city with a substantial population of elderly individuals, presents itself as a suitable choice for the adoption of sophisticated healthcare and assisted living technologies. These technologies are specifically designed to address the distinct issues that come from a progressively diverse and interconnected society. Since 2008, the Hong Kong government has implemented a healthcare system known as SMART, which is under government control. The Legislative Council has granted financing of US$90.5 million for the initial implementation of the fundamental concepts of the smart healthcare initiative in Hong Kong (Pramanik et al., 2017). It is crucial to allocate more financial resources to ensure the security and well-being of the aging population. Furthermore, according to the 2017 Hong Kong Smart City Blueprint, the need to deal with limited resources requires the promotion of sustainable economic growth and innovation in the broader framework of people-centered smart city development in Hong Kong (ITB, 2017). Several notable initiatives could be undertaken to foster sustainable economic growth and innovation in the Hong Kong region, aiming to address the challenge of resource scarcity. The aforementioned strategies encompass providing support to entrepreneurs, enhancing innovation ecosystems, and promoting investments that align with the sustainable development goals of the city. Implementing this endeavor would provide advantages for the advancement of a human-focused, intelligent urban setting in Hong Kong.

13.4 Conclusion and Future Recommendations

This chapter conducted a comprehensive analysis of the potential advantages and disadvantages linked to the adoption of a digital, intelligent, and environmentally friendly urban setting in Hong Kong. Moreover, this chapter analyzed the necessary steps needed to transform the urban area into a citizen-centric, smart municipality. The current investigation involved a thorough analysis of pertinent academic literature regarding policies and upcoming suggestions proposed by various governmental, nongovernmental, and commercial entities, as well as renowned scholars and researchers. The findings indicate that the government is actively engaged in endeavors to transform Hong Kong into a globally acknowledged smart metropolis. The integration of more than 70 initiatives within Hong Kong's smart city policy is a noteworthy achievement. The Hong Kong government places significant emphasis on prioritizing preparedness in tackling urban challenges, implementing an efficient framework for city administration, enhancing efficiency and safety measures, improving quality of life, and promoting sustainability. The Hong Kong administration has been actively involved in the development of a sustainable, technologically advanced, and globally integrated metropolitan center for the benefit of future generations. By regularly updating and deploying information and technology applications, the government may improve its evaluation of efficacy and contribute to the progress of Hong Kong as a smart city. However, despite the government's political ambitions to address these problems, certain sociocultural barriers, such as equity, justice, and public engagement, persist. In order to enhance the intellectual capacity of Hong Kong as an urban center, it is essential to implement a systematic strategy for improving the data infrastructure and ensuring its continuous upkeep. Based on current scholarly research, the government has a wide range of policy choices to select from, as illustrated in Table 13.3. This chapter offers a comprehensive examination of the benefits and drawbacks associated with the adoption of a people-centric smart city in Hong Kong. Moreover, this chapter provides a comprehensive examination of urban expansion, including in-depth scrutiny, empirical case studies, and informed policy suggestions. The primary objective of this program is to advance the growth of cities that are inclusive, sustainable, and prioritize the needs and active involvement of their residents. This resource aims to offer significant insights and information to persons engaged in urban development projects, such as decision-makers, urban planners, and other stakeholders. Implementing a people-centric approach to smart city development in Hong Kong has the potential

Table 13.3 Policy recommendations for Hong Kong

Issues	Policy options
Smart environment	– The management of environmental quality encompasses several aspects such as irrigation, waste management, photovoltaics, lighting systems, weather stations, and water supplies. – Enhancing energy efficiency is a key objective.
Smart mobility	– The topic of discussion pertains to the promotion of electric vehicle services, specifically focusing on the establishment of charging facilities. – Intelligent monitoring systems encompass a range of applications, including those designed for pedestrians, bike lanes, parking spaces, charging stations, capacity control, traffic control, and tourist saturation.
Smart living	– Intelligent fire detection systems, video monitoring, air conditioning, and management of sports facilities are key components in ensuring the efficient and secure operation of such facilities.
Smart people	– The promotion of informational MUPs, citizen cards, citizen apps, and social Wi-fi is being emphasized.
Smart education	– The incorporation of contemporary technology. – The integration of novel educational methodologies, including remote learning, mobile learning (m-learning), individualized learning, flipped and blended learning, social collaborative learning, and game-based learning.
Smart healthcare	– There is a need for increased investment in the field of geriatric healthcare in order to support the implementation of the smart healthcare plan as well as enhance the overall livability and human centeredness of smart cities.
Sustainable economic growth and innovation	– In order to alleviate the issue of resource scarcity in Hong Kong, it is imperative to provide assistance to entrepreneurs, foster innovative ecosystems, and promote investments.

tosignificantly enhance the overall well-being of its residents. By combining technology and utilizing data-driven solutions, the municipality has the ability to enhance public services, optimize transportation networks, and foster the growth of more livable and sustainable communities. An effective method for enhancing the general well-being of people in a certain community is the implementation of intelligent traffic management technologies. These technologies have the capacity to alleviate traffic congestion and reduce journey durations. The inclusion of citizens in decision-making processes within the context of people-centric smart city development enhances the facilitation of citizen empowerment and involvement. Individuals possess the ability to actively participate in the exchange of their opinions, complaints, and suggestions through digital platforms and engagement initiatives. This active participation aids in the advancement of inclusive and responsive municipal policies and services. The effective execution of Hong Kong's "Multi-functional Pilot Smart Lampposts" initiative exemplifies the significant impact of public participation in decision-making processes related to urban planning. In order to effectively address urban difficulties, it is essential to thoroughly analyze the distinct circumstances, capabilities, and opportunities that are specific to each individual region. Over several decades, Hong Kong has transformed into a dynamic experimental environment, using I&T to develop a worldwide-renowned smart city. This transition is characterized by a resilient economy and an elevated quality of life (Lai et al., 2020). The main objective of this program is to encourage active engagement from the general public in order to achieve a variety of desired outcomes. The main goal is to improve the overall well-being, cognitive capacities, and economic prosperity of individuals while simultaneously enhancing the ecological condition of the urban environment. Implementing strategies that promote the expansion of green spaces, cleanliness, livability, sustainability, resilience, and competitiveness will help achieve this. Moreover, the aim is to leverage Hong Kong's established business-friendly environment to promote innovation and build the city as a dynamic hub for experimentation and progress. Finally, the goal is to enhance the quality of care provided to the elderly and adolescents while also fostering a more unified community among the residents. The stated objectives have two main purposes: first, to enhance the digital skills and technological knowledge of organizations, individuals, and the government and second, to promote environmental sustainability in Hong Kong by reducing resource consumption while also ensuring its vibrancy, efficiency, and livability (Government of Hong Kong, 2017).

References

Abel-Smith, B. (2018). *An introduction to health: Policy, planning and financing* (1st ed.). Routledge.

Ahvenniemi, H., Huovila, A., Pinto-Seppä, I., & Airaksinen, M. (2017). What are the differences between sustainable and smart cities? *Cities, 60*, 234–245.

Albino, V., Berardi, U., & Dangelico, R. M. (2015). Smart cities: Definitions, dimensions, performance, and initiatives. *Journal of Urban Technology, 22*(1), 3–21.

Angelidou, M. (2015). Smart cities: A conjuncture of four forces. *Cities, 47*, 95–106.

Angotti, T. (2018). *Metropolis 2000: Planning, poverty and politics* (1st ed.). Routledge.

Bătăgan, L. (2011). Smart cities and sustainability models. *Informatica Economică, 15*(3), 80–87.

Brown, B. J., Hanson, M. E., Liverman, D. M., & Merideth, R. W. (1987). Global sustainability: Toward definition. *Environmental Management, 11*(6), 713–719.

Bryman, A. (2016). *Social research methods* (6th ed.). Oxford University Press.

Carr, L. T. (1994). The strengths and weaknesses of quantitative and qualitative research: What method for nursing? *Journal of Advanced Nursing, 20*(4), 716–721.

Chandra, K., Wang, J., Luo, N., & Wu, X. (2022). Asymmetry in the distribution of benefits of cross-border regional innovation systems: The case of the Hong Kong–Shenzhen innovation system. *Regional Studies*, 1–15.

Cohen, B. (2011). *Smart cities*. Co. Exist, 11.

De Jong, M., Joss, S., Schraven, D., Zhan, C., & Weijnen, M. (2015). Sustainable–smart–resilient–low carbon–eco–knowledge cities; making sense of a multitude of concepts promoting sustainable urbanization. *Journal of Cleaner Production, 109*, 25–38.

Dhingra, M., & Chattopadhyay, S. (2016). Advancing smartness of traditional settlements-case analysis of Indian and Arab old cities. *International Journal of Sustainable Built Environment, 5*(2), 549–563.

Dizdaroglu, D., & Yigitcanlar, T. (2016). Integrating urban ecosystem sustainability assessment into policy-making: Insights from the Gold Coast City. *Journal of Environmental Planning and Management, 59*(11), 1982–2006.

Gascó-Hernandez, M. (2018). Building a smart city: Lessons from Barcelona. *Communications of the ACM, 61*(4), 50–57.

Giaoutzi, M., & Scholten, H. J. (2017). A common operational picture in support of situational awareness for efficient emergency response operations. *Journal of Future Internet, 2*(1), 10–35.

Government of Hong Kong. (2017). *Hong Kong smart city blueprint Hong Kong: Office of the Government Chief Information Officer*. https://www.smartcity.gov.hk/doc/HongKongSmartCityBlueprint(EN).pdf

Government of Hong Kong. (2020). *Hong Kong Smart City blueprint 2.0*. Office of the Government Chief Information Officer. https://www.smartcity.gov.hk/modules/custom/custom_global_js_css/assets/files/HKSmartCityBlueprint(ENG)v2.pdf

Hartley, K. (2023). Public perceptions about smart cities: Governance and quality-of-life in Hong Kong. *Social Indicators Research, 166*(3), 731–753.

Herrschel, T. (2013). Competitiveness and sustainability: Can 'smart city regionalism' square the circle? *Urban Studies, 50*(11), 2332–2348.

Hiremath, R. B., Balachandra, P., Kumar, B., Bansode, S. S., & Murali, J. (2013). Indicator-based urban sustainability—A review. *Energy for Sustainable Development, 17*(6), 555–563.

Ip, K. K., & Cheng, K. C. C. (2022). Smart city development in Hong Kong: An ethical analysis. *China Perspectives, 3*, 21–30.

ITB. (2017). *Hong Kong smart city blueprint*. Innovation and Technology Bureau. https://www.smartcity.gov.hk/

ITU. (2017). *2017 Global ICT development index*. International Telecommunication Union. https://www.itu.int/net4/ITU-D/idi/2017/index.html

Kitchin, R. (2015). Making sense of smart cities: Addressing present shortcomings. *Cambridge Journal of Regions, Economy and Society, 8*(1), 131–136.

Komninos, N. (2013). *Intelligent cities: Innovation, knowledge systems and digital spaces* (1st ed.). Routledge.

KPMG. (2022). *The connected cities conference*. KPMG Advisory (China) Limited. https://kpmg.com/cn/en/home/events/2022/09/the-connected-cities-conference.html

Lai, C. S., Jia, Y., Dong, Z., Wang, D., Tao, Y., Lai, Q. H., et al. (2020). A review of technical standards for smart cities. *Clean Technologies, 2*(3), 290–310.

Leung, K. Y., & Lee, H. Y. (2023). Implementing the smart city: Who has a say? Some insights from Hong Kong. *International Journal of Urban Sciences, 27*(sup1), 124–148.

Li, V. O., Lam, J. C., Han, Y., & Chow, K. (2021). A big data and artificial intelligence framework for smart and personalized air pollution monitoring and health management in Hong Kong. *Environmental Science & Policy, 124*, 441–450.

Lytras, M., & Visvizi, A. (2018). Who uses smart city services and what to make of it: Toward interdisciplinary smart cities research. *Sustainability, 10*(6), 1998.

Ma, R., Lam, P. T., & Leung, C. K. (2018). Potential pitfalls of smart city development: A study on parking mobile applications (apps) in Hong Kong. *Telematics and Informatics, 35*(6), 1580–1592.

McLaren, D., & Agyeman, J. (2015). *Sharing cities: A case for truly smart and sustainable cities*. MIT Press.

McMichael, A. J., Butler, C. D., & Folke, C. (2003). New visions for addressing sustainability. *Science, 302*(5652), 1919–1920.

Mok, K. H., & Jiang, J. (2017). Questing for entrepreneurial university in Hong Kong and Shenzhen: The promotion of industry-university collaboration and entrepreneurship. In *The sustainability of higher education in an era of post-massification* (pp. 115–133). Routledge.

Mosannenzadeh, F., & Vettorato, D. (2014). Defining smart city. A conceptual framework based on keyword analysis. *Tema. Journal of Land Use, Mobility and Environment*.

Nam, T., & Pardo, T. A. (2011). *Conceptualizing smart city with dimensions of technology, people, and institutions*. In Proceedings of the 12th annual international digital government research conference: digital government innovation in challenging times (pp. 282–291).

Neirotti, P., De Marco, A., Cagliano, A. C., Mangano, G., & Scorrano, F. (2014). Current trends in Smart City initiatives: Some stylised facts. *Cities, 38*, 25–36.

Office of the Government Chief Information Officer. (2019). Smart city development in Hong Kong. *IET Smart Cities, 1*(1), 23–27.

Pandey, K. (2023). The path forward for Hong Kong's smart and sustainable development. JUMPSTART. https://www.jumpstartmag.com/the-path-forward-for-hong-kongs-smart-and-sustainable-development/

Pramanik, M. I., Lau, R. Y., Demirkan, H., & Azad, M. A. K. (2017). Smart health: Big data enabled health paradigm within smart cities. *Expert Systems with Applications, 87*, 370–383.

Rasinger, S. M. (2013). *Quantitative research in linguistics: An introduction* (2nd ed.). A & C Black.

Sachs, J. D. (2012). From millennium development goals to sustainable development goals. *The Lancet, 379*(9832), 2206–2211.

SCMP. (2019). *Cybersecurity—A risk that no organisation can afford to overlook*. South China Morning Post. https://www.scmp.com/presented/tech/topics/ibm-confronts-cybersecurity/article/3003189/cybersecurity-risk-no

Sharon, A. (2019). *HKIA develops digital twin*. OPEN GOV Asia https://opengovasia.com/hkia-develops-digital-twin/

Smart City. (2018). *Clean energy for all*. Smart City. https://smartcity.wien.gv.at/site/en/citizens/

Smart Nation. (2016). *Transforming Singapore through technology*. Smart Nation. https://www.smartnation.sg/

State of Green. (2016). *From challenge to opportunity*. State of Green. https://stateofgreen.com/en/

Tang, H. H. H. (2022). The strategic role of world-class universities in regional innovation system: China's Greater Bay Area and Hong Kong's academic profession. *Asian Education and Development Studies, 11*(1), 7–22.

UN. (2018). *World urbanization prospects (WUP): The 2018 revision, methodology*. United Nations, Department of Economic and Social Affairs, Population Division.

Vanolo, A. (2016). Is there anybody out there? The place and role of citizens in tomorrow's smart cities. *Futures, 82*, 26–36.

Victorian Government. (2016). *33–2016 Fact Sheet—Hong Kong (SAR of China)-14*. Victorian Government. https://www.vic.gov.au/sites/default/files/2019-08/2016-Fact-Sheet-Hong-Kong-SAR-of-China.pdf

Wang, C. C., Lin, G. C., & Li, G. (2010). Industrial clustering and technological innovation in China: New evidence from the ICT industry in Shenzhen. *Environment and Planning A, 42*(8), 1987–2010.

Yeh, A. G., Lin, C., & Yang, F. F. (2021). *Mega-city region development in China*. Routledge.

Yigitcanlar, T. (2015). Smart cities: An effective urban development and management model? *Australian Planner, 52*(1), 27–34.

Yigitcanlar, T., & Lee, S. H. (2014). Korean ubiquitous-eco-city: A smart-sustainable urban form or a branding hoax? *Technological Forecasting and Social Change, 89*, 100–114.

Kazi Humayun Kabir is a final-year Ph.D. student at the Department of Urban Planning and Design, University of Hong Kong, Hong Kong, and an Associate Professor at Development Studies Discipline, Khulna University, Bangladesh. He received his M.Sc. Degree in Geodesy and Geoinformatics (Built Environment) from KTH, Royal Institute of Technology, Sweden. He was appointed as Assistant Town Planner in the Detailed Area Planning (DAP) project of Dhaka city after his graduation. Further, he served a Danish land surveying company in Copenhagen as a project manager for the raster Geographic Information System (GIS) department. Since 2009, he has lectured extensively in the field of Urban Planning and Development Studies at Khulna University, Bangladesh, prior to his Ph.D. journey. His research interests include urban air quality, GIS, and remote sensing applications in environmental management, sustainably built environment, spatially integrated social science (SISS), cities and sustainable development, poverty and inequality analysis, smart city, and machine learning.

Md. Ayatullah Khan is a doctoral researcher at the Department of Geography, Hong Kong Baptist University, Hong Kong. He holds both master's and bachelor's degrees in Development Studies from Khulna University, Bangladesh, with an intense focus on research, monitoring, and evaluation. As part of his professional career, he has lectured at United International University, Dhaka, Bangladesh, as a part-time faculty member. In addition, he held the position of researcher at the Institute of Informatics and Development (IID) in Dhaka, Bangladesh. Also, he has had hands-on experience serving as a humanitarian and development activist at CARE Bangladesh. Furthermore, he is engaging in multiple research collaborations with faculty members from Khulna University, Bangladesh. His research interests include, but are not limited to, vulnerability assessment, climate change, climatic events and disasters, sustainable development, city sustainability, and urban development.

Chapter 14
New Urban Technology Towards Energy Harvesting and Environmental Initiatives for a Smart City

Ramesh Chandra Nayak, Arvind B. Bodhe, Jainendra Chauhan, Mahesh Vasantrao Kulkarni, Naushad Hasin Khan, Amey R. Khedikar, and Pratik C. Ghutke

Abstract The concept of smart cities in India was developed by the Government of India in June 2015 by the Honourable Prime Minister Shri Narendra Damodardas Modi. The basic aim of such a concept was to develop the country, to improve the lifestyle of individuals, and to develop the economic part of the country. The smartness of a city is identified by some parameters including technology-based manufacturing method, pollution control initiative, new energy harvesting methods, and initiative towards new job creation for the young masses by using available resources. Each year, the urban department of the Government of India announces smart city ranking by taking the data submitted by state governments. The mission of smart cities will be fulfilled if proper judgement, development, designing, and implementation happen; 45–50% of the total population of a state lives in urban areas in which maximum are common people. The lifestyle of these peoples should be cared for a smart city. In this work, some technology-based methods towards new energy creation and pollution control have been presented. A new way of energy harvesting by developing a model, which works by converting pressure energy to electric energy, and a secondary fuel concept towards zero pollution have been described in this work. Both of these models are environmentally friendly. The design and development costs for these models are very less and also easy to operate. These methods are designed to provide benefit towards the development of smart cities and to improve the lifestyle of common people in urban areas.

Keywords Energy · Path · Model · Technology · Environment · Smart city · Pollution · Urban area

---

R. C. Nayak (✉)
Department of Mechanical Engineering, Synergy Institute of Technology, Bhubaneswar, Odisha, India

A. B. Bodhe
School of Engineering and Technology, G H Raisoni University, Saikheda, Madhya Pradesh, India

S. Majumdar et al. (eds.), *Smart Cities*, S.M.A.R.T. Environments,
https://doi.org/10.1007/978-3-031-59846-3_14

14.1 Introduction

We need energy for transportation, housing, and industry. Energy is available from both renewable and non-renewable sources. Non-renewable energy sources create air pollution, but renewable energy sources are environment-friendly. Generating electricity from non-conventional sources is very beneficial to society. Solar, wind, municipal waste, and other forms of electricity generation are some of the non-conventional sources. India is the fourth country in the world in the renewable energy sector. Recently, our Prime Minister announced the development of renewable energy plants to generate electricity and suggested further research on the concept of free energy to help businesses move into the renewable energy sector. A new way of energy development method is adopted by various countries. One of the new ways to generate electricity for a smart city is presented in this chapter and a model is explained, which works to produce electricity from pressure energy; the sources of pressure energy are from the movement of human beings and vehicles. This model is presented with two different attachments, such as simple gear train attachment and epicyclic gear train attachment. Another designed model to control air pollution developed from traditional fire is presented. This model works with both primary and secondary fuels. A number of works have been conducted by various researchers towards smart city with various concepts. There are various ways to generate electricity: generating it from pressure energy is one of them. Pressure energy is developed by the movement of human beings or the movement of vehicles on a track. Each footstep and pressure on the ground by any means can generate energy. Places like railway stations, bus stops, medical, temples, and national highways have maximum movements on the ground. The pressure exerted by such movements can be converted to electricity. Ahmed et al. (2022) targeted to support promoting a mathematical representation on behalf of composite probabilistic energy elimination constraint with renewable energy units. The effort converges towards a unique matrix, supported via an algorithm called astute black window optimization (ABWO). Aspects like pollution elimination along with sustainable

J. Chauhan
Department of Hindi, Ph.D. Research Scholar, Gauhati University, Guwahati, Assam, India

M. V. Kulkarni
Department of Mechanical Engineering, Dr. Vishwanath Karad MIT World Peace University, Pune, India

N. H. Khan
Department of Mechanical Engineering, Maulana Azad College of Engineering & Technology, Patna, Bihar, India

A. R. Khedikar
Department of Civil Engineering, Tulsiramji Gaikwad Patil College of Engineering and Technology, Nagpur, Maharashtra, India

P. C. Ghutke
Department of Electrical Engineering, Tulsiramji Gaikwad Patil College of Engineering and Technology, Nagpur, Maharashtra, India

progression have been supported via renewable energy–producing sources. On passing through the ABWO approach, five observations adhere to renewable energy resources like solar, wind, etc. introduced in order to yield optimum outcomes against pollution elimination and cost. Bhutto et al. (2019) estimated the biomass assets present in Pakistan as renewable potentials which are then considered as a productive implementation of biomass on behalf of preparing food, excitation work, and a meagre decentralized electric power generation. The ultimate execution is to project modern tools in safe status, and prominent implementation of biomass potency prepares a significant plan to eliminate the universal environment and health hazards. It also assists in supporting the objectives confirmed by the Sustainable Development Goal (SDG) on behalf of ascertaining global admittance towards consistent, economic, as well as contemporary power services by 2030. Bielecki et al. (2020) confidently put forward a selective observation of the electric power generation outer influences. Environmental impact, human health status, etc. are some of the influential factors considered in this effort. The outcomes here have been appended in a chronological pattern. The diversified path leads to generating electric power; the significance of the purity aspect along with its utilization has been specifically experienced through this effort. Brook et al. (2015) to some extent have vision over the non-conventional energy sources those fundamentally diverse opinions on sustainable energy portfolio. The very intention assesses the comparative price advantages of dissimilar power holdings which is obligatory. A comparative study has been carried out over soil utilization, discharges, weather, and price inferences forecasted but divergent storylines for future energy production, and the outcome of the same is that none of which has optimal for all ecological and cost-effective standard point of frame. The author has also observed that since ecological progress has traditionally discarded atomic power alternatives, contemporary standard reactor expertise which completely reprocesses squander, including inactive security units, may determine apprehensions which ought to be further extensively implicit. Chu and Majumdar (2012) proceeded in expanding their experience into a broader perspective correlating them in the midst of an amount of transportation as well as electricity production units. Their work is of the existing power setting and discusses several observations, creativity fortunes and platform converge an affluent, prosperous and protected power prospect linked to planet. Cooper and Sehlke (2012) have successfully manipulated the latent limits from a universal aspect by reconsidering the weather wedge proposal of Pacala and Socolow (Science2004, 305 (5686), 968–972) and also estimated the potential water-use impacts of the wedges connected with energy production. In a nutshell, the author specifically points out that greenhouse gas (GHG) improvement options which regulate power alteration or exercise effectiveness can all together reduce GHG emissions, lower energy costs, and diminish energy impacts on water possessions. Donelan et al. (2008) found that substantial electricity generation with minute additional attempts formulate this process which is very much convincing for prosthetic limb power impulsion as well as diverse conveying clinical tools. Eeckhout and Hedtrich (2021) put forward a fact upon green urbanization which constrains assessments like the usage of electric power on a per capita basis keeping waste diminishing corresponding to regional mass. The observation has also confirmed that energy requisition corresponding to city output value lies near 83% which further adds total output increase via 1%, and consequently, urban energy premium appeared at 17%. The significance of energy supported via transportation, households, and industry means becomes 20, 13, and 16%, respectively. In this effort, one spatial equilibrium model which yields dependent income taxes based on city mass, and in responding to it, a greater mass of people move towards bigger cities. In a nutshell, there is not only enhancement in the monetary outcomes but also reduction in the energy usage supported under pro-urbanization counterfactual. England and Robert (2022) specifically studied the ubiquity of electrical forces within the natural environment and monitored its rise, along with a comprehension of the large number of environmental interactions that these forces may control. The work supplies the initial full spread and combination of study in this rising domain of electric ecology. The very work includes reviews of the role electricity plays in the usual ecosystem of predator-prey communications, pollination, and animal scattering, surrounded by many others, as well as the impact of anthropogenic movement on these systems. A thorough beginning to the ecology and physiology of electroreception—the accepted detection of ecologically relevant electric fields—has also been supported. Erickson and Jennings (2017) demonstrated the solar power–propelled foundation on behalf of electric transportation unit synthesis of renewable power sources to produce electric power, shaded parking zone, and other relevant issues. This effort has also pointed out that the budget concern for wind energy, solar panels, and batteries is diminishing because of modern tool introductions. The power as well as transportation segments exhibit a significant influence on health aspects. Measures like power, atmospheric standard along with its influence, transportation, and health status may prosper through solar-powered charging foundation supplements. Glazener and Khreis (2019) successfully opined upon the investments like pedestrian paths, cycling track, public transport foundations, and services. The recent evolving advance measures, e.g. electric powered autonomous transportation units not only eliminate air pollution but also put insignificant physical influence. In this work, integrated policy packages have been drawn upon in order to showcase the suitable practices of clean air, dynamic transportation, etc. Ultimately, the researcher insists that extended research must be carried out on middle and budget-class countries considering measures out of the practical region implementations, pointing air impurity reversing towards source involvement and health standards concern to transportation. Ho et al. (2014) review the accessibility and biopower potentials of the existing biomass feedstock. The author exclusively includes various nutritive ingredients classified as the foremost production feedstock, and the typical plant extracts fall in the secondary outcomes. The ecological of foremost production feedstock secured larger control upon the secondary outcomes, which the alteration expertise counters, chief limitation to complete industrial use. Invernizzi et al. (2016) established the primary substantial as well as its chemical process path attaining power production, considering electro-magnetism, piezo-electricity, and electrostatic production. The attention converges in between the characteristics of substances and the proficiency of the device. Some fresh unfamiliar procedures like tribo-electric nanogeneration (TENG) along with reverse electro-wetting (REWOD) have been meticulously addressed. Jossinet (2008) looked into the electrodynamics issue and found the underlying facts. The author confirmed the relative significance among electric characteristics via a passing means; again, their substantial importance has been justified. This work also suggested the factor accountable in the matter of living tissues concerned with the characteristics of electrical and dielectricals. This endeavour also put light upon cells that restrict the ionic surge, in particular at low frequency, keeping membranes steady on behalf of dielectric relaxation. It also emphasizes submissive reaction to feeble installed communication concern to cell film, which brings up bioelectrical bandanna explanation. In a nutshell, the work also exhibits electrical impedance spectroscopy conveniently measured on behalf of current uses. Kamal et al. (2022) observed and targeted to yield the production cost model to associate with renewable power sources for reducing the gross functional cost and diminishing greenhouse gas exhausts. In this work, particle swarm optimization (PSO) is implemented. The lower cost and exhaust scenarios are the outcome portfolios. In this observation, to get balance for yielding scarcity during the event of contamination reduction. The very observation is being made upon a micro-grid which includes diesel generator (DG), photovoltaic cell, battery, wind turbine, as well as our country's remote location load profile. The researcher here confirmed that the optimal arrangement has an absolute functional net worth of $59,195.61. The energy per unit worth appears at 0.20 $/kWh by means of an economy of emission yield of 5994 kg/year. Kluts et al. (2019) highlighted the effort which provides an outline of the present use (for bioenergy) and prospect latent of farming feedstock in support of biorefinery intentions. The work also shows that the major submission of this feedstock is due to the production of biofuels intended for street conveyance. There is an 80% biofuel produced out of biodiesel, chiefly from rapeseed oil, and the outstanding division is bioethanol out of various ingredients like beet, wheat, etc. Li et al. (2009) fabricated a wearable knee-based power-harnessing unit which yields electric power at the instant of a human walk. The effort here converges towards physiological doctrines which pave the design methods as well as project an absolute detailing of device design along with fresh observation emphasis. Consequently, the device installed exerts a single direction grab to convey knee expansion movement exclusively, another spur gear communication to intensify the angular pace, a brushless DC turning a magnetic generator to change the mechanical power into electrical power, an organized system to decide when to unlock and shut the energy production course depending on the capacity of knee angle, and a modified orthopaedic knee brace to allocate the device response torque over a bulky leg surface region. Lombardi and Carnevale (2018) focused on the ambient influence via life cycle assessment. They primarily held that the electric exchange effectiveness rises with the rise of plant area magnitude, collaborating industrial enhancements keeping economy as the scale of standard. Apart from this, the work has also estimated impact assessment consequences considering 2008 procedures. It comes out from the natural means holds affirmative during climate change indicator, which is positive when the waste to energy system functions exclusively in the energy-establishing pattern, where registered amount steps down for rising surface area. Lombardi et al. (2015) investigated and composed ideas relating with energy harnessing effectiveness estimated against diverse thermal happening along with various throwaways. It is observed that incineration is a leading thermal execution aligned with the power withdrawing technique around a steam cycle. In line with the formation of gas out of throwaway, the created syngas is usually heated in a heat exchanger, namely a steam producer for power improvement. The work is observed and executed on behalf of exclusive electric energy generation, where attainable standards powerfully rely on the resources. The superior industrial resolution implemented and continued upon budgetary frame where electric energy effectiveness will attain a magnitude of around 32–33%. Markandya and Wilkinson (2007) technically observed the diversified fuelling techniques like nuclear, fossil fuel, and most renewable energy paths, e.g. solar, wind, etc. for the generation of electricity and its impact on physical life sustenance. McKendry (2002) studied the importance of biomass renewable energy sources; they presented that the proper selection of biomass will provide maximum energy output and will save from global warming. In this work, the researchers have mentioned their three previous works related to the renewable energy sector. Finally, the third effort withholds specific gasification technologies along with its latent for biomass gasification. Nayak et al. (Nayak & Roul, 2022; Nayak et al., 2017a, b, 2018, 2021a, b, 2022) developed various innovative products for environment sustainability and natural convection heat transfer methods. The importance of solar energy and its proper utilization methods have been presented by Parlevliet and Moheimani (2014). They focused on the conversion of solar energy to chemical energy and electrical energy. In their work, they developed some new methods for maximum utilization of solar energy. Energy generated from solar by using their designed model provides better illumination as compared with other methods. Princiotta and Loughlin (2014) envisioned that the essential track will be required for radically transformed electric and mobile phone segments. The significant focus given to geo-engineering on behalf of admit increases the time period for serious emission reductions. The considerations are at the conceptual stage with a lot of unreturned charge, ecological, and opinionated issues. This endeavour results the issue that eliminate the latent for catastrophic climate change will be a typical confront, seeking the global society to alter its power system in a forceful, synchronized, and sensible manner. If this dispute is to be met, fresh technologies will have to be fabricated and employed at a faster pace. The work also includes advances in carbon arrest and storage, renewable, nuclear, and transportation technologies which are specifically significant. Radousky and Liang (2012) successfully arranged the facts concerning nanoscience adhering to energy-harnessing concepts as well as equipment. This specific approach converging towards developing new devices with enhanced efficiency shifts technological limitations commencing at cm/mm graduation to micro-nano level of electro-mechanical systems. In this effort, transportation issue, crystallography, thermodynamic verification, shifting of energy, and overall system function as well as its productivity have been carefully monitored. Ramasamy et al. (2021) observed the salinity gradients through reverse electrodialysis technology on behalf of producing electric power via diverse means. As a matter of fact, the observation has been made considering capacitive electrodes which systematically adheres to past fabrication concern to capacitive combination as well as reverse electrodialysis. In a nutshell, the observation revolves around diverse reverse electrodialyses selected out of contemporary specific methods. Rome et al. (2005) experimentally verified a rucksack of hanged forces which convert physical power. The same is carried through the upright transaction of supporting force (measuring 20–38 kg) to electrical energy for the period of usual walking (produce up to 7.4 W, else 300-fold raise over previous shoe devices (20 mW)). The author also estimated and commented that inherent metabolic reserve (against elongated sturdy rucksack support) is essential for the period of electrical energy production. Slovic et al. (2016) successfully opined a general idea upon confined air impurity restriction strategies along with the events eliminating impurity standard within megacities via absolute proof estimating their productivity. This effort also insists that the regional air contamination guidelines suit all near and far environments simultaneously. Outcomes streamline various guideline paths implemented across the regions along with the recommendation of enhanced thorough cross-city observations keeping prospective to project optimized executions in both regional and universal concerns. Tan et al. (2023) presented the water and its adjoining environment. Authors have pointed out their vision over water's rudimentary aspects and chief parameters concerns with water-hygroscopic matter linked along with energy evolvement procedure. Thereafter, the contemporary prevailing matters and prosperity in matters of evolvement have been carefully synthesized by the researchers. The dynamic procedures which are effective towards system functionality advancement work out of the entire major influential having constraint operational duration have arranged specifically. Ultimately, the leftout issues along with its upcoming perspectives of this vibrant technique have also been recommended. Tester et al. (2007) retain its intended of the board which come into view in entire form in a 2006 MIT statement; part 1 and part 2 energy types are presented as the future of geothermal energy. The endeavour constitutes a complete nationwide appraisal of US geothermal assets, assessment of drilling and reservoir expertise, and finally cost-effective modelling. This noble effort was used to assess the geologic heat energy surge on behalf of a span of geothermal potentials. It is supported via a digital demonstration of the EGS (Enhanced Geothermal System) reserve, USA. Yang et al. (2018) optimistically confirmed about the functionally coupled units, their relation with the environment potency, and consequently their expanse upon the same. These units are concerned with the issues of budget as well as facial energy feeding micro-electronics on behalf of reducing the battery control over it. Zavorka and Paar (2022) pointed out the significance of energy harnessing as well as recharging units mixing up with air standard estimation, and consequently converging towards street lighting LEDs. The absolute information gathered has been forwarded to a common server unit for observation. The implementation of contemporary tools not only saves energy degradation but also eliminates carbon dioxide emissions for most of the units. The significance of the above-mentioned tools can be introduced in both warfare, i.e. military segment and civil segment. Looking at the people priority in terms of safety measures within a region smoother logistics, noise proof, etc. has successfully projected through this effort. Zhang et al. (2019) venture the intra-corporeal energy yielding work and successfully design, equip, and harness some energy to actuate a pacemaker of potency around 8–10 μW. The very endeavour brings out a fresh paradigm of sustainable powering of implantable gadgets onto the limelight. Zhao et al. (2023) successfully considered the doctrine, methods, and implementation of piezoelectric footwear energy harvester (PFEH) technologies. At the commencement of the effort, popular piezoelectric matters were implemented, and their attributes for PEEHs have been abridged. Next to it, the force interface through the earth and active power supply upon the track as well as accelerations have been scrutinized. Ultimately, it is outlined to support the baseline, limitations, potency for PFEH modelling. Along with the same, the power surge out of the individual's walking via the PFEHs as well as the paths to enhance the energy conversion proficiency has also been projected. Zheng et al. (2021) studied and came up with China's thermoelectric generating unit's assessment over the rate of capacity utilization and over-capacity indication via stochastic frontier scrutiny. Keeping the updated cost details of fuel and electricity, the intensity of market situation is calculated by the researcher with the following aspects like among all regions, among bordering regions, and between the regions enclosed by the identical grid group. Wang et al. (2022) proposed a conceptual model concerning electric power generation out of influential parameters like droplets, natural evaporation, surge, and ambient moisture on scrutinizing primary communication at distinct liquid-matter interfaces. Key aspects in raising the output power of hydrovoltaic devices are then discussed in terms of material synthesis, structural design, and device optimization. We also provide an outlook on the potential applications of this technology ranging from sensors, power suppliers to multifunctional systems, as well as the scientific and technological challenges in transforming its potential into practical utility. The prospects of this emerging field are considered for future endeavour.

14.2 Experimental Set-Up and Observations

The work consists of two designed and tested models such as energy harvesting model and another to control air pollution. Both these models are to help common peoples in urban areas towards improvement of their lifestyle for a smart city. Principles of both models are technology based, and these models have been used in various fields and their test results have been presented. The energy-harvesting model is presented with two attachments:

(i) Simple gear train–based energy-harvesting model.
(ii) Epicyclic gear train–based energy-harvesting model.

I. *Simple Gear Train–Based Energy-Harvesting Model.*
   The purpose of this type of designed model is to generate electricity from pressure energy developed by walking on a track. Fig. 14.1 shows the simple gear

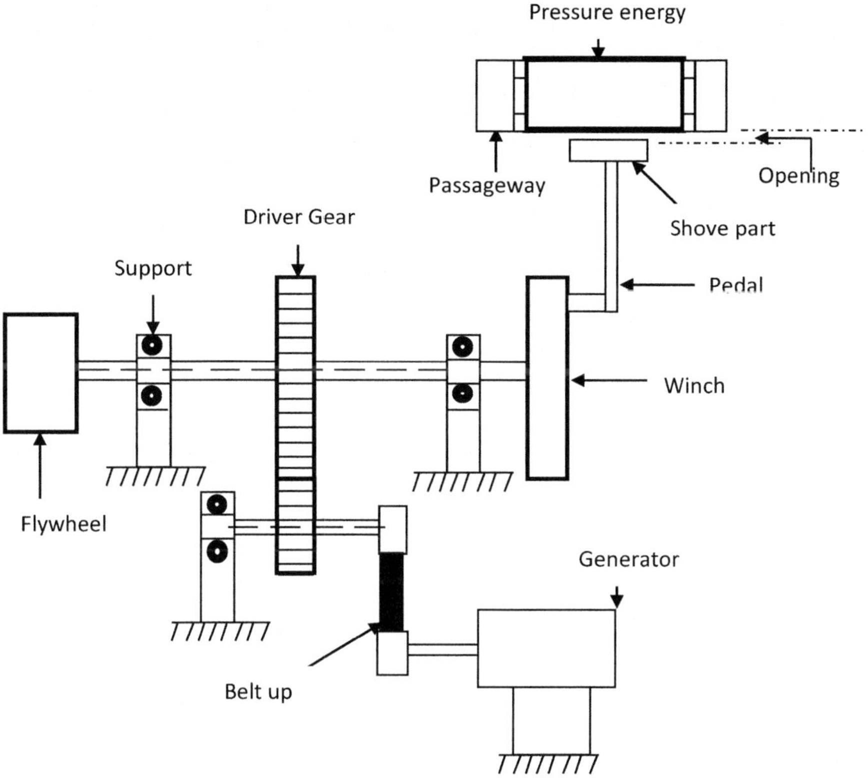

Fig. 14.1 Simple gear train–based energy harvesting model

train arrangement experimental set-up, which consists of a track, big and small gears, bearings, flywheel, belt, and motor. The dimensions of all parts are given in tabular form below. The system has been attached to a track where maximum movements are made by human beings and vehicles.

The track is attached with the walkway in various positions; when load is applied on the track, the track touches the ram of the lever, which causes to rotate the pulley; the pulley is connected with the end of the shaft. The bigger gear is placed at the middle of the shaft and the flywheel at another end of the shaft. Due to rotation of the pulley, the bigger gear rotates; a small gear is placed at another shaft, and both the gears are meshed with each other, due to rotation of the bigger gear. The smaller gear also rotates, and the rotational energy from the shaft of smaller gear shifts to the generator by belt. Due to the rotation of armature of the generator, electromotive force (emf) is developed.

Figure 14.2 shows the arrangement of the transmission system and storage system. It consists of a generator with specifications of 12 volt and 2 amp with a battery of 12 volt 7.2 amp. The distribution line is designed to supply the generated electrical energy.

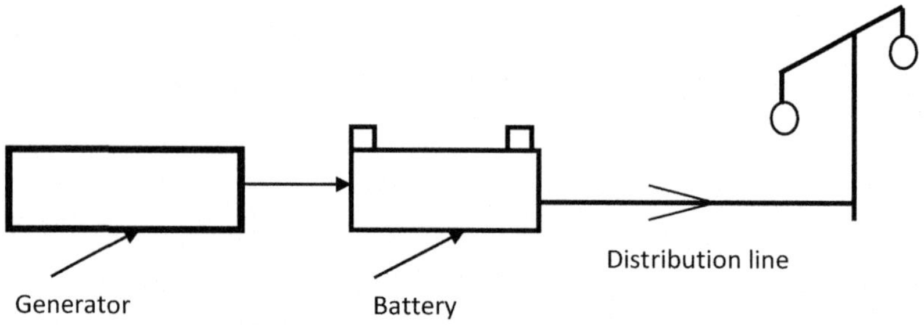

Fig. 14.2 Storage and transmission arrangement

Table 14.1 Dimensions and quantity details of parts of the developed model

Sl. no.	Parts details	Dimensions	Quantity
1	Fly wheel	20 kg	1
2	Bigger gear	75 teeth	1
3	Smaller gear	25 teeth	1
4	Belt	1387 mm	1 (flat belt)
5	Bearings	Roller bearing	6 numbers
6	Generator	12 volt, 2 amp	1
7	Track	1829 × 305 × 153 mm	1
8	Pulley	150 mm diameter	1
9	Pulley	70 mm diameter	1
10	Battery	12 volt, 2 amp	1

Dimensions of the parts in the attachment are given in Table 14.1.

The track is made of plastic; in this arrangement, acrylonitrile butadiene styrene (ABS) plastic is used for the track because of its high rigidity properties. The track dimension is taken as 6 ft. length, 1 ft. width, and 6 inch height. The track is placed under the concrete frame. The track and the complete set-up are placed with a 10 mm gap. The track is attached to the concrete frame with some spring attachment to provide only a small displacement with the application of loads on it and returns to the original position after the removal of loads from it. Figure 14.3 shows the arrangement of the position of track in the step.

From Table 14.2, it is found that one unit of the designed model is suitable to provide 60 Watt power, which helps to light a 5-watt LED bulb for 8 h.

II. *Epicyclic Gear Train–Based Energy Harvesting Model.*
   Figure 14.4 shows the designed model with an epicyclic gear train attached to generate electricity. The designed model consists of an epicyclic gear train unit, track, shove part, pedal, winch, support, and generator. The epicyclic gear train unit consists of three planet gears, one sun gear, and one ring gear. The ring gear is fixed; planet gears pass over the sun gear, and sun gear transmits power to the generator shaft.

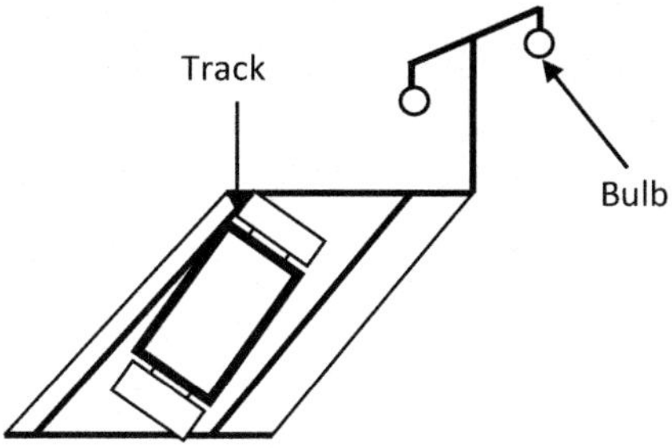

Fig. 14.3 Position of the track in the walkway

Table 14.2 Technical outputs available from the designed model

Sl. no.	Parameters	Details
1	Cost of one unit	Rs.8000/–
2	Total hits on track per day	10,000 times
3	Energy per step	3 joule
4	Total energy on the track per day	6000 joule
5	Average load in kg per step	30 kg
6	Generated voltage per one step	2.6 volt
7	Average current	1.2 m amp
8	Steps required to charge 12 volt battery	9600 steps
9	Time required for full charging of 12-volt battery	8 h

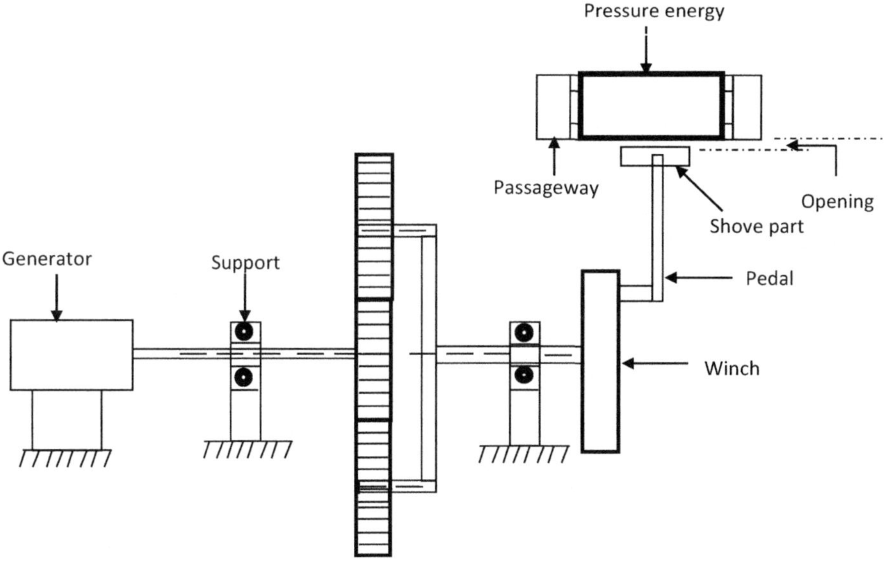

Fig. 14.4 Epicyclic gear train–based energy harvesting model

The track is fixed with the passage way, the shove part is placed at a certain gap, the shove part is attached to the pedal, and the pedal is connected with the winch eccentrically. The winch is connected with planet carrier through a shaft. When load is applied on the track, it causes to move the track slowly down, due to contact of the track with the shove during downward motion; the pedal causes to rotate the winch, as the pedal is attached with the winch eccentrically. Due to the rotation of winch, the shaft attached to it rotates; the other end of the shaft is connected with the planet carrier, and the rotation of planet carrier over the sun gear causes to transfer of rotational motion. The rotation available at the sun gear shaft causes the generator shaft to rotate, and finally, emf is developed from the generator. Table 14.3 shows the dimensions for the arrangement and details requirement.

From Table 14.4, it is found that one unit of the designed model is suitable to provide a 100-watt power, which helps to light a 10-watt LED bulb for 8 h. From the data, it is found that epicyclic arrangement provides the most advantages, because the total system provides more rotation to the generator. Due to more speed, the generator shaft is available, so for a complete charge of the battery, it requires less time and the generator provides maximum output.

Table 14.3 Dimensions and quantity details of parts for the developed model

Sl. No.	Parts details	Dimensions	Quantity
1	Sun gear	75 teeth	1
2	Planet gear	25 teeth	1
3	Ring gear	135 teeth	1
4	Bearings	Roller bearing	4 numbers
5	Winch diameter	15 inch	1
6	Generator	12 volt, 2 amp	1
7	Track	1829 × 305 × 153mm	1
8	Opening	150 mm diameter	1
9	Battery	12 volt, 2 amp	1

Table 14.4 Technical outputs available from the designed model

Sl. No.	Parameters	Details
1	Cost of one unit	Rs.9000/–
2	Total hits on track per day	10,000 times
3	Energy per step	3 joule
4	Total energy on the track per day	6000 joule
5	Average load in kg per step	30 kg
6	Generated voltage per one step	2.6 volt
7	Average current	1.2mAmp
8	Steps required to charge 12 volt battery	9600 steps
9	Time required for full charging of 12-volt battery	6 h

III. *Community Kitchen Smokeless Chulha.*
Smart cities intend to employ expertise and facts to look up the standard of life on behalf of their residents while reducing environmental influences. Nevertheless, they are not protected from pollution challenges. Some major types of pollution that can still affect smart cities include air, noise, water, light, waste, electronic waste, soil, traffic congestion, heat island effect, plastic pollution, etc. Smart cities are dynamically running on continuing with sustainable practices and technologies to tackle these pollution challenges. They may use data analytics, IoT sensors, and advanced monitoring systems to administer and diminish these issues more effectively. However, achieving pollution-free environments in urban areas becomes an enduring and multifaceted assignment.

Concerning the air contamination aspect, apart from vehicle exhaust, food preparation with some unconventional patterns like mud stoves propelled with wood/charcoal etc. is one of the significant contributors. The primary reason lying behind this specific event is because a greater food demand is sought by the budget-class people seldom touring the city. In this work, a community kitchen smokeless chulha is presented. It works with both primary and secondary fuels. The primary fuel is wood pieces and secondary fuel is superheated steam.

Figure 14.5 shows the arrangement of the experimental set-up. It consists of water tube, super heater, Insulator, support for pan, secondary fuel tank, primary fuel in unit, support, and air in unit. The dimension details for different parts of the design model are given in Table 14.5.

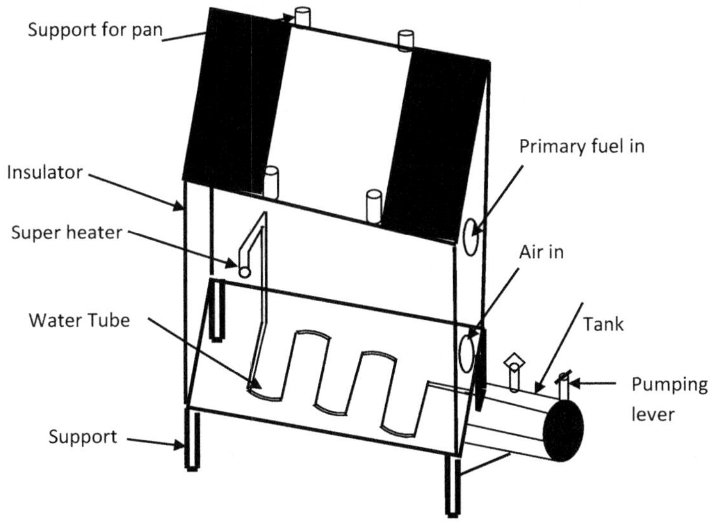

Fig. 14.5 Community kitchen of a smokeless chulha

Table 14.5 Dimension details of the set-up

Sl. no.	Parameters	Details
1	Height (h) of set-up	30 inch
2	Width (w) of set-up	24 inch
3	Depth (d) of set-up	12 inch
4	Fuel tank	4 inch × 12 inch
5	Water tube	1 inch diameter
6	Super heater dimension	3 inch height × 2 inch diameter
7	Insulator type	Glass wool
8	Air in passage diameter	2 inch
9	Primary fuel (wood) entrance diameter	3 inch
10	Primary fuel dimension	4 inch length × 2inch diameter

The designed model is suitable for community kitchen purposes, where cooking for 50–60 persons at a time is possible. The working procedure for this set-up is given below:

1. Fill water in the tank.
2. Connect a blower.
3. Close the valve of the tank.
4. Allow five to seven pumping to the water tank by pumping lever.
5. Add 200 g of wood to the chulha.
6. Add 20 ml kerosene oil to the wood.
7. Now fire the wood.
8. Allow 2–3 min of initial burning of the wood.
9. Switch on the blower for some time.
10. Due to the initial burning, water present in the tube is converted to superheated steam.
11. After some time, pump the lever.
12. The fine atomized spray of superheated steam takes place and is added with the initial burning.
13. The addition of the superheated steam with primary fuel causes to improve the flame temperature.

14.3 Observations

In a smart city, food preparation stoves can be integrated with advanced technology and smart features to enhance efficiency, safety, and sustainability in cooking processes. Stoves in a smart city can be designed with sustainability in mind, using eco-friendly materials, energy sources, and procedures. They may also have features like exhaust air purification to reduce air pollution. They can help residents/others save time and resources while reducing the city's environmental footprint.

Looking at the pattern of air pollution experienced within an advanced city, apart from the vehicle exhausting gases, it is found that the exhaust from the remotely located budget restaurant's cooking stove is one of the primary contributors. Considering the very effect on a pilot basis, one biomass-operated stove has been designed, fabricated, and observed in its functional execution. Its functional execution has been carried out under the installation/guidance of various set-ups for monitoring the pollutants in a nearby notable institution, namely IMMT, Bhubaneswar. The exhaust gas details are being successfully monitored and noted underneath for usual information and understanding. Since the result experienced is a satisfactory concern to environmental sustainable footprints, it is promoted to carry out the food preparing work in one remotely located budget restaurant. After successfully fabricating the smokeless community chulha, it has been tested with respect to various factors to confirm its suitability towards zero pollution.

Results in terms of flame properties, fuel requirements, efficiency, and calorific value are listed in Table 14.6.

Results in terms of temperature, carbon monoxide, total particulate matter, and dilution ratio are listed in Table 14.7.

The designed chulha is also compared with the traditional chulha, with respect to various factors. Table 14.8. shows the detailed comparison with respect to various factors.

From the observations, it is found that the developed chulha provides maximum flame temperature as compared with traditional chulha. Due to this maximum temperature of flame with less input primary fuel, the designed chulha is suitable for community kitchen purposes. From the environmental parameter view, it is found that the designed chulha is environment-friendly as complete burning of primary fuel is taking place. Due to this complete burning, the calorific value of the fuel increases, and as a result, there is less ash available in the chulha.

Table 14.6 Parameters related to performance

Sl. no.	Velocity of flame	Primary fuel requirement	Calorific value of primary fuel	Efficiency (%)
Observation 1	5.03 m/s	6.1 kg/h	19,430 kj/kg	38
Observation 2	5.06 m/s	6.3 kg/h	19,520 kj/kg	42
Observation 3	5.09 m/s	6.6 kg/h	19,820 kj/kg	41
Observation 4	6.21 m/s	7.2 kg/h	19,970 kj/kg	43

Table. 14.7 Parameters related to environmental factors

Sl. no.	Flame temperature	Temperature of external body	Carbon monoxide	Total particulate matter	Dilution ratio
Observation 1	800 °C	50 °C	340 ppm	135.1 mg/Mj	58.12
Observation 2	790 °C	55 °C	360 ppm	126.7 mg/Mj	54.23
Observation 3	780 °C	65 °C	395 ppm	129.8 mg/Mj	51.22
Observation 4	770 °C	60 °C	400 ppm	130.3 mg/Mj	51.34

Table. 14.8 Comparison of the designed chulha with traditional chulha

Sl. no.	Parameters	Traditional chulha	Designed community kitchen smokeless chulha
1	Maximum flame temperature	300–350 °C	700–800 °C
2	Carbon monoxide (ppm)	600 ppm	340–400 ppm
3	Calorific value	16,760 kj/kg	19,430–19,970 kj/kg
4	Operation procedure	Complicated	Easier
5	Thermal efficiency	20–25%	38–43%

14.4 Conclusion

In this work, three different models, such as harvesting energy by simple gear train arrangement, energy harvesting by epicyclic gear train arrangement, and smokeless community kitchen chulha have been presented. Both models have been tested, and their results have been represented. From the observation, it is found that:

The designed pathway model/mechanism with a simple gear train arrangement has been physically executed. The magnitudes align with the model that more than 10,000 steps during daytime are on the designed track, and 9600 steps are required to charge a 12-volt battery. It is found that one unit of the designed model is suitable to provide a 60-watt power, which helps to light a 5-watt LED bulb for 8 h.

The designed pathway model with epicyclic gear train arrangement provides a 100-watt power, which helps to light a 10-watt LED bulb for 8 h. The epicyclic gear train arrangement provides more rotation of the generator shaft every step on the track.

Though the figure drawn out through this execution is not significant, the absolute importance of clean, green sustainable harnessed becomes the crucial aspect. The practical novelty lying behind assures a definite optimistic spark inside every conscious mind.

The young mass will definitely feed/boost them on exposure to this kind of trivial execution. The model will definitely indicate the inherent potency of each manual endeavours. In a nutshell, the vision of technocrats in the domain of technicalities can be wide open, leading to the target of a sustainable earth.

The result of a smokeless community kitchen provides an advantage on environmental point of view. The exercise of long-established food preparation procedures, like open fires, as well as solid-fuel stoves, poses considerable challenges to both human health and the environment. In response to these issues, eco-friendly biofuel stoves have emerged as a sustainable solution.

Eco-friendly biofuel stoves put forward a sustainable substitute to long-established cooking techniques, addressing ecological, well-being, and economic challenges. Their implementations not only diminish greenhouse gas emissions but also get better indoor air value and facilitate local economies. Since initiatives persist to encourage their use, biofuel stoves have the potential to transform the cooking habits of communities globally, contributing to an additional sustainable and healthy opportunity.

The designed smokeless chulha produces maximum flame temperature as compared with the traditional chulha. Due to the maximum flame temperature, the designed smokeless chulha is suitable for community kitchen.

The procedure to operate the chulha is very easy, so unskilled persons can easily operate it.

References

Ahmed, I., Rehan, M., Basit, A., & Hong, K. S. (2022, July 20). Greenhouse gases emission reduction for electric power generation sector by efficient dispatching of thermal plants integrated with renewable systems. *Scientific Reports, 12*(1), 12380. https://doi.org/10.1038/s41598-022-15983-0

Bhutto, A. W., Bazmi, A. A., Karim, S., Abro, R., Mazari, S. A., & Nizamuddin, S. (2019, October). Promoting sustainability of use of biomass as energy resource: Pakistan's perspective. *Environmental Science and Pollution Research International, 26*(29), 29606–29619. https://doi.org/10.1007/s11356-019-06179-7

Bielecki, A., Ernst, S., Skrodzka, W., & Wojnicki, I. (2020, April). The externalities of energy production in the context of development of clean energy generation. *Environmental Science and Pollution Research International, 27*(11), 11506–11530. https://doi.org/10.1007/s11356-020-07625-7

Brook, B. W., & Bradshaw, C. J. (2015, June). Key role for nuclear energy in global biodiversity conservation. *Conservation Biology, 29*(3), 702–712. https://doi.org/10.1111/cobi.12433

Chu, S., & Majumdar, A. (2012, August 16). Opportunities and challenges for a sustainable energy future. *Nature, 488*(7411), 294–303. https://doi.org/10.1038/nature11475

Cooper, D. C., & Sehlke, G. (2012, March 20). Sustainability and energy development: Influences of greenhouse gas emission reduction options on water use in energy production. *Environmental Science & Technology, 46*(6):3509–3518. https://doi.org/10.1021/es201901p. Epub 2012 Mar 1.

Donelan, J. M., Li, Q., Naing, V., Hoffer, J. A., Weber, D. J., & Kuo, A. D. (2008, February 8). Biomechanical energy harvesting: Generating electricity during walking with minimal user effort. *Science, 319*(5864), 807–810.

Eeckhout, J., & Hedtrich, C. (2021, November 29). Green urbanization. *PLoS One, 16*(11), e0260393. https://doi.org/10.1371/journal.pone.0260393

England, S. J., & Robert, D. (2022, February). The ecology of electricity and electroreception. *Biological Reviews of the Cambridge Philosophical Society, 97*(1), 383–413. https://doi.org/10.1111/brv.12804

Erickson, L. E., & Jennings, M. (2017, February 16). Energy, transportation, air quality, climate change, health nexus: Sustainable energy is good for our health. *AIMS Public Health, 4*(1), 47–61. https://doi.org/10.3934/publichealth.2017.1.47

Glazener, A., & Khreis, H. (2019, March). Transforming our cities: Best practices towards clean air and active transportation. *Current Environmental Health Reports, 6*(1), 22–37. https://doi.org/10.1007/s40572-019-0228-1

Ho, D. P., Ngo, H. H., & Guo, W. (2014, October). A mini review on renewable sources for biofuel. *Bioresource Technology, 169*, 742–749. https://doi.org/10.1016/j.biortech.2014.07.022

Invernizzi, F., Dulio, S., Patrini, M., Guizzetti, G., & Mustarelli, P. (2016, October 10). Energy harvesting from human motion: Materials and techniques. *Chemical Society Reviews, 45*(20), 5455–5473. https://doi.org/10.1039/c5cs00812c

Jossinet, J. (2008). Elementary electrodynamics. *Technology and Health Care, 16*(6), 465–474. PMID: 19212042.

Kamal, M. M., Mohammad, A., Ashraf, I., & Fernandez, E. (2022, December). Rural electrification using renewable energy resources and its environmental impact assessment. *Environmental Science and Pollution Research International, 29*(57), 86562–86579. https://doi.org/10.1007/s11356-022-22001-3

Kluts, I. N., Brinkman, M. L. J., de Jong, S. A., & Junginger, H. M. (2019). Biomass resources: Agriculture. *Advances in Biochemical Engineering/Biotechnology, 166*, 13–26. https://doi.org/10.1007/10_2016_66

Li, Q., Naing, V., & Donelan, J. M. (2009 Jun). Development of a biomechanical energy harvester. *Journal of Neuroengineering and Rehabilitation, 23*(6), 22. https://doi.org/10.1186/1743-0003-6-22

Lombardi, L., & Carnevale, E. A. (2018, March). Evaluation of the environmental sustainability of different waste-to-energy plant configurations. *Waste Management, 73*, 232–246. https://doi.org/10.1016/j.wasman.2017.07.006

Lombardi, L., Carnevale, E., & Corti, A. (2015, March). A review of technologies and performances of thermal treatment systems for energy recovery from waste. *Waste Management, 37*, 26–44. https://doi.org/10.1016/j.wasman.2014.11.010

Markandya, A., & Wilkinson, P. (2007, September 15). Electricity generation and health. *Lancet, 370*(9591), 979–90. https://doi.org/10.1016/s0140-6736(07)61253-7. PMID: 17876910.

McKendry, P. (2002, May). Energy production from biomass (part 1): Overview of biomass. *Bioresource Technology, 83*(1), 37–46. https://doi.org/10.1016/s0960-8524(01)00118-3

Nayak, R. C., & Roul, M. K. (2022). Technology to develop a smokeless stove for sustainable future of rural women and also to develop a green environment. *Journal of The Institution of Engineers (India): Series A, 103*(1), 97–104.

Nayak, R. C., Roul, M. K., & Sarangi, S. K. (2017a). Experimental investigation of natural convection heat transfer in heated vertical tubes. *International Journal of Applied Engineering Research, 12*, 2538–2550.

Nayak, R. C., Roul, M. K., & Sarangi, S. K. (2017b). Experimental investigation of natural convection heat transfer in heated vertical tubes with discrete rings. *Experimental Techniques, 41*(6), 585–603.

Nayak, R. C., Roul, M. K., & Sarangi, S. K. (2018). Natural convection heat transfer in heated vertical tubes with internal rings. *Archives of Thermodynamics, 39*(4), 85–111.

Nayak, R. C., Roul, M. K., Sarangi, S. K., & Sarangi, A., Sarangi, A. (2021a). Enhancement of natural convection inside vertical tubes using internal obstacles as rings with different arrangements. In *Current advances in mechanical engineering* (Lecture Notes in Mechanical Engineering) (Vol. 52, pp. 191–201). Springer.

Nayak, R. C., Roul, M. K., Sarangi, S. K., Sarangi, A., & Sarangi, A. (2021b). Forced draft and superheated steam for design and development of community smoke less chulha to help women in rural areas. In *Advanced manufacturing systems and innovative product design* (Lecture Notes in Mechanical Engineering) (pp. 93–102). Springer.

Nayak, R. C., Roul, M. K., & Roul, P. D. (2022). Design and development of smokeless stove for a sustainable growth. *Archives of Thermodynamics, 43*(1), 109–125.

Parlevliet, D., & Moheimani, N. R. (2014, June 11). Efficient conversion of solar energy to biomass and electricity. *Aquatic Biosystems, 10*, 4. https://doi.org/10.1186/2046-9063-10-4

Princiotta, F. T., & Loughlin, D. H. (2014, September). Global climate change: The quantifiable sustainability challenge. *Journal of the Air & Waste Management Association (1995), 64*(9), 979–994. https://doi.org/10.1080/10962247.2014.923351

Radousky, H. B., & Liang, H. (2012, December 21). Energy harvesting: An integrated view of materials, devices and applications. *Nanotechnology, 23*(50), 502001. https://doi.org/10.1088/0957-4484/23/50/502001

Ramasamy, G., Rajkumar, P. K., & Narayanan, M. (2021, December). Generation of energy from salinity gradients using capacitive reverse electro dialysis: A review. *Environmental*

Science and Pollution Research International, 28(45), 63672–63681. https://doi.org/10.1007/s11356-020-12188-8. Epub 2021 Jan 5.

Rome, L. C., Flynn, L., Goldman, E. M., & Yoo, T. D. (2005, September 9). Generating electricity while walking with loads. *Science, 309*(5741), 1725–1728. https://doi.org/10.1126/science.1111063

Slovic, A. D., de Oliveira, M. A., Biehl, J., & Ribeiro, H. (2016 Feb). How can urban policies improve air quality and help mitigate global climate change: A systematic mapping review. *Journal of Urban Health, 93*(1), 73–95. https://doi.org/10.1007/s11524-015-0007-8

Tan, J., Wang, X., Chu, W., Fang, S., Zheng, C., Xue, M., Wang, X., Hu, T., & Guo, W. (2023 Jan). Harvesting energy from atmospheric water: Grand challenges in continuous electricity generation. *Advanced Materials, 27*, e2211165. https://doi.org/10.1002/adma.202211165

Tester, J. W., Anderson, B. J., Batchelor, A. S., Blackwell, D. D., DiPippo, R., Drake, E. M., Garnish, J., Livesay, B., Moore, M. C., Nichols, K., Petty, S., Toksoz, M. N., Veatch, R. W., Baria, R., Augustine, C., Murphy, E., Negraru, P., & Richards, M. (2007 Apr 15). Impact of enhanced geothermal systems on US energy supply in the twenty-first century. *Philosphical Transactions of Series A Mathematical Physical and Engnierring Sciences, 365*(1853), 1057–1094. https://doi.org/10.1098/rsta.2006.1964

Wang, X., Lin, F., Wang, X., Fang, S., Tan, J., Chu, W., Rong, R., Yin, J., Zhang, Z., Liu, Y., & Guo, W. (2022, June 20). Hydrovoltaic technology: From mechanism to applications. *Chemical Society Reviews, 51*(12), 4902–4927. https://doi.org/10.1039/d1cs00778e

Yang, X., Cheng, M., Zhang, L., Zhang, S., Liu, X., & Shi, F. (2018, July 20). Electricity generation through light-responsive diving-surfacing locomotion of a functionally cooperating smart device. *Advanced Materials*, e1803125. https://doi.org/10.1002/adma.201803125. Epub ahead of print.

Zavorka, R., & Paar, M. (2022, October 15). A focus on electromobility within smart city solutions-charging stations, renewable energy, and air quality monitoring. *Sensors (Basel), 22*(20), 7841. https://doi.org/10.3390/s22207841

Zhang, L., Cheng, M., Luo, H., Zhang, H., Ju, G., Liu, P., Zhou, Y., & Shi, F. (2019, April). Mini-generator based on reciprocating vertical motions driven by intracorporeal energy. *Advanced Healthcare Materials, 8*(8), e1900060. https://doi.org/10.1002/adhm.201900060

Zhao, B., Qian, F., Hatfield, A., Zuo, L., & Xu, T. B. (2023, June 23). A review of piezoelectric footwear energy harvesters: Principles, methods, and applications. *Sensors (Basel), 23*(13), 5841. https://doi.org/10.3390/s23135841

Zheng, X., Wu, C., & He, S. (2021, February 1). Impacts of market segmentation on the over-capacity of the thermal electricity generation industry in China. *Journal of Environmental Management, 279*, 111761. https://doi.org/10.1016/j.jenvman.2020.111761. Epub 2020 Dec 15.

Dr. Ramesh Chandra Nayak is presently working as the Professor and Head of mechanical Engineering Department at Synergy Institute of Technology, Bhubaneswar, Odisha, India. He has 16 years of teaching, research, and administrative experience. He has done his B. Tech. and M. Tech. in Mechanical Engineering from BPUT, Odisha, and Ph.D. in Mechanical Engineering with a specialization in Thermal Engineering from SOA University, Bhubaneswar, Odisha, India.

Dr. Arvind B. Bodhe is working as a Dean in the School of Engineering and Technology at G H Raisoni University, Saikheda. He graduated in Mechanical Engineering at RTMNU, Nagpur University, Nagpur, and Maharashtra, India. He secured a Master of Technology in Production Technology from Yeshwantrao Chauhan College of Engineering, Nagpur, Maharashtra, India. He secured Ph.D. in Mechanical Engineering from RTMNU, Nagpur University, Nagpur, Maharashtra, India. He is in the field of Production Technology and Industrial Engineering. He has been in teaching profession for more than 20 years. He has keen interest in environmental sciences and its applications. He filed 17 patents.

Jainendra Chauhan has an M.A., B.Ed, M.Phil., Assamese Language Education, Diploma in Communicative Assamese, NET, SLET, TET, Ph.D. research scholar. He was a contractual Assistant Professor for about 2 years. He received the Indian Excellence Award and many national-level honours.

Dr. Mahesh Vasantrao Kulkarni holds a master's degree in Mechanical (Design) Engineering and a doctorate degree in Mechanical Engineering from Savitribai Phule Pune University. He is currently employed as an Assistant Professor in the School of Mechanical Engineering at Dr. Vishwanath Karad MIT World Peace University in Pune, India. He has a total of 25 years of experience, including 16 years of teaching and 9 years of industrial experience. He has over 33 peer-reviewed international journal papers and conference proceedings to his credit.

Naushad Hasin Khan serves as an Assistant Professor in the Mechanical Engineering Department at Maulana Azad College of Engineering & Technology. He earned his doctoral degree from the National Institute of Technology, Jamshedpur, focusing on specialized areas including complex fluid flow and heat transport phenomena. His primary research involves exploring the rheology and hydrothermal behaviour of complex fluids and nanoparticle.

Dr. Amey R. Khedikar is presently working in the capacity of Registrar at Tulsiramji Gaikwad-Patil College of Engineering and Technology, Wardha Road, Mohgaon, Nagpur, (an autonomous institute, affiliated to RTM Nagpur University, Nagpur Accredited by NAAC A+ Grade and accredited by NBA [ME&EE]), Maharashtra (India). He is the former Head of Civil Engineering Department at the same institute. He completed M. Tech in Structural Engineering, from RTM Nagpur University, Nagpur and has more than 12 years of experience in teaching, research, and administration.

Dr. Pratik C. Ghutke. He is working as an Assistant Professor in the Department of Electrical Engineering Tulsiramji Gaikwaad, Patil College of Engineering and Technology Nagpur Maharashtra, India, since 2015. He received a Bachelor of Technology degree in Electrical Engineering from Dr. Babasaheb Ambedkar Technological University Lonere District, Raigad, Maharashtra, India, in 2011. He received a Master of Technology degree in Control System Engineering from Walchand College of Engineering, Sangli, Maharashtra, India, in 2015.

Chapter 15
Shaping Intuitive E-Governance: Perspectives from a Smart City in a Developing Nation

Debjeet Ganguly, Amit Kundu, and Chandan Kumar Bhattacharyya

Abstract This chapter investigates the quality variables that affect the perceptions of the users of the e-governance information system (e-Gov IS) implemented by the Newtown Kolkata Development Authority, the urban local body of Newtown Kolkata, a Smart City under the Government of India's Smart Cities Mission. Utilizing econometrics tools, the study explores the framework for developing an intuitive e-governance system, taking into consideration the peculiarities of a context that is as varied as India's. A comprehensive e-government system success model is developed and validated, ultimately refined to five pivotal factors, namely, system usability and satisfaction, operational efficiency of E-service, system service support, information integrity and relevance, and perceived economic returns of e-governance. All factors, except system service support, exhibit a significant influence on intuitive e-governance adoption among the users. The chapter concludes by underscoring the importance of comprehending these determinants as India advances its e-governance initiatives. While core aspects like system usability, operational efficiency, economic returns, and information quality are crucial, the undervaluation of system service support prompts a call for further research.

Keywords E-governance · Information systems · Smart cities · Digital governance · Quality of information systems

D. Ganguly (✉)
Research Scholar, Sister Nivedita University, Kolkata, West Bengal, India

A. Kundu
Department of Management, North Eastern Hill University, Shillong, Meghalaya, India
e-mail: amitkundu@nehu.ac.in

C. K. Bhattacharyya
LMS, Techno India Group, Kolkata, West Bengal, India

S. Majumdar et al. (eds.), *Smart Cities*, S.M.A.R.T. Environments,
https://doi.org/10.1007/978-3-031-59846-3_15

15.1 Introduction

The advent of information technology has sparked a worldwide transformation, and India is no exception. Smart government services are one area where information technology is leveraged, whereby it is seen to be a useful tool for improving a public entity's ability to communicate with and provide services to groups of people as well as business organizations (Ameen et al., 2019). Cities are changing rapidly in the Global South and smart government services, which make use of information technology, and are becoming more vital and widespread. And thus, one of the important goals confronting scholars and policymakers is to figure out how to successfully create, utilize, and implement information systems in urban settings, with unpredictable appearances and a kaleidoscope of stakeholders, and how to construct integration between them (Ganguly, 2022). In a nation like India, where digital divisions may deepen pre-existing socio-economic gaps, this is an especially challenging undertaking. Moreover, many parameters applicable in the developed world might not hold true for the contexts set in the Southern Hemisphere. Consequently, it is imperative to formulate a comprehensive strategy that ensures the optimal utilization of e-resources by all sections, within a robust and inclusive quality framework.

15.2 Smart Urban Governance: The Problem of Digital Divide

Smart governance is an integral element of the contemporary smart city development process because a smart governance paradigm is deployed by the administration of a smart city to implement the necessary standards required to achieve its governance goals. Human smart cities need novel governance mechanisms in which public bodies "hear and talk" with citizens, form policies, and support services to render local administration more accessible, participatory, efficient, and reflective of the residents' preferences (Lopes, 2017). Smart urban governance, which includes the incorporation of digital technologies and data-driven methods, is a revolutionary concept for advancing urban development and augmenting the quality of life in cities. However, the widespread issue of the digital divide poses an important hurdle to the effective implementation of smart initiatives, especially in developing nations (Deursen & Helsper, 2015) and perpetuates disparities in access to information, services, and opportunities in urban environments (Castells, 2009). The lack of access to digital infrastructure and technological resources disproportionately affects marginalized communities, which are frequently households with low incomes, remote regions, and underprivileged neighbourhoods (Graham & Aurigi, 1997). The impact of the digital divide on smart city governance is multidimensional. The exclusion of some parts of the population from technological advances might impede the equal distribution of benefits provided by smart initiatives (Bibri,

2018). Therefore, efforts are needed to design intelligent and intuitive initiatives with inclusiveness in consideration, taking into account the diverse requirements and capacities of various communities (Taewoo & Pardo, 2011).

15.3 Intuitive E-Governance

E-governance represents a significant advancement in the field of public administration, offering the potential for a smooth and efficient digital transformation of governmental services. The notion "intuitive" emphasizes the commitment to not just digitize but also simplify e-governance processes (Heeks, 2003). Whereas the expansion of e-governance presents a paradigm shift in the delivery of public services, the true advancement lies in the integration of "intuitiveness" within this framework (Ministry of Housing and Urban Affairs, 2015). The concept of intuitive e-governance encompasses more than just converting traditional procedures into digital formats. It involves enhancing these processes to become notably more user-friendly, beneficial, and responsive to the wants and expectations of those who use them. The significance of intuitive e-governance in India's smart cities is crucial due to several factors:

Enhancing digital accessibility: One notable characteristic of intuitive e-governance is its prioritization of user-friendliness. Simplified user interfaces, accompanied by clear instructions and simple navigation, serve to accommodate a broader demography, therefore mitigating the perceived complexity of the digital experience for those who are new to it. This aspect has great importance in addressing the prominent digital disparity in India, as it facilitates seamless access to e-services for both urban technology enthusiasts and rural individuals who are less experienced in this domain.

Achieving optimal benefits for citizens: The design of intuitive e-governance prioritizes user centricity. Streamlining digital interactions enables individuals to make their jobs easier and to fully capitalize on the advantages offered by smart city developments (Ahvenniemi et al., 2017). This encompasses expedited service delivery, clear procedural frameworks, real-time feedback tools, and enhanced direct contact channels with municipal authorities.

Enhancing trust and participation: The use of an easy platform consistently improves trust levels. When individuals are able to easily access and beneficially use digital government services, concerns related to bureaucratic difficulties diminish. The inherent simplicity of this approach fosters a broader acceptance and engagement.

Enhancing digital literacy: The use of user-friendly e-governance systems may serve as effective mechanisms for promoting digital literacy. The comparatively beneficial and user-friendly characteristics of digital technologies have the potential to motivate those with little digital experience to explore the digital realm more extensively, hence promoting widespread digital empowerment at a national level.

Optimizing resources: The optimization of resources is a crucial consideration for government entities. One effective approach is the use of user-friendly platforms, which may significantly minimize the need for substantial user training and continuous maintenance. Not only does this practice contribute to resource conservation, but it also guarantees a more streamlined and effective provision of public services (United Nations, 2020).

Reducing digital intimidation: The persistent obstacle in the rollout of electronic governance is the perceived intricacy of digital platforms, which leads to digital intimidation. The use of intuitive e-governance system leads to a significant reduction in intimidation. Individuals are more motivated and inclined to adopt and place confidence in a platform that exhibits user-friendly navigation, hence facilitating a seamless and efficient shift to digital services.

The deployment of intuitive e-governance in India's smart cities extends beyond the simple process of digitalization. This entails the integration of advanced technology and user-centric design to establish an ecosystem where digital governance is not only intelligent but also accessible, advantageous, and inclusive for all individuals involved.

15.4 E-Gov IS Success and Role of Motivators

A sustainable, inclusive digital society and successful digital economy need not just have the technology components of the digital society at its core but also a wide range of "analogue components" (World Bank, 2016). French et al. identified four main groups of people who are not using the Internet. These groups include those who think the Internet is not for them, those who do not have the right support to access the Internet, those who find it too complicated, and those who cannot afford to go online (French et al., 2019). Different people in these groups require different kinds of support to become Internet users, each with their own specific needs and barriers. It hence becomes critical for a researcher examining the acceptability of e-governance information systems (e-Gov IS) in digitally divided countries such as India to investigate the impact of motivators in this context.

While much effort has been put to understand the role of socio-economic factors as predictors of information system (IS) use, yet motivational determinants maintain their importance in this question (Scheerder et al., 2017). This line of research treats motivation as an independent variable that can affect Internet usage and attitudes. This has been reinforced by various other studies too. Wang et al., in their study on the effect of motivation on the digital divide and whether it has any moderating effect, demonstrated that motivators are the salient predictors of Internet usage. They postulated that the digital divide should not be merely considered as a fallout of socio-economic disparities but that the different sections of society have different motivational factors towards using the Internet (Wang & Liu, 2022). Motivators are also important in shaping people's attitudes and behaviours towards adopting and utilizing e-governance systems (Venkatesh et al., 2003). Under such a context, understanding and addressing the motivators that might overcome these inequalities become critical in a nation like India, which is characterized by enormous disparities in digital literacy, financial condition levels, as well as access to technology.

15.4.1 ROI as a Motivator for E-Governance IS Success

Since in developing nations like India, obtaining equitable access to online information systems is hampered by the digital divide, the potential for bridging this divide thus hinges on the ability to encourage individuals, often without technical expertise or means, to adopt and utilizes these systems. According to research, perceived time and cost advantages impact citizens' adoption of e-governance efforts (Dwivedi et al., 2015; Rana et al., 2019). This leads us to believe that the concept of return on investment (ROI) can be a powerful motivator that has the potential to be used to address this issue in digitally divided nations like India, and the importance of return on investment (ROI) and cost savings as incentives for the adoption of e-governance information systems cannot be overstated.

15.5 Additional Factors that Contribute Towards a Successful Information System

15.5.1 System Quality

An information system's system quality is critical for maintaining its dependability, performance, and overall effectiveness. System quality is described as the system's appropriateness, dependability, and stability in terms of use, functionality, adaptability, understandability, and reliability (Petter & McLean, 2009; Issac et al., 2019). It has been considered as one of the mainstays in the search for factors responsible for the success of an IS (Delone & McLean, 1992). Weill and Ross (2004) define system quality as the extent to which an information system fulfils stated criteria and standards. It indicates the degree to which the system satisfies certain requirements concerning its functionality, usability, efficiency, reliability, maintainability, security, etc. A high standard of system quality ensures that the information system serves the organization's procedures effectively, improves user satisfaction, and minimizes risks and disruptions.

15.5.2 Quality of Service

The term "service quality" refers to the comprehensiveness of a service provider's assistance (Delone & McLean, 2003). According to Parasuraman et al. (1988), it includes the dimensions of tangibility, dependability, responsiveness, assurance, and empathy, among others (Parasuraman et al., 1988; Delone & McLean, 2003). The promptness and readiness of the system to give help and resolve user inquiries or issues is referred to as responsiveness. The consistency and correctness of the system's performance, as well as the dependability of its services, are all aspects of reliability. Assurance denotes the system's support staff's expertise, knowledge, and trustworthiness. Empathy refers to a system's capacity to comprehend and handle the requirements, problems, and preferences of its users. Tangibles are the physical or palpable features of the system and its services. Service quality is critical for an information system's performance since it directly affects user satisfaction, trust, and loyalty. According to research by Parasuraman et al. (1988), service quality is a crucial factor of user happiness and loyalty in information systems. User engagement and system adoption are also influenced by quality of service. A system that delivers good support services and successfully handles user demands encourages users to actively interact with the system, discover its features, and maximize its possibilities.

15.5.3 Quality of Information

It refers to the properties of information that influence its value and utility for decision-making and other organizational tasks, such as accuracy, completeness, reliability, relevance, and timeliness (Petter & McLean, 2009; Seddon, 1997). Users depend on high-quality information to make educated choices, facilitate efficient operations, and plan strategically. According to Wang and Strong (1996), one of the major elements of information quality is accuracy, which is critical to the success of an information system. It denotes the accuracy and precision with which the system provides information. They also stress the importance of completeness, arguing that incomplete or missing information may lead to biased judgements and unproductive actions. Gorla et al. (2010) in their work emphasize the importance of reliability in information quality, highlighting the need for accurate information for making good decisions and organizational performance, since it reflects the consistency and dependability of the system's information. Redman (1998) highlighted the fact that incorrect information may lead to bad judgements and destroy faith in the system. The degree to which information is topical and suitable to the unique context and requirements of the users is referred to as relevance. According to Eppler and Wittig (2000), relevance is important because information systems should deliver customized information that suits users' individual demands and goals. Timeliness is another important aspect of information quality since it indicates how quickly information is made accessible to consumers. Wang and Strong stress the importance of timeliness in information quality, claiming that timely information is critical for successful decision-making and organizational responsiveness (Wang & Strong, 1996). High-quality information guarantees that users can depend on the system to make educated choices and run effective operations.

15.5.4 User Satisfaction

In modern studies, user satisfaction remains a critical criterion for assessing information systems (IS). According to Banoet al., user happiness has a substantial impact on the success and efficacy of information systems by influencing users' attitudes, intentions, and actual system use (Bano et al., 2017). They emphasize the significance of building information systems with a focus on user satisfaction, since this leads to increased user acceptance and adoption rates. Furthermore, Ahmed et al.'s study emphasizes the relevance of user satisfaction in determining user loyalty and continuing the use of IS. They discovered that greater levels of user satisfaction were connected with higher levels of user loyalty and a lower propensity to migrate to alternative systems (Ahmad et al., 2022). Recent research has related user happiness to the idea of user experience throughout information systems. According to DeLone and McLean, one of the major characteristics of the new DeLone and McLean IS success model is user satisfaction, demonstrating its lasting relevance in assessing IS performance.

15.5.5 Use/Intention to Use

Within the scope of the information system success models, the dimension of system use and intention to use is extremely significant. According to DeLone and McLean's IS Framework, it is a crucial variable in figuring out the efficacy of any information systems (Delone & McLean, 2003). System usage denotes users' actual usage of the system, while intention to use denotes users' desire and intentions to adopt and use the system in the future. According to the model, higher system use and positive intention to use are related to users' evaluations of system efficacy and overall satisfaction. These factors interact with other dimensions to determine the success of information systems, such as system quality, information quality, and service quality.

15.5.6 Perceived Usefulness

An information system's perceived utility plays a critical role in establishing its acceptance and efficacy. Research has demonstrated that users` perceived usefulness is favourably linked to their intention to use e-governance information systems (Al- Gahtani, 2011; Liang et al., 2007). As per Davis's (1989) technology acceptance model (TAM), perceived usefulness is "the degree to which a person thinks that using a given technology would enhance his or her job performance" (Davis, 1989). Multiple investigations indicate that an information system's perception of utility amongst its users is a reliable indicator of users' intention to utilize it (Agarwal & Prasad, 1999; Venkatesh et al., 2003). In a research on the spread of online shopping, Venkatesh and Davis discovered that perceived utility had a substantial beneficial influence on consumers' intent to use an Internet shopping system (Venkatesh & Davis, 2000). Also, it has been noted that a key factor in deciding how an information system is actually employed in practice is perceived utility. In a study of electronic health record systems, Gajanayake et al. found that perceived usefulness had a significant positive effect on actual system usage (Gajanayake et al., 2014). It has been found to be positively associated with users' satisfaction with e-governance information systems (Chiang, 2013; Zhou et al., 2010). Users who perceive the system as useful are more likely to be satisfied with its performance and to continue using it in the future. Furthermore, the perceived usefulness and value of e-governance information systems have been seen to yield a positive impact on citizens' trust in government (Parent et al., 2005). When individuals think that the e-governance system is beneficial, they are more inclined to trust the government and see its services favourably.

15.5.7 Objective of the Study

This study enquires into the influence and implications of diverse factors shaping satisfaction levels among end users interfacing with the e-government information system (e-Gov IS) within a smart city framework, contextualized in a modern city of a developing nation like India, with its peculiar challenges and user biases and predispositions. Against the backdrop of a heightened momentum towards digitalization of governance, especially in the regions characterized by pronounced digital divides and socio-economic inequities, the research delves into the interplay between technological determinants and their analogous non-digital counterparts in modulating user perspectives on such systems. Central to this exploration is the role of the motivational drivers, in the adoption and consistent engagement with the e-Gov IS. The culminating objective is to gauge the resultant impact on the perception of intuitive e-governance, determining its stride towards a more user-centric governance paradigm.

15.6 Research Methodology and Instrument Development

The reference e-governance information system of our study was the online e-Gov IS of Newtown Kolkata Development Authority, the Urban Local Body of Newtown Kolkata, a designated smart city under the Smart Cities Mission of the Government of India. According to past studies, we created 16 measures (Table 15.1) to evaluate the satisfaction level of end users with different attributes of the software application. The accompanying questionnaire asked respondents to rate their satisfaction with each measure on a 5-point scale (1- being extremely unsatisfied and 5- being extremely satisfied).

The survey was conducted from March 2023 to May 2023, in both online and offline methods (via survey forms at the facilitation kiosks). A total of 505 responses were received which we consider to be sufficient for the study, out of which 418 were considered to be fit for use.

The demographics of the respondents are mentioned in Figs. 15.1, 15.2, and 15.3.

15.7 Research Methodology

15.7.1 Instrument Reliability

Cronbach's alpha (α) is a widely used measure in psychometrics to evaluate the internal consistency of a questionnaire or scale. It assesses the extent to which a set of items in a questionnaire are correlated and measure a single underlying construct reliably.

In the provided SPSS output for a questionnaire comprising 16 items and 418 responses, Cronbach's alpha is computed to be 0.959 (Ref Table 15.2). This high coefficient signifies a strong internal consistency among the items, indicating that the questionnaire items are closely related and collectively measure the intended construct effectively. Such robust reliability ($\alpha > 0.9$) suggests that the questionnaire is dependable for measuring the targeted concept within the research domain.

15.7.2 Factor Analysis

15.7.2.1 Validation

The data's factor analysis was carried out on IBM SPSS Statistics 27 to examine the validity of the proposed model and reveal the underlying structure of the observed variables. To ascertain the number of components to extract, various criteria were taken into account, including eigenvalues, cumulative percentage of variance explained, and theoretical interpretability.

Table 15.1 Survey instrument

Indicators	Parameter	Definition of the parameter	Source
IQ1	The information, data, and records that the Newtown Kolkata development authority (NKDA)'s online e-governance system presents are error-free	Precision	Doll and Torkzadeh (1988)
IQ2	For doing my job, the information, data, and records available at the NKDA's online e-governance system are enough	Sufficiency	
IQ3	The information, data, and records available on the NKDA's online e-governance system is always current and complete	Relevancy	
SYS1	I consider the online e-governance system of NKDA to be friendly to users	User-friendliness	Doll and Torkzadeh (1988)
SYS2	I think it is really simple to use the NKDA's online e-governance system	Ease of use	
SYS3	The system is flexible and versatile enough for me to work with	Flexibility	Hossein (2015)
SYS4	It does not take me long to log into the system	Speed	Iivari (2005)
SV1	Every time I have a system issue, the NKDA online e-government system genuinely strives to find a solution	Sincerity	Wang and Tang (2003)
SV2	I receive personalized attention and care from the NKDA's online e-government system, whenever so needed	Responsiveness	
EX1	Usage of this online e-service has enabled me to save time	Time cost savings	Osman et al. (2014)
EX2	Usage of this online e-service has enabled me to save on my money	Economic cost savings	
UX1	Overall, I am pleased with the NKDA's online e-governance system	Satisfaction	Doll and Torkzadeh (1988), Palvia (1996), Rai et al. (2002)
UX2	My expectations have been achieved by the NKDA's online e-governance technology	Expectancy/ prospect	
UF1	With the NKDA e-governance system, I can complete tasks more quickly compared to a manual method	Better than alternative	Nugraheni and Bayastura (2021)
UF2	The fact that the system may be used from wherever one wants makes me think it is effective	Mobility	
IEGV	My work has become easier and quicker thanks to the NKDA's online e-governance technology	Intuitive e-governance	Etezadi-Amoli and Farhoom (1996)

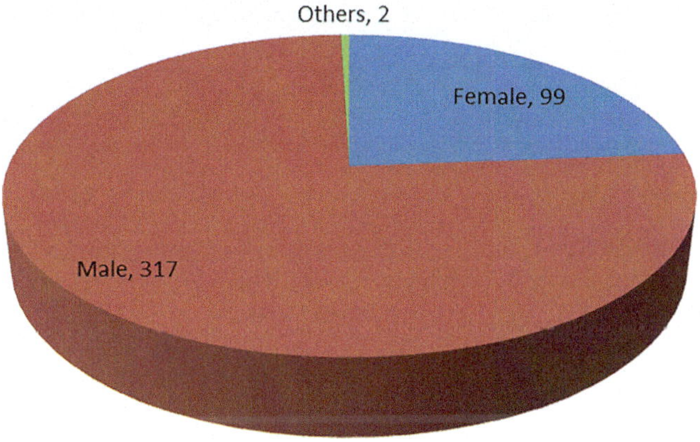

Fig. 15.1 Gender demography

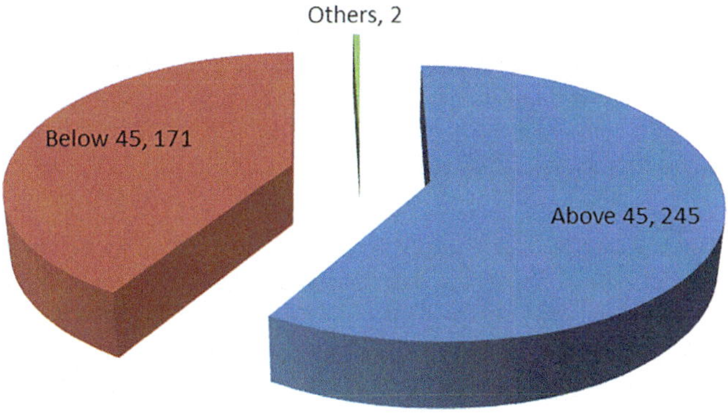

Fig. 15.2 Age demography

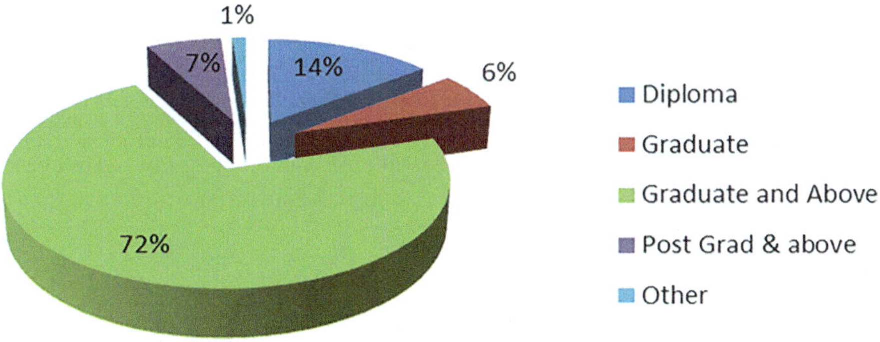

Fig. 15.3 Educational qualification

Table 15.2 Cronbach's alpha

Reliability statistics	
Cronbach's alpha	N of items
0.959	16

Table 15.3 KMO and Bartlett's test

KMO and Bartlett's test		
Kaiser-Meyer-Olkin measure of sampling adequacy		0.947
Bartlett's test of sphericity	Approx. Chi-square	5574.127
	Df	105
	Sig.	0.000

The Kaiser-Meyer-Olkin (KMO) measure of sampling adequacy yielded a value of 0.947 (Ref Table 15.3), surpassing the recommended threshold of 0.6, indicating the appropriateness of the dataset for factor analysis. The Bartlett's test of sphericity yielded a statistically significant result (Approx. Chi-square = 5574.127, df = 105, Sig. = 0.000), supporting the suitability of the data for factor analysis.

15.7.2.2 Output of Factor Analysis

From the extracted components, five were selected based on a combination of well-established guidelines, considering eigenvalues, cumulative percentage of variance explained, and meaningful theoretical interpretation. The combination of these components collectively capture a substantial portion of the observed variance in the data. The cumulative percentage of variance (83.5%) explained by these components also supports their meaningful inclusion in the analysis (Table 15.4).

The five numbers of factors that have emerged after factor analysis along with their respective loadings have been represented in Table 15.5.

To assign theoretical meaning to the components, variables with the highest loadings in each component were identified. The first factor (F_1) included variables related to system usability and satisfaction (UX1, UX2, SYS2, SYS1, SYS4, and SYS3). The second factor (F_2) represented operational efficiency of e-service (UF2, EX1, and UF1). The third factor (F_3) centred on system service support (SV2 and SV1). The fourth factor (F_4) was associated with information integrity and relevance (IQ2, IQ1, and IQ3) and the fifth factor (F_5) represented perceived economic returns of e-governance (PEX) (Ref Table 15.6).

Table 15.4 Total variance explained

Total variance explained									
	Initial eigenvalues			Extraction sums of squared loadings			Rotation sums of squared loadings		
Component	Total	% of Variance	Cumulative %	Total	% of Variance	Cumulative %	Total	% of Variance	Cumulative %
1	9.375	62.501	62.501	9.375	62.501	62.501	3.463	23.090	23.090
2	1.241	8.276	70.777	1.241	8.276	70.777	3.144	20.960	44.049
3	0.837	5.579	76.356	0.837	5.579	76.356	2.596	17.304	61.354
4	0.627	4.180	80.536	0.627	4.180	80.536	2.261	15.073	76.427
5	0.450	2.999	83.535	0.450	2.999	83.535	1.066	7.108	83.535
6	0.367	2.444	85.980						
7	0.353	2.356	88.336						
8	0.323	2.156	90.492						
9	0.301	2.006	92.498						
10	0.267	1.777	94.275						
11	0.239	1.594	95.869						
12	0.209	1.391	97.260						
13	0.166	1.104	98.364						
14	0.137	0.911	99.275						
15	0.109	0.725	100.000						

Extraction method: principal component analysis

Table 15.5 Rotated component matrix

Rotated component matrix[a]					
	Component				
	1	2	3	4	5
SYS2	***0.856***				
SYS1	***0.852***				
SYS4	***0.677***				
SYS3	***0.612***				
UX2	***0.514***				
UX1	***0.508***				
UF2		***0.854***			
EX1		***0.808***			
UF1		***0.720***			
SV2			***0.837***		
SV1			***0.830***		
IQ2				***0.768***	
IQ1				***0.720***	
IQ3				***0.585***	
EX2					***0.738***

Extraction method: principal component analysis
Rotation method: Varimax with Kaiser normalization
[a]Rotation converged in six iterations

Table 15.6 Factor output with the associated variables and the respective factor loadings

Factor no.	Name of the factors	Underlying variable name of the variables with codes	Factor loadings
F1	System usability and satisfaction (SUUX)	Ease of use [SYS2]	0.856
		User friendliness [SYS1]	0.852
		System speed [SYS4]	0.677
		System flexibility [SYS3]	0.612
		User expectancy [UX2]	0.514
		User satisfaction [UX1]	0.508
F2	Operational efficiency of E-service (OEES)	System mobility [UF2]	0.854
		Time costs [EX1]	0.808
		Better than alternative [UF1]	0.720
F3	System service support (Ssv)	Responsiveness [SV2]	0.837
		Sincerity [SV1]	0.83
F4	Information integrity and relevance (II&R)	Precision [IQ1]	0.720
		Sufficiency [IQ2]	0.768
		Relevancy [IQ3]	0.585
F5	Perceived economic returns of e-governance (PEX)	Economic cost savings [EX2]	0.738

15.7.3 Regression Analysis

The factors that have now emerged were subsequently put to regression analyses, to understand the impact of the five emerged factors on intuitive e-governance (IEGV) (Ref Table 15.7). The analysis reveals some interesting findings which as expected are peculiar to the context of the study. Firstly, the model's R-value, standing at 0.813, indicates a pronounced linear relationship between the predictors and intuitive e-governance. This is bolstered by the R square value of 66.1%, which suggests that a substantial portion of the variability in intuitive e-governance is accounted for by the predictors included in the model (ref Table 15.8). The robustness of the model has also been established (level of significance 0.000, ref. Table 15.9).

It has been observed that four factors, namely, system usability and satisfaction (F_1), operational efficiency of e-service (F_2), information integrity and relevance (F_4), and perceived economic returns of e-governance (F_5) have become significant, and all of them have a positive influence on the intuitive e-governance (IEGV) (ref Table 15.10). Operational efficiency of e-service has turned out to be the most influencing one (high standard coefficient value of 0.371) followed by system usability and satisfaction (coefficient value of 0.261) (ref Table 15.10). The output of the regression analysis has made substantial contributions to justify the stated objectives of the research study and the resultant impact of the four factors on the perception of intuitive e-governance and has established for determining its stride towards a more user-centric governance paradigm.

Table 15.7 Details of the variables entered for the regression analysis—*model summary*

Variables entered/removed[a]			
Model	Variables entered	Variables removed	Method
1	II & R, PEX, SSv, OEES, SUUx[b]	.	Enter

[a]Dependent variable: IEGV
[b]All requested variables entered

Table 15.8 Regression model output

Model summary				
Model	R	R square	Adjusted R square	Std. error of the estimate
1	0.813[a]	0.661	0.656	0.555

[a]Predictors: (constant), II & R, PEX, SSv, OEES, SUUx

Table 15.9 Regression output and level of significance

Model		Sum of squares	Df	Mean square	F	Sig.
1	Regression	244.934	5	48.987	158.828	0.000[a]
	Residual	125.838	408	0.308		
	Total	370.773	413			

[a]Predictors: (constant), II & R, PEX, SSv, OEES, SUUx

Table 15.10 Regression model relationship with the variables

Model		Unstandardized coefficients B	Std. error	Standardized coefficients Beta	T	Sig.
1	(Constant)	0.308	0.148		2.087	0.038
	PEX	0.152	0.035	0.181	4.395	0.000
	SUUx	0.256	0.055	0.261	4.690	0.000
	OEES	0.416	0.051	0.371	8.128	0.000
	SSv	0.004	0.038	0.005	0.115	0.908
	II&R	0.118	0.054	0.111	2.187	0.029

15.8 Interpretation of the Results

The regression analysis reveals crucial insights when applied to a developing nation like India, which has been emphasizing digital transformation and e-governance over the past few years.

The significant positive influence of perceived economic returns of e-governance (Factor 5) indicates that the perceived economic benefits from e-governance play a pivotal role in its intuitive adoption. In the Indian context, where resource allocation is often under scrutiny and there's a constant push to maximize returns on investments, demonstrating and confirming that clear economic benefits from e-governance projects could significantly drive their adoption and acceptance.

System usability and satisfaction (factor 1) and operational efficiency of e-service (factor 2) have also emerged as key determinants. Given India's diverse demographic and varied digital literacy levels, user-friendliness and operational efficiency are paramount. E-services that are easily navigable, understandable, and efficient cater to a wider audience, ensuring inclusivity, a primary goal in India's digital initiatives.

Furthermore, information integrity and relevance (factor 4) being significant, albeit with a smaller coefficient, underlines the importance of trustworthiness and context relevance of e-governance platforms. For a populous nation like India, where misinformation can rapidly spread and where regional and cultural nuances matter, the integrity and local relevance of information on e-governance platforms are crucial.

System service support (factor 3) not emerging as a significant determinant is intriguing. One hypothesis, especially in an Indian context, could be the relative novelty and unfamiliarity of many users with advanced system service supports. It would also be prudent to mention that though the strong e-governance systems have been endeavoured to be developed, training and skill development measures at the implementing level should also be improved for achieving significant improvements in human responsiveness to make the projects viable. Then, only a significant portion of the population may be motivated by the skilled human workforce to work around the basic usability and functionality of the e-governance systems.

Moreover, given the challenges of connectivity, infrastructure, and digital literacy in many parts of India, the emphasis might naturally shift towards ensuring basic services work seamlessly rather than sophisticated system support. The non-significance could also hint at a gap in awareness; users might not fully recognize or utilize advanced system support, thereby not perceiving them as critical for intuitive e-governance.

15.9 Conclusion and Limitations

The online e-Gov IS (e-governance information system) has been developed by the Newtown Kolkata Development Authority (NKDA), the Urban Local Body of Newtown Kolkata, a designated smart city under the Smart Cities Mission of the Government of India. The assessment of the mentioned programme has been undertaken by the present researchers on the basis of subjective analyses of the perceived notions of the citizens of NKDA on the basis of the identified variables. The outcomes of the analysis, using econometrics tools such as factor and regression analysis, have made insightful analysis of the challenges of the implementation of smart city projects in the Indian context. We have developed and verified a thorough, multifaceted model of e-government system success in our work, which initially considered 16 variables, and then, by factor analysis, it was reduced to 5 factors, namely, system usability and satisfaction (F_1), operational efficiency of e-service (F_2), system service support (F_3), information integrity and relevance (F_4), and perceived economic returns of e-governance (F_5). All of the factors except system service support (F_3), significantly or marginally influenced the adoption of e-governance system among the citizens of the smart city. We consider that the outputs of this investigation have several significant ramifications for e-governance research and practice.

In conclusion, as India steers its path towards more integrated and comprehensive e-governance, understanding these determinants and their implications is crucial. While core aspects like usability, efficiency, economic returns, and information integrity remain pivotal, it might also be worthwhile for policymakers and implementers to investigate and address the apparent undervaluation of system service support, ensuring a holistic and inclusive digital governance model.

References

Agarwal, R., & Prasad, J. (1999). Are individual differences germane to the acceptance of new information technologies? *Decision Sciences, 30*(2), 361–391. https://doi.org/10.1111/j.1540-5915.1999.tb01614.x

Ahmad, F., Mustafa, K., Hamid, S. A. R., Khawaja, K. F., Zada, S., Jamil, S., Qaisar, M. N., Vega-Muñoz, A., Contreras-Barraza, N., & Anwer, N. (2022, July 28). Online customer experience leads to loyalty via customer engagement: Moderating role of value co-creation. *Frontiers in Psychology*, (13), 897851. https://doi.org/10.3389/fpsyg.2022.897851. PMID: 35967631; PMCID: PMC9369676.

Ahvenniemi H, Huovila A, Pinto-Seppä I, Airaksinen M (2017) What are the differences between sustainable and smart cities? *Cities*, *60*(Part A): 234–245. ISSN0264 2751. https://doi.org/10.1016/j.cities.2016.09.009. https://www.sciencedirect.com/science/article/pii/S0264275116302578

Al-Gahtani, S. S. (2011). Modeling the electronic transactions acceptance using an extended technology acceptance model. *Applied Computing and Informatics, 9*(1), 47–77. https://doi.org/10.1016/j.aci.2009.04.001

Ameen, A., Alfalasi, K., Gazem, N. A., & Isaac, O. (2019). *Impact of system quality, information quality, and service quality on actual usage of smart government*. In 2019 1st International conference of intelligent computing and engineering: toward intelligent solutions for developing and empowering our societies, ICOICE 2019, 0–5. https://doi.org/10.1109/ICOICE48418.2019.9035144

Bano, M., Zowghi, D., & da Rimini, F. (2017). User satisfaction and system success: An empirical exploration of user involvement in software development. *Empirical Software Engineering, 22*. https://doi.org/10.1007/s10664-016-9465-1

Bibri, S. E. (2018). The IoT for smart sustainable cities of the future: An analytical framework for sensor-based big data applications for environmental sustainability. *Sustainable Cities and Society, 38*, 230–253.

Castells, M. (2009). *Communication power*. Oxford University Press.

Chiang, Li. (2013). Exploring the effects of trust, perceived risk, and E-services systems on public services in E-Government, 102–118. https://doi.org/10.4018/978-1-4666-4313-0.ch008

Davis, F. D. (1989). Perceived usefulness, perceived ease of use, and user acceptance of information technology. *MIS Quarterly, 13*(3), 319–340. https://www.jstor.org/stable/249008

DeLone, W. H., & McLean, E. R. (1992). Information systems success: The quest for the dependent variable. *Information Systems Research, 3*(1), 60–95.

Delone, W., & McLean, E. (2003). The DeLone and McLean model of information systems success: A ten-year update. *Journal of Management Information Systems, 19*, 9–30. https://doi.org/10.1080/07421222.2003.11045748

Deursen, A. J. A. M. & Helsper, E. (2015). The third-level digital divide: who benefits most from being online? https://doi.org/10.1108/S2050-206020150000010002

Doll, W. J., & Torkzadeh, G. (1988). The measurement of end-user computing satisfaction. *MIS Quarterly, 12*, 259–272. https://doi.org/10.2307/248851

Dwivedi, Y. K., Rana, N. P., Jeyaraj, A., Clement, M., & Williams, M. D. (2015). Re-examining the unified theory of acceptance and use of technology (UTAUT): Towards a revised theoretical model. *Information Systems Frontiers, 17*(6), 1183–1208.

Eppler, M., & Wittig, D. (2000). Conceptualizing information quality: A review of information quality frameworks from the last ten years, pp. 83–96.

Etezadi-Amoli, J., Farhoom, A. F. (1996). A structural model of end user computing satisfaction and user performance. *Information & Management, 30*(2), 65–73. ISSN 0378-7206. https://doi.org/10.1016/03787206(95)00052-6

French, T., Quinn, L., & Yates, S. (2019). *Digital motivation: Exploring the reasons why people are offline*. Good Things Foundation. https://www.goodthingsfoundation.org. 2019/02.

Gajanayake, R., Sahama, T., & Iannella, R. (2014). The role of perceived usefulness and attitude on electronic health record acceptance. *International Journal of E-Health and Medical Communications, 5*, 108–119. https://doi.org/10.4018/ijehmc.2014100107

Ganguly, D. (2022). Quality of e-governance in smart cities : A qualitative review of literature. *International Journal of Mechanical Engineering*, *7*(1), 2441–2447. https://kalaharijournals.com/resources/281-300/IJME_Vol7.1_292.pdf

Gorla, N., Somers, T., & Wong, B. (2010). Organizational impact of system quality, information quality, and service quality. *The Journal of Strategic Information Systems, 19*, 207–228. https://doi.org/10.1016/j.jsis.2010.05.001

Graham, S., & Aurigi, A. (1997). Virtual cities, social polarisation, and the crisis in urban public space. *Journal of Urban Technology, 4*(1), 109–132. https://doi.org/10.1080/10630739708724546

Heeks, R. (2003). *Most eGovernment-for-development projects fail: How can risks be reduced.*

Hossein, M. (2015). Investigating users' perspectives on e-learning: An integration of TAM and IS success model. *Computers in Human Behavior, 45*, 359–374., ISSN 0747-5632. https://doi.org/10.1016/j.chb.2014.07.044

Iivari, J. (2005). An empirical test of the delone-mclean model of information system success. *ACM SIGMIS Database, 36*, 8–27. https://doi.org/10.1145/1066149.1066152

Isaac, O., Bhaumik, A., Ameen, A., Alameria, M., & Ameenc, A. (2019). Antecedent and consequence of internet usage: An empirical study among government employees in UAE. *International Journal on Emerging Technologies, 10*(1a), 41–49. https://www.researchgate.net/publication/335465796

Liang, H., Saraf, N., Hu, Q., & Xue, Y. (2007). Assimilation of enterprise systems: The effect of institutional pressures and the mediating role of top management. *MIS Quarterly, 31*(1), 59–87.

Lopes, N. V. (2017). *Smart governance: A key factor for smart cities implementation.* In 2017 IEEE international conference on smart grid and smart cities (ICSGSC) (pp. 277–282).

Ministry of Housing and Urban Affairs, Government of India. (2015). *Smart cities: Mission statement & guidelines.*

Nugraheni, D. M. K., & Bayastura, S. F. (2021). Analysis of factors that influence satisfaction and usefulness for attendance system with the Delone & McLean model (Case study: Attendance system at Diponegoro University). *Journal of Physics: Conference Series, 1943*(1). https://doi.org/10.1088/1742-6596/1943/1/012108

Osman, I. H., Anouze, A. L., Irani, Z., et al. (2014). COBRA framework to evaluate e-government services: A citizen-centric perspective. *Government Information Quarterly, 31*(2), 243–256.

Palvia, P. C.(1996). A model and instrument for measuring small business user satisfaction with information technology. *Information & Management, 31*(3), 151–163. ISSN 0378-7206. https://doi.org/10.1016/S0378-7206(96)01069-5

Parasuraman, A. P., Zeithaml, V. A., & Berry, L. L. (1988). SERVQUAL: A multiple-item scale for measuring consumer perceptions of service quality. *Journal of Retailing, 64*(1), 12–40.

Parent, M., Vandebeek, C. A., Gemino, A. C (2005). Building citizen trust through e-government. *Government Information Quarterly, 22*(4), 720–736. ISSN 0740-624X. https://doi.org/10.1016/j.giq.2005.10.001. https://www.sciencedirect.com/science/article/pii/S0740624X05000869

Petter, S., & McLean, E. R. (2009). A meta-analytic assessment of the DeLone and McLean IS success model: An examination of IS success at the individual level. *Information and Management, 46*(3), 159–166. https://doi.org/10.1016/j.im.2008.12.006

Rai, A., Lang, S. S., & Welker, R. B. (2002). Assessing the validity of is success models: An empirical test and theoretical analysis. *Information Systems Research, 13*(1), 50–69. https://doi.org/10.1287/isre.13.1.50.96

Rana, N. P., Dwivedi, Y. K., Lal, B., & Williams, M. D. (2019). Examining citizens' adoption behavior in e-government: Identifying the critical role of privacy concern, social influence and benefits perception. *Information Systems Frontiers, 21*(2), 363–381.

Redman, T. C. (1998). The impact of poor data quality on the typical enterprise. *Communications of the ACM, 41*(2), 79–82.

Scheerder, A., van Deursen, A., & van Dijk, J. (2017). Determinants of internet skills, use and outcomes: A systematic review of the second- and third-level digital divide. *Telematics and Informatics, 34*(8), 1607–1624. https://doi.org/10.1016/j.tele.2017.07.007

Seddon, P. B. (1997). A respecification and extension of the DeLone and McLean model of IS success. *Information Systems Research, 8*(3), 240–253. https://doi.org/10.1287/isre.8.3.240

Taewoo, N., & Pardo, T. A. (2011). *Conceptualizing smart city with dimensions of technology, people, and institutions*. In Proceedings of the 12th annual international digital government research conference: Digital government innovation in challenging times (dg.o' 11) (pp. 282–291). Association for Computing Machinery. https://doi.org/10.1145/2037556.2037602

United Nations. (2020). *UN E-government survey 2020*. United Nations Department of Economic and Social Affairs.

Venkatesh, V., & Davis, F. D. (2000). A theoretical extension of the technology acceptance model: Four longitudinal field studies. *Management Science, 46*(2), 186–204. https://www.jstor.org/stable/2634758

Venkatesh, V., Morris, M. G., Davis, G. B., & Davis, F. D. (2003). User acceptance of information technology: Toward a unified view. *MIS Quarterly, 27*(3), 425–478. https://www.jstor.org/stable/30036540

Wang, L., & Liu, C. (2022). The role of motivation in the digital divide: The moderating effect of device access. *Mobile Media & Communication, 10*(3), 406–420. https://doi.org/10.1177/20501579211060621

Wang, R. Y., & Strong, D. M. (1996). Beyond accuracy: What data quality means to data consumers. *Journal of Management Information Systems, 12*(4), 5–33.

Wang, Y.-S., & Tang, T.-I. (2003). Assessing customer perceptions of website service quality in digital marketing environments. *JOEUC, 15*, 14–31. https://doi.org/10.4018/joeuc.2003070102

Weill, P., & Ross, J. (2004). *IT Governance: How top performers manage IT decision rights for superior results*.

World Bank. (2016). *World development report 2016: Digital dividends*. http://www.worldbank.org/en/publication/wdr2016

Zhou, T., Lu, Y., & Wang, B. (2010). Integrating TTF and UTAUT to explain mobile banking user adoption. *Computers in Human Behavior, 26*(4), 760–767. https://doi.org/10.1016/j.chb.2010.01.013

Debjeet Ganguly is a research scholar under the School of Business at Sister Nivedita University, Kolkata, India. The area of his research spans total quality management, smart cities, and electronic governance. He is also a civil service officer in the State of West Bengal, India, and in his career, he has shouldered varied administrative responsibilities across multiple roles and domains, both in far-flung marginal areas and in urban settings. He is presently functioning as Senior Deputy Secretary and Officer on Special Duty in the Department of Industries, Commerce and Enterprises in Government of West Bengal. His academic credentials include the attainment of a Master of Business Administration degree with specialization in Total Quality Management (TQM) from Sikkim Manipal University. Additionally, he has also completed Postgraduate Diploma in Personnel Management (PGDPM, recognized by AICTE) from the National Institute of Personnel Management (NIPM), Kolkata.

Dr. Amit Kundu is presently an Associate Professor, Department of Management, North Eastern Hill University (NEHU), Shillong, Meghalaya, where he has been since 11 October 2023. From May 2018 to 10 October 2023, he served as Professor and Dean, Schools of Management, Techno India Group. He has also functioned as Dean, School of Business, Sister Nivedita University (Under Techno India Group), from 1 August 2022 to 10 October 2023. He was appointed as convenor of the PhD programme of Sister Nivedita University from its inception (2019), and has discharged the duties as PhD convenor till the end of his tenure at Techno India Group. He served as HoD, School of Management, Techno Main, Salt Lake, from 2006 to 2018. He worked at Consulting Engineering Services (India) Private Limited in New Delhi, in the capacity of Business Development Executive from 2001 to 2003 and was also associated with them from 2007 to 2012 as Consultant—Market Analyst and Planner. He received his B. Tech from Jadavpur University in 1999 and MBA from Indian Institute of Engineering Science and Technology (IIEST), Shibpur, Howrah, in 2001. He received his PhD in Business Management from the University of Calcutta. His research interests span the areas of marketing management, strategic management, and entrepreneurship development. Five scholars have already been awarded PhD under his sole co-supervision from the University of Calcutta, Jadavpur University, and Sister Nivedita University. Amit has made over 60+ scholarly contributions, including nearly 40 peer-reviewed research papers. He has presented more than 100 research papers both in India and abroad He received the Outstanding Research Award in 2013 for his scholarly research paper presentation on his research work in the USA. He has given numerous invited talks and tutorials and acted as a visiting professor at esteemed business schools, such as ASCI (Hyderabad), IIFT (Kolkata), IISWBM (Kolkata), Department of Business Management, University of Calcutta, to name a few.

Prof Dr. Chandan Kumar Bhattacharyya is currently heading the Academic Quality Assurance Cell (AQAC) of Techno India Group Colleges for synergy in academic excellence among group colleges. Before that, he was heading the ERP project of digital transformation of the entire academic and allied activities of Techno India Group in different verticals. During his journey towards academic excellence, Prof Bhattacharyya executed as Dean of Sister Nivedita University (SNU) and Dean Academic, Techno International NewTown (TINT) Engineering College at Kolkata, West Bengal. He was also the former Principal of Techno Main Salt Lake [TMSL] Engineering College at Salt Lake, Kolkata. Dr. Bhattacharyya earned his M. Tech degree in Computer Science and Engineering from the University of Calcutta, MBA degree (Topper in Systems as specialization) from the Indian Institute of Social Welfare and Business Management (IISW&BM), University of Calcutta, and PhD in Computer Science and Engineering from Jadavpur University.

Chapter 16
Green Buildings and Smart Cities: A Perfect Harmony of Sustainability and Progress

Prasanna Goge, Snil Saspara, and Jiten Shah

Abstract The urgent global focus on sustainable development has driven a paradigm shift in the construction industry, emphasizing the need for a balance between economic growth, environmental preservation, and social well-being. Buildings, significant contributors to energy consumption and environmental degradation, require transformative approaches to ensure sustainability. This research explores the integration of green building principles within the framework of smart cities, aiming to redefine urban development standards and achieve a sustainable future. The chapter emphasizes the collaboration required among various stakeholders, including architects, engineers, contractors, governments, and the public, across the planning, design, and construction phases. It delves into innovative strategies and technologies underpinning green buildings and smart cities, highlighting barriers and opportunities for successful integration. The study recognizes the historic role of the construction industry in unsustainable development, making the emergence of green building practices crucial. Various green building rating systems, like Leadership in Energy and Environmental Design (LEED) and Building Research Establishment Environmental Assessment Method (BREEAM), have been introduced to encourage eco-friendly constructions, and their critical analysis is a part of this research. The chapter underscores the importance of collaboration, education, incentives, and effective project management in promoting sustainable construction practices. It emphasizes the potential of green building initiatives to significantly enhance energy efficiency, reduce greenhouse gas emissions, and contribute to a more sustainable future. The integration of green buildings within the framework of

---

P. Goge · S. Saspara · J. Shah (✉)
Department of Civil Engineering, Institute of Infrastructure Technology, Research and Management (IITRAM), Ahmedabad, Gujarat, India
e-mail: prasanna.goge.20c@iitram.ac.in; snil.saspara.20c@iitram.ac.in; jitenshah@iitram.ac.in

S. Majumdar et al. (eds.), *Smart Cities*, S.M.A.R.T. Environments,
https://doi.org/10.1007/978-3-031-59846-3_16

smart cities is seen as an imperative, requiring alignment of policies, education, research, and societal involvement to unlock their potential and drive the transformation toward a greener and smarter urban landscape.

Keywords Green buildings · Smart cities · Sustainable development · Construction · Planning · Design

16.1 Introduction

The urgent global focus on sustainable development, emphasizing a balance between economic growth, environmental preservation, and social well-being (Ahn et al., 2016), has catalyzed a paradigm shift in the construction industry, a significant contributor to energy consumption and environmental degradation (Hoffman & Henn, 2008; Olubunmi et al., 2016). Buildings, responsible for a substantial portion of energy consumption and carbon dioxide emissions, necessitate transformative approaches in design, construction, and operation to ensure sustainability (Yong Han Ahn et al., 2016, Olubunmi et al., 2016). This shift is exemplified by the integration of green building principles, promoting resource efficiency, improved indoor environmental quality, and sustainability throughout a building's life cycle (Ahn et al., 2016., Olubunmi et al., 2016). Concurrently, the emergence of smart cities, leveraging advanced technologies to optimize resource use and enhance infrastructure, offers a powerful approach to urban development (Ahn et al., 2016). Integrating green building principles within the smart city framework presents a compelling synergy, aligning sustainability and technological advancement to redefine urban development standards (Ahn et al., 2016, Darko et al., 2017). This research endeavors to explore the integration of green buildings and smart city initiatives, emphasizing the collaboration required among various stakeholders—architects, engineers, contractors, governments, and the public—across the planning, design, and construction phases (Ahn et al., 2016, Hoffman & Henn, 2008, Darko et al., 2017). The study delves into innovative strategies and technologies underpinning green buildings and smart cities, striving for a cohesive and sustainable urban future. Furthermore, the chapter examines barriers and opportunities for successful integration, providing insights into future perspectives that can drive advancements in this transformative approach to urban development. The construction industry, a pivotal force shaping the natural and built environment with substantial global economic and social impacts (Darko et al., 2017), has historically driven unsustainable development, contributing significantly to greenhouse gas emissions and waste generation (Darko & Chan, 2016, 2017). In response, the green building movement has emerged, aiming to create resource-efficient and ecologically sound constructions (Doan et al., 2017). This movement has gained prominence in the context of sustainable global development (Doan et al., 2017). Despite progress, barriers to the widespread adoption of green building practices persist (Darko et al., 2017). Various green building rating systems like Leadership in Energy and Environmental Design (LEED), Building Research Establishment Environmental Assessment Method (BREEAM), and Green Star have been introduced to encourage energy-efficient and environmentally friendly constructions (Doan et al., 2017). However, a systematic review of their criteria and updated processes is lacking (Doan et al., 2017). Moreover, green building practices have evolved from a buzzword to an economically significant aspect of the construction industry, especially in the context of energy-efficient construction (Zuo & Zhao, 2014). Both public and private sector investments in green building projects have been substantial. This chapter seeks to explore the harmonious integration of green buildings within the context of smart cities, offering a comprehensive analysis of barriers, critical comparisons of green building rating systems, the status and future agenda of green building research, and the economics of green building (Darko et al., 2017; Zuo & Zhao, 2014). This multifaceted investigation is essential for the development of smart cities and the future of urban living, providing valuable insights for stakeholders, policymakers, and researchers alike. In recent years, the imperative for sustainable urban development has gained significant traction, accentuated by increasing urbanization and its environmental impacts. Central to achieving sustainability is the seamless integration of green building principles within the framework of smart cities. This integration encapsulates environmental consciousness and propels both progress and economic growth. This chapter delves into the symbiotic relationship between green buildings and smart cities, illustrating how their integration can pave the way for a sustainable and progressive urban future. Green building, a prominent facet of sustainable architecture, embraces environmentally friendly materials and energy-efficient design strategies. Emphasizing the necessity of incorporating green architectural design into building practices, studies have underscored its pivotal role in achieving sustainable development and enhancing people's quality of life (Zheng, 2021). Furthermore, the discourse extends to green building literacy, highlighting the significance of educating the masses about the benefits of green building in fostering a culture of sustainability (Cole, 2019). Considering that buildings constitute a substantial portion of energy consumption, knowledge dissemination becomes vital. Additionally, the chapter discusses the sustainability considerations of green buildings, viewing them as a potential solution to address the environmental, economic, and social challenges posed by traditional buildings (Liu et al., 2022). A bibliometric review of green building research offers insights into its trajectory and identifies emerging trends and areas necessitating further exploration (Zhao et al., 2019). Considering these scholarly contributions, this chapter integrates insights from research papers to guide the discourse on how green buildings, emphasizing sustainability, can seamlessly blend with the innovative framework of smart cities, ultimately fostering a sustainable and progressive urban landscape. The ensuing discussion unveils the synergies between green buildings and smart cities, illustrating their potential to mitigate environmental impact while propelling urban development.

16.2 Literature Review

16.2.1 Unifying Green Building Principles and Smart City Objectives

Green buildings and smart cities represent fundamental paradigms crucial for achieving sustainable urban development. The integration of green building principles within smart city frameworks is imperative, given the significant environmental impacts associated with construction activities over their life cycle (Ahn et al., 2016; Darko & Chan, 2016; Liu et al., 2022). Green buildings, characterized by environmentally conscious design and resource efficiency, play a pivotal role in mitigating these impacts (Ahn et al., 2016; Liu et al., 2022). The construction industry, a significant economic driver, provides a fundamental context for implementing green building methods (Samari et al., 2013; Zheng, 2021). Understanding the incentives that drive green building adoption is vital to promote sustainability within the construction sector (Olubunmi et al., 2016). Combining insights from the construction industry, green building incentives, and overarching objectives of smart cities, this study aims to explore the seamless integration of green building principles within smart city frameworks, harmonizing sustainability with urban progress (Ahn et al., 2016; Samari et al., 2013; Olubunmi et al., 2016). This chapter offers an integrated overview of green building principles, their role in the construction industry, and the importance of incentives while highlighting their interconnectedness in the context of smart cities, as guided by the specified research methodology. The development and integration of green buildings within the broader framework of smart cities signify a pivotal paradigm for sustainable urbanization. To comprehensively explore this integration, a systematic literature review was conducted, drawing insights from various scholarly works that illuminate the barriers, challenges, and dimensions of sustainability associated with green buildings in the context of smart cities (Darko & Chan, 2016). The literature review encompassed peer-reviewed academic journal articles published between 2000 and 2015, aimed at ensuring the inclusion of contemporary and pertinent sources (Darko & Chan, 2016).

16.2.2 Assessing the Synergy of Green Buildings and Smart Cities

The study delved into existing green building assessment tools such as LEED, BREEAM, and others, shedding light on their structures and roles in promoting sustainable building practices (Doan et al., 2017). These assessment tools were recognized for their significance in driving the adoption of green building strategies and evaluating sustainability outcomes. The integration of findings from the literature review demonstrated the interplay of environmental, social, and economic sustainability within the context of smart cities and green building initiatives, underscoring the need for a holistic approach to achieve sustainability and progress in urban environments (Doan et al., 2017). This extensive literature review laid the groundwork for a thorough investigation into the integration of green buildings and smart cities, offering a comprehensive understanding of the challenges and opportunities for sustainable urban development in the modern era (Darko & Chan, 2016; Doan et al., 2017). Green buildings, being a pivotal component of sustainable construction, emphasize energy and resource efficiency, with minimal environmental impact throughout their life cycle (Hwang & Tan, 2010). The integration of green building principles into urban development is fundamental, considering the substantial growth of the construction sector and the ensuing environmental impact (Hwang & Tan, 2010). Such initiatives are bolstered by legislative and economic drivers. This study draws on the methodologies of both green building practices and rating systems (Energy Star and LEED) to assess the harmonious integration of green buildings into smart cities (Hwang & Tan, 2010). By analyzing data from these rating systems and incorporating principles from the methodology of green building, we aim to comprehensively evaluate the potential synergy between green buildings and smart cities in promoting sustainability, energy efficiency, and overall urban progress. The overarching goal is to underscore the significance of incorporating green building practices into the fabric of smart cities to achieve a sustainable and progressive urban future (Hwang & Tan, 2010).

16.2.3 Holistic Approach of Green Buildings in Urban Sustainability

In the quest for sustainable urban development, the concept of green buildings and their integration into the fabric of smart cities has gained prominence (Zheng, 2021). Green buildings epitomize structures designed to minimize environmental harm, optimizing the use of natural resources without disrupting the ecological balance (Zheng, 2021). They encompass energy efficiency, land, water, and material conservation, aiming to reduce waste and pollution while providing inhabitants with healthy, efficient, and applicable living spaces (Zheng, 2021). In parallel, the broader theme of green building design is viewed through the lenses of environmental literacy and science literacy, with the core idea that a green building literate individual possesses foundational knowledge from both environmental/sustainability and science education (Cole, 2019). This study's methodology intricately weaves these concepts into a unified framework, exploring the seamless integration of green buildings within smart cities (Zheng, 2021). Through an interdisciplinary literature review, the foundational dimensions of green building literacy were adapted, refined, and harmonized with the goals of sustainable development within smart cities (Cole, 2019). This refined framework encompassed key dimensions such as energy efficiency, resource optimization, environmental impact reduction, and human-centric design. Real-world case studies were then presented to exemplify successful integration, followed by expert validation to enhance the framework's robustness and applicability (Cole, 2019; Zheng, 2021). Through this integrated approach, this research aims to shed light on how the harmonization of green buildings and smart cities can pave the way toward a sustainable and progressive urban future. Future directions are discussed, encompassing potential trends, innovations, and advancements in this realm (Cole, 2019; Zheng, 2021). The concept of green buildings, often described as energy-saving, ecological, or sustainable structures, is fundamental to sustainable urban development (Liu et al., 2022). The green building concept encompasses a holistic approach that goes beyond the building itself, incorporating considerations for people, society, the economy, and the environment throughout the building's life cycle (Liu et al., 2022). This comprehensive perspective necessitates collaborative efforts from all involved sectors to design green buildings successfully (Liu et al., 2022). Despite an increasing number of research papers on green buildings, a notable research gap exists, with many studies focusing on specific aspects of green building areas (Liu et al., 2022). Notably, a systematic review encompassing a comprehensive understanding of green building areas is lacking, underscoring the significance of addressing this gap (Liu et al., 2022). To bridge this research gap and gain a nuanced understanding of the barriers, progress, and future research directions for sustainable development of green buildings, this study adopts a research methodology involving systematic literature review, bibliometric analysis, and visualization techniques (Liu et al., 2022).

16.2.4 Balancing Sustainability and Progress

The integration of green building principles within smart city frameworks is a topic of growing importance, as sustainability becomes a paramount concern in urban development (Ahn et al., 2016; Liu et al., 2022). The harmonious alignment of green buildings and smart cities holds the promise of creating urban environments that are ecologically responsible, economically viable, and socially equitable (Ahn et al., 2016; Liu et al., 2022). However, this integration presents its own set of challenges and complexities that need to be understood and addressed to achieve a seamless fusion of sustainability and progress. This chapter serves as a foundational exploration into this crucial intersection of sustainability and urbanization, paving the way for further research and initiatives that can reshape our cities into hubs of sustainability and innovation. In this pursuit of sustainability, green buildings have emerged as a central pillar of eco-friendly and resource-efficient construction practices. Green buildings not only reduce negative environmental impacts but also enhance the overall quality of life for the occupants. These structures are designed with a focus on energy efficiency, water conservation, use of sustainable materials, waste reduction, and improved indoor air quality (Ahn et al., 2016). The integration of green building principles into the broader context of smart cities adds another layer of complexity and opportunity, aiming to create urban spaces that are technologically advanced, sustainable, and inclusive. The construction industry, being a significant contributor to environmental degradation, has recognized the urgency of incorporating sustainable practices. Green building initiatives have gained traction due to their potential to reduce the industry's carbon footprint and decrease its overall environmental impact (Samari et al., 2013). Sustainable construction practices not only benefit the environment but also lead to economic advantages, including lower operating costs and increased property values. These economic incentives provide a strong motivation for both developers and investors to embrace sustainable building approaches (Olubunmi et al., 2016). Additionally, governments and regulatory bodies are increasingly advocating for green building certifications and incentives to encourage the adoption of sustainable practices in the construction sector (Samari et al., 2013). One of the most notable green building certification systems is the Leadership in Energy and Environmental Design (LEED), which sets the global standard for sustainable and high-performing buildings. LEED certification provides a framework for designing, constructing, operating, and maintaining buildings with minimal environmental impact. Similarly, the Building Research Establishment Environmental Assessment Method (BREEAM) is another widely recognized certification that focuses on assessing the environmental sustainability of buildings. These certifications act as guiding tools, promoting sustainable construction practices and aiding in the seamless integration of green buildings into the fabric of smart cities (Doan et al., 2017). The synergy between green buildings and smart cities lies in their mutual goal of sustainability. Smart cities utilize advanced technologies and data-driven solutions to optimize resource use, enhance infrastructure, and improve the overall quality of life for their citizens (Ahn et al., 2016). By incorporating green building principles, smart cities can further amplify their sustainability efforts. For instance, smart technologies can be employed to monitor and optimize energy consumption within green buildings, making them even more efficient (Ahn et al., 2016). This integration ensures a holistic approach to urban development, where technology and sustainability work hand in hand to create environmentally sound and technologically advanced cities. Efforts to integrate green buildings into the smart city landscape require a multidimensional approach. Policymakers, urban planners, architects, engineers, and stakeholders from various sectors must collaborate to establish a framework that facilitates the seamless fusion of green building principles into smart city initiatives. This collaboration should involve setting clear sustainability goals, adopting, and refining existing assessment tools, incentivizing green building practices, and promoting public awareness and education about the benefits of green buildings within the context of smart cities (Ahn et al., 2016; Doan et al., 2017). In conclusion, the integration of green buildings and smart cities presents a promising path toward achieving sustainable urban development. Green buildings, with their emphasis on energy efficiency and environmental responsibility, align seamlessly with the objectives of smart cities, which leverage technology for sustainable and efficient urban management. The literature reviewed herein underscores the interconnectedness of these paradigms and emphasizes the need for further research and concerted efforts to realize the full potential of this integration. Moving forward, a deeper understanding of the challenges and opportunities in harmonizing green building principles within smart city frameworks will pave the way for urban environments that are not only progressive but also environmentally and socially responsible.

16.3 Major Findings

This research endeavors to shed light on the crucial integration of green buildings within the ambit of smart cities, aiming for a harmonious blend of sustainability and progress. Through a meticulous analysis, the study has unearthed several major findings, providing a roadmap for the successful implementation of green building initiatives within smart cities.

16.3.1 Integrated Construction Process (ICP) Model for Green Buildings

The cornerstone of achieving sustainable construction lies in a well-structured approach. In line with this notion, the research introduces the integrated construction process (ICP) model, a pioneering framework that addresses the limitations inherent in existing project delivery methods (Ahn et al., 2016). This model operates across four distinct layers, meticulously outlining the stages of green construction activities, delineating stakeholder roles, and providing practical examples for a comprehensive understanding and seamless execution (Ahn et al., 2016). Within this structured approach, each green construction activity is carefully detailed, encompassing content, goals, and procedural guidelines, thus facilitating a holistic and methodical pathway toward achieving sustainability objectives in construction projects (Ahn et al., 2016). The ICP model stands as a testament to the necessity of a structured framework, acting as a catalyst for the successful integration and operationalization of green construction technologies and strategies within the context of smart cities. Within the ambit of green building development, barriers often obstruct the smooth trajectory toward sustainability. Recognizing these barriers is a pivotal step toward surmounting them effectively. The research underscores the significance of strategies to overcome these impediments, emphasizing the necessity of altering prevailing mindsets to promote behavioral shifts toward sustainable practices (Hoffman & Henn, 2008). Strategies such as framing green buildings attractively, targeting specific demographics, and integrating environmental literacy into various training systems within the building sector are identified as effective tools for stimulating the necessary change (Hoffman & Henn, 2008). Demographic characteristics are acknowledged as influential factors shaping attitudes toward sustainability. Factors such as gender, age, education, income, and political affiliation play critical roles in determining individual receptiveness to green technologies (Hoffman

& Henn, 2008). Tailoring marketing strategies to align with the environmental values of specific demographics emerges as a potent tool for fostering acceptance of green technologies (Hoffman & Henn, 2008). Moreover, restructuring and incentivizing the adoption of green building practices are recognized as fundamental enablers (Hoffman & Henn, 2008). Implementing a more collaborative and integrated approach in construction processes, accompanied by contractual relationships that encourage shared risk and reward, is deemed crucial for successful implementation (Hoffman & Henn, 2008). Incentivizing risk reduction and indemnification strategies are proposed to address subcontractors' concerns and foster the adoption of new technologies (Hoffman & Henn, 2008). It is also highlighted that the evolution of green building standards must align with advancements in the field, ensuring a comprehensive promotion of sustainability (Hoffman & Henn, 2008). Government policies, especially tax reforms, are recommended to stimulate demand for green buildings and components (Hoffman & Henn, 2008). The research underscores the integral role of addressing barriers and leveraging strategic initiatives to promote sustainability and progress within urban environments.

16.3.2 Demographic Influences on Green Building Adoption

Understanding the dynamics of individual preferences and attitudes toward sustainability is pivotal in promoting the adoption of green building practices. Demographic characteristics have been identified as significant determinants in shaping these attitudes. Gender, age, education, income, and political affiliation were identified as influential factors that significantly impact how individuals perceive and embrace green building technologies (GBTs) (Hoffman & Henn, 2008). Targeting specific demographics that align with environmental values can substantially facilitate the acceptance and adoption of green technologies. It is imperative to tailor marketing and educational efforts to resonate with the distinct preferences and predispositions of various demographic segments. By doing so, we can present green buildings as attractive options that complement the values and lifestyles of these demographics, consequently fostering a higher degree of acceptance and integration within the market.

16.3.3 Structural Changes and Incentive Reforms

Overcoming barriers to the widespread adoption of green building practices necessitates substantial structural changes within the construction industry. One of the fundamental structural shifts proposed by the research is the integration of a more collaborative and integrated approach to construction processes (Hoffman & Henn, 2008). This approach entails fostering contractual relationships that encourage shared risk and reward among stakeholders involved in a construction project. By promoting a collaborative atmosphere where risks and benefits are shared, concerns and apprehensions regarding the adoption of green technologies can be effectively mitigated, promoting a smoother transition toward sustainability. Moreover, incentivizing risk reduction and indemnification strategies were identified as crucial tools in encouraging the adoption of new technologies within the construction sector (Hoffman & Henn, 2008). These incentives act as catalysts, mitigating concerns and promoting a favorable environment for the integration of green technologies. The role of government policies in incentivizing the adoption of green building practices cannot be understated. Tax reforms were highlighted as a viable means to stimulate the demand for green buildings and their components (Hoffman & Henn, 2008). By offering tax incentives, governments can significantly influence investment choices, making green alternatives more appealing and financially viable for potential consumers. This financial encouragement not only facilitates the adoption of green practices but also helps drive the growth of the green building market, promoting sustainability at a broader scale.

16.3.4 Evolution of Green Building Standards

Green building standards play a critical role in ensuring that sustainable practices are consistently upheld across the construction industry. However, for these standards to remain relevant and effective, they must evolve in alignment with advancements in the field of sustainable construction. The research emphasizes the necessity of updating and refining green building standards to keep pace with emerging technologies and evolving best practices (Hoffman & Henn, 2008). This ensures that the standards remain comprehensive and relevant, effectively promoting sustainability while accommodating the dynamic nature of the construction industry. The continual enhancement of these standards is crucial for fostering a culture of sustainability, promoting innovation, and setting a benchmark for environmentally conscious construction practices. Recognizing demographic influences, advocating for structural changes, incentivizing adoption through policy measures, and evolving standards stand as essential steps toward achieving a harmonious fusion of sustainability and progress within urban environments.

16.3.5 Understanding Barriers to Green Building Development

Barriers to green building development represent significant hurdles that hinder the widespread adoption of sustainable practices. A critical barrier identified in the research is the lack of expertise in green building development within the construction industry (Samari et al., 2013). This dearth of knowledge and skill prolongs project timelines and inhibits the effective implementation of green building concepts. Overcoming this barrier necessitates an emphasis on education and training programs targeted at equipping professionals with the requisite expertise in sustainable construction practices. Bridging this knowledge gap is instrumental in accelerating the adoption of green building technologies and promoting sustainability within the industry. Government involvement and strategies in promoting green building initiatives also play a pivotal role, but their effectiveness may vary based on the context (Samari et al., 2013). The research highlights the importance of evaluating and refining existing government strategies to ensure their efficiency and alignment with the unique dynamics of the construction industry in a particular region. Moreover, understanding the reservations and perceived risks of construction companies toward engaging with green projects is vital (Samari et al., 2013). Addressing these concerns, perhaps through tailored incentives or targeted support, can encourage construction firms to actively participate in green building initiatives.

16.3.6 Incentives for Promoting Green Building

Incentives are fundamental tools for promoting the adoption of green building practices. The research underscores the critical role of both financial and nonfinancial incentives, emphasizing the need to increase the contribution of the private sector in providing these incentives (Olubunmi et al., 2016). Financial incentives could include tax benefits, grants, or subsidies that make green building projects financially viable and appealing for investors and developers. On the other hand, nonfinancial incentives may encompass certifications, recognition, or fast-track permitting, providing reputational benefits to encourage adoption. Tailoring incentives to local economic and environmental conditions was identified as a strategic approach to enhance their effectiveness (Olubunmi et al., 2016). It ensures that incentives are aligned with the specific needs and circumstances of a region, making them more appealing and impactful. Designing incentives with a long-term perspective is crucial, considering that the benefits of green building, such as energy savings and reduced maintenance costs, accrue over an extended period (Olubunmi et al., 2016). Hence, incentivizing a long-term commitment to sustainability is key to driving adoption and fostering a sustainable built environment. Addressing the issue of split incentives, where stakeholders have differing motivations for green building, was deemed crucial to improve the effectiveness of incentives (Olubunmi et al., 2016). Strategies need to be devised to align the interests of various stakeholders, ensuring that incentives benefit all parties involved. Moreover, increasing awareness of incentives among stakeholders, particularly building owners, is vital for encouraging their adoption and promoting green building (Olubunmi et al., 2016). Effective communication and education campaigns can bridge the information gap and maximize the impact of incentives on the adoption of green building practices. The barriers and incentives highlighted in this major theme provide critical insights into the dynamics of green building adoption. By understanding these challenges and leveraging the power of incentives, policymakers and industry stakeholders can formulate targeted strategies to drive sustainable practices and foster a more environmentally conscious construction industry.

16.3.7 Overcoming Barriers to Green Building Adoption

The research underscores the necessity of overcoming barriers that impede the successful uptake of green building practices to achieve sustainable development in the construction industry (Darko et al., 2017). These barriers were categorized into five main challenges, shedding light on the multifaceted nature of obstacles faced in promoting sustainable building practices. First, a lack of information, education, research, knowledge, awareness, and expertise was identified as a fundamental challenge hindering the adoption of green building practices (Darko et al., 2017).

Bridging this knowledge gap through educational initiatives, awareness campaigns, and targeted research efforts is critical in surmounting this barrier. The higher costs associated with green building technologies compared to conventional building methods represent another significant challenge (Darko et al., 2017). To address this, financial incentives and subsidies may be pivotal in offsetting these additional costs and promoting the adoption of sustainable construction practices (Darko et al., 2017; Olubunmi et al., 2016). Additionally, strategic partnerships between public and private sectors could be established to share the financial burden and encourage the implementation of sustainable technologies. The absence of incentives and support, particularly from government authorities, was identified as a substantial barrier to the adoption of green building practices (Darko et al., 2017). Government involvement through policy formulation, incentives, and regulation is crucial in creating a supportive environment for sustainable construction initiatives. Public-private collaborations and stakeholder engagement could enhance the effectiveness of these government-driven efforts (Darko et al., 2017). Insufficient interest and demand from stakeholders present another barrier (Darko et al., 2017). Raising awareness about the benefits of green building practices and fostering a cultural shift toward sustainability is vital in generating interest and demand. Educational campaigns targeting various stakeholders, including consumers, developers, and policymakers, can effectively address this challenge. Lastly, the absence of mandatory green building codes and regulations was recognized as a significant barrier (Darko et al., 2017). Strengthening and enforcing building codes that prioritize sustainability and green practices are imperative for fostering a culture of environmental consciousness within the construction industry. Collaboration between regulatory bodies, industry associations, and educational institutions can play a pivotal role in achieving this objective (Darko et al., 2017).

16.3.8 Key Drivers and Promotion Strategies for Green Building Adoption

Greater energy and water efficiency emerged as prominent drivers for the adoption of green building technologies (Darko & Chan, 2016). Given the increasing concerns about resource scarcity and climate change, showcasing how green buildings significantly contribute to energy and water conservation is crucial in encouraging adoption. Enhanced company image and reputation constitute another influential driver (Darko & Chan, 2016). Companies increasingly value their public image and perceive sustainable buildings to demonstrate corporate social responsibility and environmental stewardship. Incorporating sustainability in their built environment adds to their brand value and appeal. Improved occupants' well-being and comfort are compelling reasons for the adoption of green building technologies (Darko & Chan, 2016). Creating a conducive and healthy indoor environment through sustainable building practices appeals to both building owners and occupants. This driver emphasizes the tangible benefits that green buildings can offer in terms of enhanced living and working spaces. Reduced environmental impact and life cycle costs were identified as drivers that resonate with both environmentally conscious consumers and cost-conscious developers (Darko & Chan, 2016). Demonstrating how green buildings are not only environmentally friendly but also economically viable in the long run is critical for driving adoption. To further promote green building adoption, various promotion strategies are suggested. Financial and market-based incentives were highlighted as effective strategies to encourage uptake (Darko & Chan, 2016). Financial incentives, such as tax benefits or grants, can directly reduce the cost burden associated with green building practices, making them more attractive to investors and developers. Market-based incentives, on the other hand, create a competitive advantage for green buildings, encouraging their adoption through market forces. Better information dissemination regarding costs and benefits was recognized as a vital promotion strategy (Darko & Chan, 2016). Providing clear, accurate, and accessible information on the costs and benefits associated with green building practices enables stakeholders to make informed decisions. Dispelling misconceptions and demonstrating the long-term advantages of sustainability can significantly influence adoption rates. Green labeling, such as certifications or eco-labels, was proposed as a powerful strategy to influence consumer choices (Darko & Chan, 2016). Certifications like LEED and Energy Star act as visible indicators of a building's sustainability, instilling confidence in prospective occupants and investors. Mandatory green building codes were identified as a forceful strategy to drive adoption, ensuring that a minimum level of sustainability is mandated across all construction projects (Darko & Chan, 2016). The enforcement of such codes creates a level playing field and establishes sustainability as a nonnegotiable aspect of construction. Moreover, intensified research, education, and training in green building technologies were advocated as strategies to enhance knowledge and awareness within the construction industry (Darko & Chan, 2016). Providing educational resources and training opportunities ensures that professionals are well equipped to embrace and implement green building practices effectively. Research on barriers, drivers, and promotion strategies provide a valuable roadmap for policymakers, stakeholders, and industry professionals to accelerate the adoption of green building technologies and practices, ultimately fostering sustainability within the construction industry (Darko et al., 2017; Darko & Chan, 2016). Overcoming barriers and leveraging the identified drivers through effective promotion strategies are essential steps toward achieving sustainable building practices and fostering a more environmentally conscious construction industry. These findings underscore the need for structured approaches, mindset changes, and supportive policies to promote the successful integration of green buildings in the context of smart cities.

16.3.9 Global Trends in Green Building Research

The landscape of green building research has witnessed a significant surge in published papers, particularly after the 2000s (Doan et al., 2017). This upsurge signifies the growing emphasis on sustainable construction and the increasing awareness of the need for environmentally conscious building practices. Prominent green building rating systems have garnered substantial attention, with LEED standing out as the most recognized and widely adopted, followed by BREEAM and CASBEE (Doan et al., 2017). The global reach of these rating systems highlights the international recognition and applicability of sustainable building standards. Contributions to this research have been widespread, with key nations such as the USA, Canada, the UK, China, and Australia emerging as major contributors (Doan et al., 2017). The international collaboration in green building research underlines the global consensus and concerted efforts to drive sustainability in the construction sector. A comparative analysis of assessment criteria and point allocation in new construction manuals reveals common emphasis areas, including indoor environment quality, energy, and materials (Doan et al., 2017). This commonality in focus areas across various rating systems reflects the fundamental importance of these aspects in promoting sustainability. However, notable variations exist in the approach to evaluation, with CASBEE adopting a unique level-based approach, distinguishing itself from the point-based systems seen in BREEAM, LEED, and Green Star NZ (Doan et al., 2017). This diversity in assessment methodologies reflects the adaptability of sustainability principles to different contexts and the innovative approaches employed to assess and promote green building practices globally (Table 16.1).

Table 16.1 Criteria checklist for green buildings

LEED green building rating system checklist	Possible points
Modules	110
sustainable sites	26
Water efficiency	10
Energy and atmosphere	35
Materials and resources	14
Indoor environmental quality	15
Innovation in design	6
Regional priority	4

16.3.10 Critical Success Factors for Green Buildings

Achieving green buildings involves understanding and optimizing critical success factors across technological, managerial, and behavioral dimensions (Zuo & Zhao, 2014). Technological advancements stand as a pivotal dimension, with the utilization of renewable energy technologies playing a central role in reducing energy consumption and emissions (Zuo & Zhao, 2014). Incorporating these technologies effectively into building designs and systems is crucial for achieving energy efficiency and sustainability goals. Additionally, effective waste management throughout the construction and demolition stages is identified as another critical aspect that significantly impacts green building objectives (Zuo & Zhao, 2014). Life cycle assessment (LCA) is highlighted as a popular method for assessing the environmental impacts of buildings, analyzing material and energy flows across various stages of the building life cycle (Zuo & Zhao, 2014). Implementing LCA provides valuable insights into the environmental footprint of buildings and aids in making informed decisions to minimize negative impacts. At the managerial level, project management practices and organizational commitment are identified as key drivers for successful green building initiatives (Zuo & Zhao, 2014). Efficient project management, underpinned by a strong commitment to sustainability, ensures that green building projects stay on track and meet their sustainability objectives. Public policies play a significant role in influencing the green building market, with initiatives like the Commercial Building Disclosure Program incentivizing developers and building owners to invest in high-performance buildings (Zuo & Zhao, 2014). Government interventions and policies create an enabling environment for the proliferation of sustainable building practices, amplifying the impact of individual efforts. Behavioral and cultural factors are recognized as critical success factors, highlighting the need to enhance stakeholder awareness and understanding of sustainability concepts (Zuo & Zhao, 2014). By fostering a culture that values sustainability and environmental consciousness, stakeholders are more likely to actively participate in and support green building initiatives. Addressing these critical success factors is essential for promoting sustainability within the construction industry and accelerating the adoption of green building practices globally. These major findings provide critical insights into the trajectory, approaches, and factors shaping the domain of green building practices and technologies, shedding light on the importance of technology adoption, effective waste management, life cycle assessment, managerial strategies, policy frameworks, and behavioral aspects (Doan et al., 2017; Zuo & Zhao, 2014). Understanding these dynamics and integrating them into strategies and initiatives is pivotal in driving sustainable building practices and fostering a more environmentally conscious construction industry. These insights contribute to the broader goal of achieving a sustainable built environment and align with the overarching theme of integrating green buildings within smart cities for a harmonious blend of sustainability and progress.

16.3.11 Economic Implications of Sustainable Building Practices

A critical aspect of promoting sustainable building practices is understanding the economic implications associated with such endeavors (Hwang & Tan, 2010). Cost emerges as a significant challenge in the adoption of green building projects, often cited as a barrier to widespread implementation (Hwang & Tan, 2010). However, it is crucial to recognize that the initial investment in sustainable construction is an investment in the future. Government incentives, subsidies, and tax breaks can play a pivotal role in offsetting the higher upfront costs, making green building projects financially viable and attractive for developers and investors (Hwang & Tan, 2010). By effectively communicating the long-term economic benefits, such as reduced energy consumption and lower maintenance expenses, stakeholders can appreciate the financial advantages of sustainable practices and overcome cost-related concerns.

16.3.12 Incentivizing Sustainable Construction

Effective communication and collaboration within project teams are emphasized, with a need for a higher level of engagement compared to conventional projects (Hwang & Tan, 2010). Integrating green management goals from top-down and involving green building experts in the project team can ensure that sustainability objectives are woven into every stage of the construction process. Moreover, a lack of credible research on the benefits of green buildings inhibits market demand (Hwang & Tan, 2010). Government subsidies for research and development in green building systems are proposed to bridge the gap between research and practical implementation (Hwang & Tan, 2010). These strategies aim to foster a deeper understanding of the economic benefits associated with sustainable building practices and drive increased interest and investment in green building projects.

16.3.13 Economic Value of Energy Efficiency

Energy efficiency is highlighted as a significant factor influencing economic value in the real estate market (Eichholtz et al., 2010). Buildings certified for energy efficiency under programs like Energy Star experience higher effective rents, demonstrating the financial benefits of reduced energy usage (Eichholtz et al., 2010). Similarly, sustainability attributes, evaluated through certifications like LEED, contribute to higher effective rents (Eichholtz et al., 2010). This implies that sustainability features in buildings are valued by occupants, leading to enhanced market value. Consequently, the market is increasingly recognizing the economic value derived from energy-efficient and sustainable buildings.

16.3.13.1 Economic Benefits of Green Buildings

In the real estate sector, green buildings, whether Energy Star or LEED certified, enjoy higher rents and asset prices compared to conventional office spaces (Eichholtz et al., 2010). This not only underscores the economic benefits of sustainable building practices but also highlights the growing market awareness and preference for environmentally friendly and energy-efficient buildings (Eichholtz et al., 2010). Property investors are increasingly attributing a lower risk premium to energy-efficient and sustainable commercial spaces, indicating a shift in market perception toward recognizing the long-term economic viability of green buildings (Eichholtz et al., 2010). Sustainable building practices, therefore, hold significant economic promise and potential for long-term financial gains, aligning with the global movement toward sustainability and environmental responsibility. These collective findings underscore the economic benefits and market premiums associated with sustainable building practices, emphasizing the need for supportive policies and certifications to propel sustainability in the real estate sector. Encouraging the adoption of sustainable construction practices not only contributes to a more environmentally conscious future but also offers significant economic advantages, enhancing property value, attracting tenants, and reducing operational costs in the long run.

16.3.13.2 Applications of Green Building Concepts in Housing Design

Optimization of building layout is identified as a pivotal consideration, emphasizing efficient orientation and design to harness solar energy and enhance indoor comfort (Zheng, 2021). Rational utilization of resources and energy is emphasized, highlighting the importance of clean energy and renewable materials to reduce waste and minimize environmental impact (Zheng, 2021). Selection of green and environment-friendly building materials is underscored, emphasizing their role in creating a healthier living environment and reducing pollution (Zheng, 2021). Moreover, greening design according to local conditions is emphasized, stressing the need to tailor architectural designs based on local natural environments and climate (Zheng, 2021). These insights into the practical applications of green building concepts in housing design emphasize the need for a holistic approach that encompasses both technological advancements and thoughtful design strategies. Integrating these concepts into the early stages of design and construction allows for the creation of sustainable living spaces that promote environmental conservation and enhance the quality of life for occupants (Figs. 16.1, 16.2, and 16.3).

16.3.14 Educational Frameworks for Advancing Sustainability

The development of green building knowledge categories is deemed crucial, providing effective principles and illustrations for integrating sustainability practices in educational contexts (Cole, 2019). These knowledge categories are refined based on feedback, ensuring alignment with prominent standards for green building design

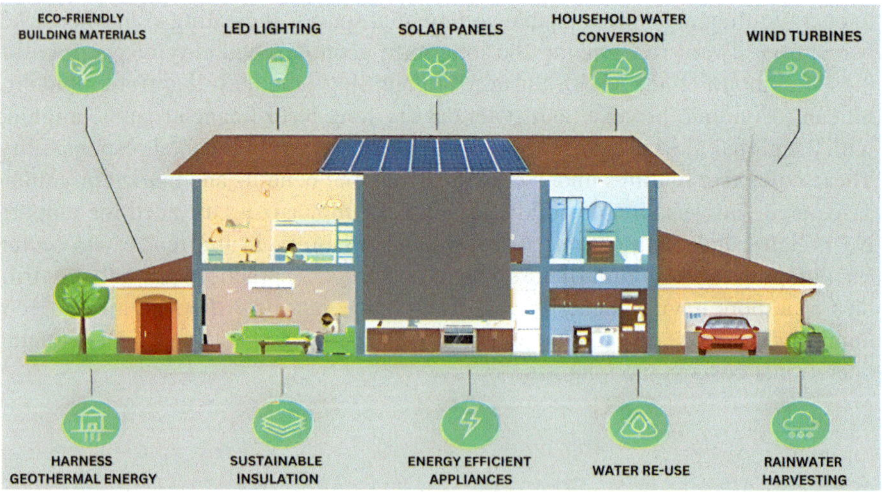

Fig. 16.1 Green building techniques

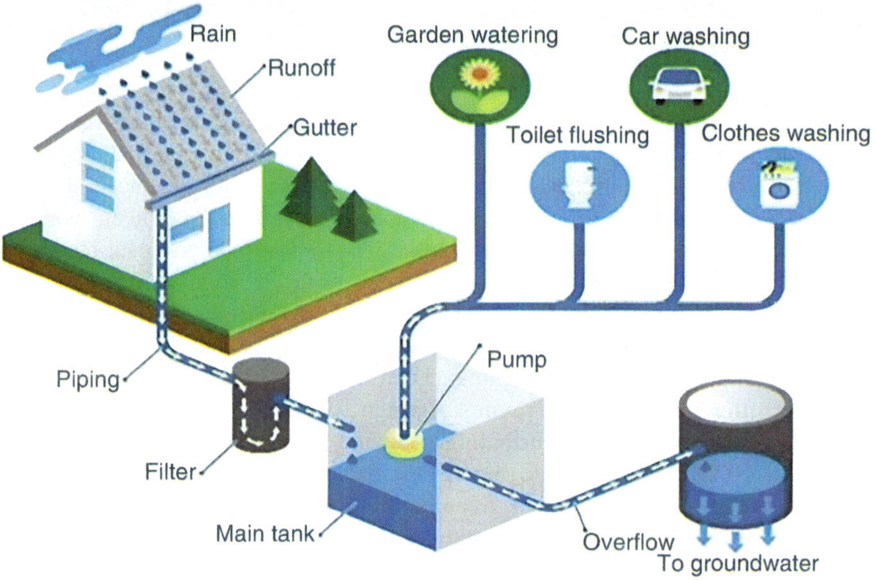

Fig. 16.2 Rainwater harvesting

Fig. 16.3 Solar panel system

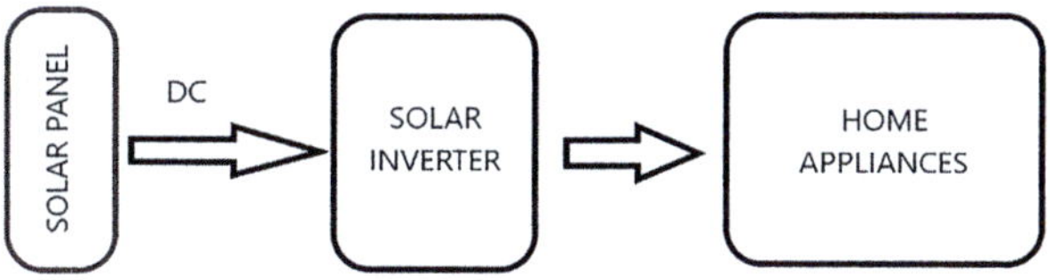

(Cole, 2019). Furthermore, the research underscores the practical implementation of green building literacy frameworks within educational settings (Cole, 2019). Alignment between green building literacy and science standards is identified as essential for promoting adoption in K-12 classrooms (Cole, 2019). This integration ensures that sustainability concepts are ingrained in the educational system, fostering a new generation of environmentally conscious individuals. Promoting green building education across various school systems, both with and without green buildings, is advocated, emphasizing inclusivity and access to resources (Cole, 2019). This approach ensures that sustainability education is not limited to specific regions or demographics but is accessible to all, creating a more informed society with a strong foundation in sustainable practices. In summary, the research findings underscore the significance of green building applications in housing design and the critical role of educational frameworks in advancing sustainability and environmental consciousness (Cole, 2019; Zheng, 2021). These insights provide a foundation for integrating green building principles into architectural design, contributing to a more sustainable and eco-conscious future.

16.3.15 Multidimensional Aspects of Sustainability in Green Building Implementation

Green building implementation is underscored by pivotal aspects of environmental, economic, and social sustainability (Liu et al., 2022). Recognizing and addressing these multidimensional aspects of sustainability is essential for the successful integration of green building practices into the construction industry. The environmental benefits of green buildings are substantial, encompassing significant reductions in energy consumption, CO_2 emissions, and overall environmental impact (Liu et al., 2022). Sustainable building practices, such as the utilization of renewable energy sources and efficient waste management, play a fundamental role in mitigating environmental harm and promoting a healthier ecosystem. Economically, green buildings offer long-term cost savings through reduced energy consumption and maintenance expenses, thus contributing to economic growth and investment returns (Liu et al., 2022). This economic dimension of sustainability emphasizes the financial viability and attractiveness of green building projects, making them appealing to both investors and developers. Moreover, green buildings reflect corporate social responsibility, positively impacting a company's image and indoor environmental quality, consequently enhancing the well-being of occupants (Liu et al., 2022). Prioritizing social sustainability ensures that green buildings provide safe, healthy, and conducive spaces for people to live and work, thus nurturing a thriving community. However, green building implementation is not without challenges, and addressing these challenges is pivotal to advancing sustainability. Balancing energy efficiency and thermal comfort while optimizing natural ventilation poses an environmental challenge (Liu et al., 2022). Striking this balance is crucial in ensuring that green buildings are not only energy-efficient but also comfortable and habitable for occupants. Economically, higher initial investment costs are a hurdle, requiring a balanced perspective to elucidate the long-term benefits, particularly in terms of energy savings (Liu et al., 2022). Overcoming this economic challenge involves effective communication of the return on investment and the enduring financial advantages of sustainable construction. Socio-culturally, limited awareness and understanding of green building concepts among consumers and stakeholders hinder widespread adoption, necessitating improved public participation and a better understanding of sustainable development (Liu et al., 2022). Bridging this awareness gap involves educational initiatives, awareness campaigns, and community engagement efforts that focus on promoting a culture of sustainability.

16.3.16 Trends in Green Building Research

The trends in green building research reveal a multidisciplinary landscape, covering subjects such as engineering, environmental sciences, construction, and energy (Zhao et al., 2019). This multidisciplinary approach highlights the collaborative and comprehensive nature of research in the field of green building, recognizing that addressing sustainability necessitates input and expertise from various domains. Advanced technologies like building information modeling (BIM) and the Internet of Things (IoTs) have gained attention, signifying the direction of research toward smart green buildings (Zhao et al., 2019). The integration of these cutting-edge technologies into green building practices reflects the industry's recognition of the potential for technology to enhance sustainability and efficiency. Key research areas identified include green and cool roof systems, water efficiency, occupants' comfort and satisfaction, financial benefits of green building, and life cycle assessment (LCA) integrated with rating systems (Zhao et al., 2019). These research areas underscore the industry's focus on critical aspects such as resource efficiency, occupant well-being, economic viability, and holistic sustainability assessment. Green building implementation stands as a linchpin for sustainable development, showcasing substantial environmental, economic, and societal benefits (Liu et al., 2022). The multidisciplinary nature of research, the integration of advanced technologies, and the identification of key research clusters indicate a promising trajectory toward sustainable and innovative green building practices (Zhao et al., 2019). Balancing the benefits of sustainability with the challenges posed by implementation remains a critical focus for the future of green building development (Liu et al., 2022).

16.3.17 Synthesis of Major Findings

The research has uncovered a multitude of critical findings, shedding light on the integration, challenges, opportunities, and implications of green building practices within the context of smart cities. The major findings can be synthesized into a comprehensive narrative that encapsulates the essence and significance of each discovery. The research initiated by delineating a structured approach, the integrated construction process (ICP) model, is crucial for successful integration of green buildings within the framework of smart cities (Ahn et al., 2016). This model, characterized by its four-layer structure, not only addresses the existing limitations in project delivery methods but also provides a systematic pathway for achieving sustainability objectives in construction projects (Ahn et al., 2016). Through this structured approach, the research emphasized the importance of organized procedures and stakeholder involvement to drive successful integration. Subsequently, the research underscored the necessity of overcoming barriers hindering the widespread adoption of green building practices. Strategies ranging from mindset reframing to structural changes were identified to encourage behavioral shifts toward sustainability (Hoffman & Henn, 2008). Demographics, such as age, gender, education, and income, were recognized as pivotal influencers in shaping attitudes and behaviors toward sustainability (Hoffman & Henn, 2008). Furthermore, incentivizing risk reduction and evolution of green building standards were proposed as effective strategies to promote sustainable practices (Hoffman & Henn, 2008). These strategies shed light on the need for comprehensive approaches involving both psychological and structural aspects to drive sustainability. The research delved into the dimensions of expertise, governmental involvement, and incentive effectiveness to promote green building initiatives (Samari et al., 2013). Lack of expertise emerged as a significant barrier, emphasizing the need for educational programs and continuous learning to bridge the knowledge gap (Samari et al., 2013). Moreover, the economic implications of green building practices, including enhanced property values and higher effective rents, were highlighted to underline the financial attractiveness of sustainable construction (Hwang & Tan, 2010; Eichholtz et al., 2010). Further, the research broadened its perspective to encompass sustainable building practices and their economic implications. The economic benefits of green buildings, including long-term cost savings and higher asset prices, were highlighted, emphasizing the positive correlation between sustainability and financial gains (Eichholtz et al., 2010). Additionally, the integration of green building literacy within educational frameworks emerged as a critical aspect, ensuring a well-informed and sustainability-conscious future generation (Cole, 2019; Zheng, 2021). The multidimensional nature of sustainability was illuminated, emphasizing the need for a comprehensive approach encompassing environmental, economic, and social aspects (Liu et al., 2022). Challenges, ranging from striking a balance between energy efficiency and thermal comfort to overcoming initial investment costs, were recognized as key hurdles in achieving sustainability (Liu et al., 2022). However, the research noted that technological advancements and the integration of smart technologies like BIM and IoTs are propelling the green building industry toward innovative and sustainable solutions (Zhao et al., 2019). The major findings converge on a central theme: the imperative to promote sustainability within the construction industry, particularly in the context of smart cities. From structured approaches to mindset changes, economic incentives to educational initiatives, the research emphasizes a multifaceted strategy to foster sustainability. Green building practices, when harmoniously integrated into the fabric of smart cities, not only mitigate environmental impact but also yield economic and social benefits. Encouraging sustainable practices in construction is a fundamental step toward creating a more environmentally conscious and prosperous future.

16.4 Conclusion

In the pursuit of sustainable urban development, the integration of green building principles within the context of smart cities emerges as a fundamental and imperative endeavor (Ahn et al., 2016). Green building practices encompass a holistic approach to sustainability, addressing energy, water, and material conservation alongside enhancements in indoor air quality (Ahn et al., 2016). This synthesis advocates for a comprehensive framework, known as the integrated construction process (ICP), to guide stakeholders, including architects, engineers, and contractors, in effectively implementing green building strategies and technologies, ultimately simplifying the certification process (Ahn et al., 2016). The ICP ensures systematic and thorough documentation of green construction activities, facilitating a streamlined transition toward sustainable construction practices (Ahn et al., 2016). A critical aspiration is for green building practices to become mainstream, obviating the necessity for certification systems such as LEED (Hoffman & Henn, 2008). Overcoming social and psychological barriers holds paramount importance in this transition, necessitating an identification and subsequent mitigation of these obstacles to seamlessly integrate green building practices into every facet of the building process (Hoffman & Henn, 2008). Structural shifts within organizational systems and alignments with societal value systems are imperative for this transformative change to materialize (Hoffman & Henn, 2008). Furthermore, recognizing how prevailing cognitions and routines influence decisions is crucial for meaningful and sustainable transformation (Hoffman & Henn, 2008). Governmental roles and incentives emerge as pivotal drivers in advancing green building development, offering structural incentives, subsidies, tax incentive schemes, and market assistance to eliminate barriers hindering the adoption of green building practices (Samari et al., 2013). Both external and internal incentives, intrinsic and extrinsic in nature, play a vital role in motivating stakeholders to embrace green building practices (Olubunmi et al., 2016). Enhancing these incentives and fostering collaboration between the government and private sector are essential strategies to propel the adoption of green building practices on a broader scale (Olubunmi et al., 2016). To anticipate a surge in the adoption of green building technologies (GBTs) within the construction industry, understanding the influencing factors is imperative (Darko et al., 2017). Key barriers include resistance to change stemming from conventional technologies and a lack of knowledge about GBTs, highlighting the necessity for comprehensive awareness and education initiatives (Darko et al., 2017). Critical drivers encompass heightened energy and water efficiency, underlining the importance of resource conservation in promoting GBTs (Darko et al., 2017). Financial incentives and readily available information emerge as significant strategies for the promotion of GBTs, underscoring the importance of tailored policies and dissemination of knowledge for successful implementation (Darko et al., 2017).

The integration of green building principles and GBTs within the broader framework of smart cities stands as a cornerstone for achieving sustainable urban development. Overcoming barriers and leveraging incentives emerge as critical strategies in promoting the seamless amalgamation of these practices within the fabric of modern cities. Continued research and collaborative efforts among stakeholders remain essential in advancing the field and realizing a sustainable and progressive urban future (Ahn et al., 2016; Darko et al., 2017). The extensive literature reviewed across multiple research papers offers a comprehensive understanding of the barriers, advancements, and economic implications pertaining to green building practices and sustainable construction, shedding light on critical aspects including barriers to adoption, evaluation of green rating systems, economic considerations, and project management challenges within the realm of sustainable construction. The identified barriers to the adoption of green building practices encompass a diverse array of factors, ranging from a lack of information and cost implications to the absence of incentives and support, inadequate interest and demand, and a deficiency in relevant codes and regulations. Addressing these barriers necessitates a collaborative approach, involving government bodies, industry associations, developers, and contractors. Key recommendations include augmented government involvement, information dissemination, incentives, and educational initiatives to boost awareness and stimulate interest in green building practices (Darko & Chan, 2016). The research focusing on green rating systems like BREEAM, LEED, CASBEE, and Green Star NZ illuminates their evolution, impact, and the imperative for continuous improvement. These rating systems play a pivotal role in promoting sustainability by evaluating buildings based on various sustainability parameters. A thorough analysis of their similarities, differences, strengths, and weaknesses provides valuable insights for practitioners and researchers, aiding them in comprehending these systems and advocating for comprehensive sustainability assessments in future building projects (Doan et al., 2017). Delving into the economic implications of sustainable building practices, the research reveals that green buildings can yield economic premiums. Despite the initial higher costs associated with green building construction, these investments result in higher rents and asset prices, showcasing their financial viability. The findings underscore the need for incentives, public education, and effective communication to drive the demand for green buildings and further encourage their adoption within the real estate sector (Zuo & Zhao, 2014). Project management within the context of green building projects presents unique challenges compared to conventional projects. The research underscores that effective communication and seamless integration of building design and construction are pivotal for successful project management. Additionally, the high costs associated with green building projects pose a major obstacle, underscoring the importance of fiscal incentives, public education, and communication-conducive delivery systems to promote green building initiatives (Hwang & Tan, 2010). Considering the urgent need for sustainability and the global impact of climate change, the economic and environmental benefits of green building practices are increasingly being recognized. The research emphasizes the potential of green building initiatives to significantly enhance energy efficiency and reduce greenhouse gas emissions, aligning with global efforts to combat climate change. The findings further stress the significance of policy measures, market awareness, and sustainable construction practices in driving the green building revolution (Eichholtz et al., 2010).

In conclusion, the combined insights from these research papers underscore the critical role of collaboration, education, incentives, and effective project management in promoting sustainable construction practices. Green building initiatives not only offer environmental benefits but also present viable economic opportunities. Encouraging the adoption of green building practices through appropriate policies and promoting awareness among stakeholders can significantly contribute to a more sustainable future (Darko & Chan, 2016; Eichholtz et al., 2010). The integration of green buildings within the overarching framework of smart cities represents a pivotal and indispensable endeavor for achieving sustainable urban development (Zheng, 2021). The construction industry, particularly construction enterprises, should proactively engage in green construction practices, with concerted efforts involving laws, policies, management, standards, supervision, and technical research, all coordinated to attain effective green outcomes (Zheng, 2021). Education emerges as a critical facilitator in promoting green building practices, and its integration into schooling is essential to enhance awareness and understanding of green principles (Cole, 2019). To ensure public comprehension of green building concepts, a comprehensive framework outlining diverse educational content and outcomes is imperative (Cole, 2019). This framework should emphasize knowledge, affect, behavior, and active participation in green building practices, fostering a society that not only inhabits but actively contributes to green building sustainability (Cole, 2019). The significance of green building initiatives in achieving Sustainable Development Goals (SDGs) is profound (Liu et al., 2022). Green buildings significantly contribute to energy conservation and the reduction of CO_2 emissions, aligning with the global objectives for a sustainable future (Liu et al., 2022). Despite facing economic and social challenges during development, green building initiatives offer substantial potential to mitigate energy consumption and reduce the greenhouse effect (Liu et al., 2022). Bibliometric reviews of global green building research highlight its economic, social, and environmental sustainability, emphasizing its potential in reducing greenhouse gas emissions from buildings (Zhao et al., 2019). The research trends underscore evolving areas of focus, such as automation and control systems and computer science, and emphasize critical research topics like performance, sustainability, and building systems (Zhao et al., 2019). Understanding these trends and knowledge gaps can guide future research, addressing areas such as corporate social responsibility integration, real performance validation, and leveraging technology to enhance green building practices (Zhao et al., 2019). In the pursuit of sustainable urban futures, the integration of green buildings within smart cities stands as an imperative. It necessitates the alignment of policies, education, research, and societal involvement to unlock the potential of green building initiatives, mitigating environmental impact and fostering progress toward a sustainable world. The challenges are acknowledged, but the collective efforts and advancements in knowledge will drive the transformation toward a greener and smarter urban landscape (Zheng, 2021).

References

Ahn, Y. H., Jung, C. W., Suh, M., & Jeon, M. H. (2016). Integrated construction process for green building. *Procedia Engineering, 145*, 670–676.

Cole, L. B. (2019). Green building literacy: A framework for advancing green building education. *International Journal of STEM Education, 6*(18). https://doi.org/10.1186/s40594-019-0171-6

Darko, A., & Chan, A. P. (2016). Review of barriers to green building adoption. *Sustainable Development, 25*(3), 167–179.

Darko, A., Chan, A. P. C., Ameyaw, E., He, B. J., & Olanipekun, A. O. (2017). Examining issues influencing green building technologies adoption: The United States green building experts' perspectives. *Energy and Buildings, 144*, 320–332.

Doan, D. T., Ghaffarianhoseini, A., Naismith, N., Zhang, T., Ghaffarianhoseini, A., & Tookey, J. (2017). A critical comparison of green building rating systems. *Building and Environment., 123*, 243.

Eichholtz, P., Kok, N., & Quigley, J. M. (2010). *The economics of green building*. Berkeley Program on Housing and Urban Policy. Retrieved from https://escholarship.org/uc/item/3k16

Hoffman, A., & Henn, R. (2008). Overcoming the social and psychological barriers to green building (*Ross School of Business Working Paper Series Working Paper No. 1106*).

Hwang, B. G., & Tan, J. S. (2010). Green building project management: obstacles and solutions for sustainable development. *Sustainable Development, 20*(5), 335–349.

Liu, T., Chen, L., Yang, M., Sandanayake, M., Miao, P., Shi, Y., & Yap, P.-S. (2022). Sustainability considerations of green buildings: A detailed overview on current advancements and future considerations. *Sustainability, 14*, 14393. https://doi.org/10.3390/su142114393

Olubunmi, O. A., Xia, P. B., & Skitmore, M. (2016). Green building incentives: A review. *Renewable and Sustainable Energy Reviews., 59*, 1611.

Samari, M., Godrati, N., Esmaeilifar, R., Olfat, P., & Mohd Shafiei, M. W. (2013). The investigation of the barriers in developing green building in Malaysia. *Modern Applied Science, 7*(2), 1.

Zhao, X., Zuo, J., Wu, G., & Huang, C. (2019). A bibliometric review of green building research 2000–2016. *Architectural Science Review, 62*(1), 74–88. https://doi.org/10.1080/00038628.2018.1485548

Zheng, L. (2021). Research on the application of green building in building design. *IOP Conference Series: Earth and Environmental Science, 783*, 012160.

Zuo, J., & Zhao, Z. Y. (2014). Green building research–current status and future agenda: A review. *Renewable and Sustainable Energy Reviews, 30*, 271–281.

Prasanna Goge, B. Tech student of the Institute of Infrastructure Technology Research and Management (IITRAM) with expertise in construction management techniques including BIM and sustainable development. He has an 8.6 CPI with an honors qualification in civil engineering. He has secured the third position in poster making and presentation for the Alumni cell and is the winner of pitching presentation of Industry Academia Conclave (IAC), IITRAM. He has completed projects on software-based learning in various domains in civil engineering. He also has industrial experience in project management and real estate valuation. He has also completed a training workshop on structural audit and nondestructive testing at All India Institute of Local Self-Government (AIILSG), Mumbai.

Snil Saspara is a B. Tech student of the Institute of Infrastructure Technology Research and Management (IITRAM) with expertise in construction management techniques including BIM and sustainable development. He has a 7.7 CPI in civil engineering and has cleared the GATE examination. He has completed projects on software-based learning in various domains in civil engineering. He also has industrial experience in project management. He has worked on infrastructure projects for Indian Railways.

Dr. Jiten Shah, Ph.D., has expertise in vehicular and human behavior, traffic safety, transportation planning, smart cities, geometric design, and traffic flow modeling including nonmotorized traffic. Three Ph.D. and fourteen M. Tech students work under his guidance. He was involved in various consultancy work like junction improvement, road safety audit, traffic survey, pavement design, proof checking, etc. He has several publications to his credit in reputed international and national journals. He is a recipient of the gold medal from the Gujarat Institute of Civil Engineers and Architects (GICEA). He is a reviewer/editor for journal/s such as the *American Society of Civil Engineers* (ASCE), *Transportation Letters*, *Taylor & Francis*, *Journal of Traffic and Transportation Engineering*, *Transportation Research Record*, *TRB*, and *World Review of Intermodal Transportation Research*. He is also a member of renowned societies/institutes including Transportation Research Group of India (TRG), Institute of Urban Transportation (India), and Institute of Engineers (India).

Chapter 17
Evolving Paradigms of Smart Resilient Cities: Strategic Integration of Smart Technologies

Apurvaa Trivedi and Hardik Shah

Abstract The chapter presents a comprehensive analysis of the collaborative relationship between smart technologies and resilience strategies in urban planning. It delves into how these technologies bolster a city's preparedness, response, and recovery capabilities in the face of disasters. It explores the evolution of smart resilient cities, emphasizing the integration of technological innovation with urban resilience. It also addresses the critical role of governance models and policy frameworks in implementing smart resilience strategies, highlighting the importance of collaborative efforts among multiple stakeholders, including human resources. The chapter also presents global examples showcasing practical applications of these strategies. Emerging technologies like 5G and edge computing are discussed for their potential in revolutionizing urban resilience, while future directions focus on anticipating and planning for disasters with advanced predictive analytics and autonomous systems. The importance of integrating smart technologies with resilience planning to create disaster-resistant urban environments is also discussed.

Keywords Smart resilient cities · Urban planning · Disaster management · Smart technologies · Urban resilience

---

A. Trivedi
Himalayan School of Management Studies, Swami Rama Himalayan University, Dehradun, India

H. Shah (✉)
Institute of Management, Nirma University, Ahmedabad, Gujarat, India
e-mail: hardik@nirmauni.ac.in

S. Majumdar et al. (eds.), *Smart Cities*, S.M.A.R.T. Environments,
https://doi.org/10.1007/978-3-031-59846-3_17

17.1 Introduction

This chapter explores the intersection of smart city technologies and resilience strategies in urban planning to combat disasters. It begins with an overview of how smart technologies enhance a city's ability to prepare for, respond to, and recover from various disasters.

The rapid urbanization of the twenty-first century, coupled with the increasing threats of climate change, has necessitated a transformative approach to urban development. This chapter introduces the concept of "smart resilient cities"—urban environments that leverage smart technologies to enhance their ability to withstand, respond to, and recover from disasters. The integration of resilience into smart city frameworks represents a convergence of urban planning, technology, and disaster management disciplines, aiming to create cities that are not only technologically advanced but also robust against various emergencies.

Resilience in urban planning has evolved from being a reactive measure to a proactive strategy. Godschalk (2003) defines urban resilience as "the ability of an urban system to maintain continuity through all shocks and stresses while positively adapting and transforming toward sustainability." This definition underscores the importance of resilience as an ongoing process rather than a static goal (Godschalk, 2003).

The role of smart technologies in enhancing urban resilience is multifaceted. Technologies such as the Internet of Things (IoT), artificial intelligence (AI), and big data analytics have been instrumental in advancing predictive capabilities, improving resource management, and facilitating efficient disaster response. Chourabi et al. (2012) describe how smart city initiatives utilize these technologies to integrate and analyze data from various sources, thereby enhancing decision-making processes and emergency responses (Chourabi et al., 2012).

However, the integration of smart technologies into urban resilience planning is not without challenges. Issues such as data privacy, the digital divide, and the need for significant infrastructural investments pose considerable hurdles. Kitchin (2014) discusses the complexities of data management and the ethical considerations surrounding privacy and surveillance in smart cities (Kitchin, 2014).

The governance models and policy frameworks also play a pivotal role in the implementation of smart resilience strategies. As noted by Meijer and Bolívar (2016), effective governance in smart cities requires the collaboration of multiple stakeholders, including government, private sector, academia, and citizens, to ensure comprehensive and inclusive planning (Meijer & Bolívar, 2016). This chapter sets the stage for a detailed exploration of how smart city technologies can be strategically integrated into urban planning to enhance resilience against disasters. The following sections delve into the specific technologies, strategies, and case studies that exemplify the concept of smart resilient cities.

17.1.1 Foundations of Smart Resilient Cities

17.1.1.1 Concept and Evolution of Smart Resilient Cities

The concept of smart resilient cities represents an innovative fusion of urban resilience with smart technologies. This integration targets enhancing a city's ability to prepare for, respond to, and recover from various disasters and crises. The evolution of smart resilient cities can be traced back to the early 2000s, when the notion of "smart cities" began to gain prominence. Initially focused on technological advancements and digitalization, the concept gradually expanded to incorporate resilience, especially in the wake of increased urban vulnerabilities to climate change and disasters (Hollands, 2008; Caragliu et al., 2011).

Caragliu et al. (2011) provide a comprehensive definition of smart cities, emphasizing the role of human capital, social and relational capital, environmental interest, and information and communication technology (ICT) in urban development (Caragliu et al., 2011). The addition of resilience to this framework signifies a shift toward a more holistic approach to urban development, where the emphasis is not just on technological advancement but also on sustainability and disaster preparedness (Coaffee, 2008).

17.1.2 Theoretical Frameworks in Smart Urban Resilience

Urban resilience involves a range of theoretical frameworks that guide the planning, development, and management of cities. One key framework is the "systems theory," which views a city as a system comprising various interconnected components. Meijer and Bolívar (2016) discuss how systems theory can be applied to understand the complexities and interdependencies within smart cities, thus aiding in the development of more resilient urban environments (Meijer & Bolívar, 2016).

Another important framework is the "urban resilience model," which focuses on a city's capacity to absorb, recover, and prepare for future shocks. Ahern (2011) emphasizes the importance of designing resilient urban landscapes that can adapt and transform in response to changing conditions, thus ensuring long-term sustainability (Ahern, 2011).

17.1.2.1 Role of Smart Technologies in Enhancing Urban Resilience

Smart technologies are increasingly recognized for their pivotal role in enhancing urban resilience. This section explores how the integration of technologies like the Internet of Things (IoT), artificial intelligence (AI), and big data analytics contribute significantly to strengthening cities against disasters.

17.1.2.2 Benefits of Smart Technologies in Urban Resilience

1. *Real-Time Data Collection and Analysis*: Technologies like IoT and AI enable the collection and analysis of real-time data, facilitating predictive and adaptive decision-making in urban management. This aspect is crucial for preparing and responding to disasters, as noted by Batty et al. (2012), who underscore the importance of urban informatics and real-time data in improving decision-making processes, thereby bolstering a city's resilience (Batty et al., 2012).
2. *Enhanced Situational Awareness*: Smart technologies significantly enhance situational awareness through improved sentient capacity, communication, connectivity, user interfaces, and data analytics. This enhancement is critical in disaster scenarios, as articulated by Bellini and Nesi (2018), who discuss the role of smart technologies in improving a city's capacity to cope with emerging threats (Bellini & Nesi, 2018).
3. *Efficient Resource Management*: The use of AI and big data analytics leads to more efficient management of resources, contributing to sustainable urban development. This efficiency is vital for ensuring rapid recovery from disasters, as highlighted by Xiong et al. (2022) in their discussion on the role of smart technologies in urban operations (Xiong et al., 2022).

17.1.2.3 Contribution of Smart Technologies in Urban Resilience

1. *Data Management and Public Participation*: Smart technologies facilitate the creation, sustenance, and enhancement of resilience within smart cities by managing citywide geodata and fostering public participation (Samarakkody et al., 2023).
2. *Resilience Characteristics Enhancement*: These technologies enhance various resilience characteristics like robustness, modularity, and agility, applicable in sectors like early warning systems, social media, and transportation (Sharifi & Allam, 2023).
3. *Efficient Disaster Response*: Townsend (2013) highlights how smart city technologies improve disaster response and recovery efforts by facilitating effective communication and coordination during emergencies (Townsend, 2013).

17.1.2.4 Challenges in Implementing Smart Technologies

1. *Understanding Smartness-Resilience Relationship*: The relationship between "smartness" and "resilience" is complex and requires further exploration, particularly in how digital technologies aid resilience-building processes (Cañavera-Herrera et al., 2022).

2. *Need for Interdisciplinary Research*: There is a pressing need for interdisciplinary research to better understand the interplay between smartness and resilience, extending to physical, social, and environmental systems (Cañavera-Herrera et al., 2022).
3. *Context-Specific Technology Utilization*: The effective use of technological innovations in resilience-building necessitates an understanding of their current usage and application in various contexts (Samarakkody et al., 2023).

17.1.3 Smart Technologies in Disaster Resilience

17.1.3.1 Role of IoT and AI in Disaster Management

The integration of the Internet of Things (IoT) and artificial intelligence (AI) in disaster management has opened new avenues for effectively monitoring, predicting, and responding to urban disasters. This section expands upon the roles of IoT and AI in disaster management by incorporating additional academic studies and real-world applications.

1. *Comprehensive Monitoring and Communication*: IoT's capabilities extend beyond data collection and include facilitating communication between professionals. This aspect plays a vital role in environmental management, first aid, people tracking, and the distribution of supplies and medicines (Bail et al., 2021).
2. *Disaster Mitigation, Rescue, and Recovery*: IoT technologies are instrumental in all phases of disaster management, including mitigation, rescue, response, and recovery. They enhance the efficiency and effectiveness of these operations (Bail et al., 2021).
3. *Early Warning Systems*: IoT is pivotal in developing early-warning systems for potential disasters such as fires and earthquakes. Such systems are essential for timely evacuations and mitigating disaster impacts (Sharma et al., 2021).

17.1.3.2 Contributions of AI in Disaster Management

1. *Data-Driven Insights and Forecasting*: AI leverages diverse data types, including geospatial, social media, and wireless network sensor data, to enhance understanding and forecasting of natural disasters. This broad spectrum of data enhances the precision of AI predictions and humanitarian assistance (Seng et al., 2022).
2. *AI in All Phases of Disaster Management*: AI applications are not limited to a single phase of disaster management but span across mitigation, preparedness, response, and recovery. AI processes disaster-related data to support informed decision-making and develops AI-based decision support tools (Sun et al., 2020).

17.1.3.3 Synergistic Integration of IoT and AI

The combination of IoT and AI technologies provides a more holistic approach to disaster management. IoT offers real-time data collection, while AI analyzes this data for predictive insights. This synergy not only enhances the speed and efficiency of emergency response but also improves decision-making processes and contributes significantly to disaster resilience (Shahat et al., 2020; Munawar et al., 2022).

17.1.3.4 Challenges and Future Directions

Implementing IoT in disaster management faces challenges such as the need for robust information and communication technology (ICT) systems and overcoming barriers to utilizing disruptive technologies. The integration of AI with IoT, through AI's analysis of IoT-generated data, promises faster and more efficient responses to disasters. However, addressing these challenges is crucial for the successful implementation and maximization of these technologies' potential (Munawar et al., 2022). The IoT's diverse applications in disaster management, ranging from communication facilitation to early-warning systems, combined with AI's capacity to leverage various data types for improved decision-making, underscore their vital roles in enhancing disaster management strategies. The challenges present in their implementation highlight the need for continued development and integration of these technologies to achieve optimal disaster resilience.

17.1.3.5 Application of Transformative Technologies in Disaster Preparedness and Response in Smart Cities

The advent of disruptive technologies, notably data analytics and predictive modeling, has significantly influenced the way smart cities approach disaster resilience. This section delves into the pivotal role these technologies play in forecasting, mitigating, and managing disasters in urban environments. Innovative technologies are increasingly recognized as game changers in developing strategies for disaster resilience in urban settings. The application of modern technology in analyzing risks associated with natural disasters like floods, earthquakes, and bushfires is critical. These technologies facilitate the development of effective contingency plans and risk management policies essential for disaster relief (Li et al., 2020).

17.1.3.6 Integration of AI, IoT, and Big Data

The integration of artificial intelligence (AI), the Internet of Things (IoT), and big data has emerged as a cornerstone in understanding and managing disaster risks. By utilizing historical data from previous disasters, these systems provide a comprehensive analysis of disaster risks and public safety. AI, in particular, has been instrumental in analyzing earthquake fault lines, weather modeling, and environmental chemical exposures, thus enabling the creation of a substantial data repository concerning disaster risks (Munawar et al., 2022; Syifa et al., 2019; Ovesný et al., 2014).

17.1.3.7 Application of Disruptive Technologies in Disaster Management

Disruptive technologies play a crucial role in disaster mitigation, preparedness, response, and recovery. The focus is on leveraging these technologies for emergency evacuation, rescue operations, and post-disaster management. The combination of disruptive technology and disaster risk understanding is key to overcoming challenges in disaster management (Shah et al., 2019; Vorbach et al., 2017; Zhang et al., 2021; Alahakoon et al., 2023).

17.1.3.8 Artificial Intelligence in Disaster Management

AI‘s capabilities in learning, predicting, and providing timely information are invaluable during all phases of disaster management. It has been applied for flood risk control, remote sensing data analysis, and infrastructure damage assessment. Techniques like random forest, neural networks, and support vector machines (SVM) have been employed for efficient damage detection. AI’s potential in enhancing weather predictions and providing early warnings for disaster preparation is noteworthy (Xie et al., 2019; Choubin et al., 2019; Nabian & Meidani, 2018; Reilly et al., 2013; Fang, 2015).

17.1.3.9 Big Data Analysis in Disaster Management

The big data paradigm involves the management of large data sets for disaster evaluation. This paradigm, which encompasses data collection, storage, processing, and visualization, is pivotal in enhancing decision-making during disasters. Big data analytics have been crucial in assessing the economic impact of disasters and identifying affected communities. However, this paradigm also raises concerns regarding data privacy and security (Ichimura et al., 2022; Reilly et al., 2013).

Data analytics and predictive modeling are integral to the resilience of smart cities against disasters. By effectively managing and analyzing vast amounts of data, these technologies enable informed decision-making, risk reduction, and efficient post-disaster recovery, thereby fortifying urban areas against the adverse impacts of disasters.

17.1.4 Challenges and Solutions in Smart Resilient Urban Planning

17.1.4.1 Integrating Smart Technologies with Urban Infrastructure

The integration of smart technologies into urban infrastructure is a transformative process that redefines the urban landscape. This integration, as Neirotti et al. (2014) point out, is not just about the technological advancements but also about the innovative application of these technologies to enhance urban life. Central to this transformation is the deployment of the Internet of Things (IoT) and sensor networks. As Zanella et al. (2014) elaborate, IoT devices play a pivotal role in collecting real-time data, which is essential for the dynamic management of city resources and services. This data becomes the backbone of urban planning and management, contributing to areas such as traffic control, public safety, and environmental monitoring.

The utilization of big data analytics, as discussed by Batty et al. (2012), is another crucial aspect. Through sophisticated data analysis, city planners can decipher patterns and trends from the vast amounts of data generated, leading to more informed decision-making. This approach not only enhances operational efficiency but also contributes to the strategic planning of urban development.

Moreover, the incorporation of smart grids, as described by Gungor et al. (2013), revolutionizes urban energy systems. These grids offer a more sustainable and reliable energy distribution, accommodating renewable energy sources and facilitating real-time energy management. The concept of digital twins, as introduced by Kitchin (2014), further extends the capabilities of smart urban planning. By creating virtual replicas of physical spaces, urban planners can simulate and predict various urban dynamics, thus enabling more effective planning and problem-solving.

However, these technological advancements bring forth significant challenges. Data privacy and security emerge as paramount concerns, as highlighted by Meijer and Bolívar (2016). Protecting the sensitive information collected from urban sensors and IoT devices is critical to maintaining public trust and ensuring the ethical use of technology. Additionally, the issue of interoperability, as Angelidou (2015) notes, presents a significant challenge. The diverse range of smart technologies and platforms requires a unified approach to ensure seamless communication and data exchange. This interoperability is essential for the holistic integration of smart technologies into urban infrastructure, enabling different systems to work together efficiently and effectively.

17.1.4.2 Balancing Technological and Human Factors

While the integration of technology is crucial, achieving a balance between technological and human factors is equally important for the success of smart cities. Hollands (2008) emphasizes that smart urban planning should not only focus on technological advancements but also consider the social, cultural, and ethical implications of these technologies. It is essential to adopt a human-centric approach to urban planning, which involves understanding and addressing the needs, values, and aspirations of the urban population.

Public engagement is a key component of this approach. As Chourabi et al. (2012) suggest, involving citizens in the planning and implementation of smart city initiatives ensures that these projects are aligned with the community's needs and preferences. This participatory approach fosters a sense of ownership among residents and leads to more sustainable and accepted urban solutions.

Furthermore, ethical considerations, as discussed by Kitchin (2016), play a critical role in smart city development. Issues related to surveillance, data privacy, and potential misuse of technology need to be addressed to maintain public trust and acceptance. Additionally, incorporating cultural sensitivity into urban planning is vital. Allwinkle and Cruickshank (2011) highlight the importance of respecting local cultures and traditions in the implementation of technological solutions. This sensitivity ensures that these solutions are not only effective but also resonate with the cultural context of the city, thus enhancing their acceptance and impact.

However, the integration of technology in urban planning is not without challenges. The digital divide, as discussed by Vanolo (2014), is a significant concern. Ensuring equitable access to the benefits of smart city technologies is essential to prevent the exacerbation of existing social and economic inequalities. This includes providing access to digital resources and training for all segments of the population, thereby ensuring that everyone can benefit from the advancements of a smart city.

In addition to bridging the digital divide, addressing socioeconomic factors is crucial. Neirotti et al. (2014) argue that smart city initiatives must consider the diverse socioeconomic backgrounds of urban residents to avoid deepening existing inequalities. This includes designing inclusive and accessible smart solutions that cater to the varied needs of different social groups, thereby ensuring that the benefits of smart urban planning are distributed equitably across the entire urban population.

The integration of smart technologies with urban infrastructure and balancing these technological advancements with human factors are critical components in the development of resilient and sustainable smart cities. These approaches require careful consideration of technological capabilities, ethical implications, cultural sensitivities, and socioeconomic factors to ensure the successful implementation of smart city initiatives.

17.1.5 Policy and Governance in Smart Resilient Cities

17.1.5.1 Regulatory Frameworks for Smart Resilience

In the context of smart resilient cities, regulatory frameworks play a crucial role in guiding and shaping the development and implementation of smart city initiatives. These frameworks provide the legal and ethical basis for integrating smart technologies into urban infrastructure while ensuring the protection of citizens' rights and data privacy. As noted by Chourabi et al. (2012), effective regulatory frameworks must address issues related to technology deployment, data governance, and public-private partnerships. These frameworks should be flexible enough to adapt to rapid technological advancements while maintaining robust protections for individual and communal rights.

An important aspect of these frameworks is their role in facilitating innovation. As emphasized by Kitchin (2016), regulations should not stifle technological innovation but rather encourage responsible and ethical development of smart solutions. This involves creating policies that balance the need for data collection and analysis with privacy and security concerns.

17.1.5.2 Human Governance Models and Stakeholder Collaboration

Governance models in smart cities are evolving to accommodate the multifaceted nature of smart resilient urban planning. These models often involve a combination of government, private sector, and community stakeholders working collaboratively. As Hollands (2008) asserts, effective governance in smart cities requires the active participation of all stakeholders to ensure that the development is inclusive and beneficial for the entire community. This collaborative approach fosters transparency, accountability, and shared responsibility in decision-making processes.

Stakeholder collaboration is particularly vital in addressing complex urban challenges. As Meijer and Bolívar (2016) point out, collaboration enables the pooling of resources, expertise, and perspectives, leading to more comprehensive and effective solutions. It is essential for governance models to facilitate such collaboration through clear communication channels, shared goals, and equitable participation.

17.1.6 Global Examples of Smart Resilient Strategies

17.1.6.1 Tokyo's Earthquake-Resistant Infrastructure

Tokyo, located in one of the world's most earthquake-prone areas, has implemented advanced technologies for earthquake resistance. The city's approach includes the use of seismic isolation structures and early warning systems that utilize IoT technologies for real-time monitoring and response. Tokyo's resilience strategy also involves community-based disaster risk reduction programs, enhancing the city's preparedness for earthquakes. This comprehensive approach integrates smart technology with community resilience, setting a model for earthquake-prone cities globally (Edgington, 2010).

17.1.6.2 Copenhagen's Climate Adaptation Plan

Copenhagen's Climate Adaptation Plan exemplifies a resilient strategy against climate change and rising sea levels. The city incorporates smart technologies in its urban infrastructure, including green roofs, rainwater harvesting systems, and climate-resilient urban development. These initiatives are coupled with real-time data monitoring systems for water levels and weather conditions, enabling proactive responses to climate threats. Copenhagen's integrated approach to climate adaptation and smart technology usage serves as a blueprint for other coastal cities facing similar challenges (City of Copenhagen, 2011).

17.1.6.3 San Francisco's SMART Program for Disaster Preparedness

San Francisco, known for its susceptibility to earthquakes and wildfires, has implemented the SMART (strategic management and response team) program. This initiative leverages smart technologies such as geographic information systems (GIS), cloud computing, and mobile applications to enhance disaster preparedness and response. The program focuses on real-time data analysis, public communication, and emergency response coordination, integrating these elements into the city's broader resilience strategy (San Francisco, 2022). San Francisco's SMART Program, particularly the SF72 initiative, exemplifies an innovative approach to disaster preparedness. Developed by San Francisco's Department of Emergency Management in collaboration with IDEO, SF72 focuses on creating a culture of positive preparedness in the community. It emphasizes the use of technology and community engagement to prepare for disasters like earthquakes and wildfires.

17.1.7 Future Directions in Smart Urban Resilience

17.1.7.1 Emerging Technologies and Their Potential

The advent of emerging technologies such as 5G and edge computing is set to revolutionize urban resilience. 5G technology, with its high-speed and low-latency capabilities, enables more efficient data transmission, crucial for real-time monitoring and response in urban environments (Alsharif et al., 2020). This can significantly enhance emergency services and disaster response capabilities. Edge computing, on the other hand, processes data closer to where it is generated, reducing reliance on central data centers and improving response times in critical situations (Shi et al., 2016). These technologies also support advanced IoT applications, making cities more responsive and adaptive to emergency scenarios.

17.1.7.2 Anticipating and Planning for Future Disasters

As cities evolve, anticipating and planning for future disasters becomes increasingly crucial. The evolution of smart city technologies could play a pivotal role in addressing future challenges. Advanced predictive analytics, powered by AI and machine learning, can analyze vast data sets to forecast disasters more accurately (Meier, 2015). The integration of autonomous systems and drones for rapid response and relief operations could also become standard in disaster management (Erdelj et al., 2017). These advancements, coupled with continuous innovation in communication and sensor technologies, can significantly enhance urban resilience against future disasters.

17.2 Conclusion

Integrating resilience and smart technologies in urban environments marks a transformative shift in disaster management and urban planning. This integration paves the way for cities that are not only technologically advanced but also robust against various emergencies. With the rapid urbanization and increasing threats of climate change, this approach is more relevant than ever. Utilizing smart technologies such as IoT, AI, and big data analytics, cities can enhance their predictive capabilities, improve resource management, and facilitate efficient disaster responses. However, challenges like data privacy, digital divide, and infrastructural investments are significant. Successful examples from Tokyo, Copenhagen, and San Francisco demonstrate the practical application of these concepts. Looking ahead, emerging technologies like 5G and edge computing offer new possibilities for urban resilience. Anticipating and planning for future disasters with these technologies will be crucial in evolving smart city strategies. The chapter illustrates the importance of a holistic approach, combining technological innovation with ethical and cultural considerations, to create sustainable and disaster-resilient urban futures.

References

Ahern, J. (2011). From fail-safe to safe-to-fail: Sustainability and resilience in the new urban world. *Landscape and Urban Planning, 100*(4), 341–343. https://doi.org/10.1016/j.landurbplan.2011.02.021

Alahakoon, D., Nawaratne, R., Xu, Y., De Silva, D., Sivarajah, U., & Gupta, B. (2023). Self-building artificial intelligence and machine learning to empower big data analytics in smart cities. *Information Systems Frontiers*, 1–20.

Allwinkle, S., & Cruickshank, P. (2011). Creating smarter cities: An overview. *Journal of Urban Technology, 18*(2), 1–16. https://doi.org/10.1080/10630732.2011.601103

Alsharif, M. H., Kelechi, A. H., Albreem, M. A., Chaudhry, S. A., Zia, M. S., & Kim, S. (2020). Sixth generation (6G) wireless networks: Vision, research activities, challenges and potential solutions. *Symmetry, 12*(4), 676. https://doi.org/10.3390/sym12040676

Angelidou, M. (2015). Smart cities: A conjuncture of four forces. *Cities, 47*, 95–106. https://doi.org/10.1016/j.cities.2015.05.004

Bail, R. D. F., Kovaleski, J. L., da Silva, V. L., Pagani, R. N., & Chiroli, D. M. D. G. (2021). Internet of things in disaster management: Technologies and uses. *Environmental Hazards, 20*(5), 493–513.

Batty, M., Axhausen, K. W., Fosca, G., Pozdnoukhov, A., Bazzani, A., Wachowicz, M., et al. (2012). Smart cities of the future. *The European Physical Journal Special Topics, 214*(1), 481–518. https://doi.org/10.1140/epjst/e2012-01703-3

Bellini, E., & Nesi, P. (2018). Exploiting smart technologies to build smart resilient cities. In *Routledge handbook of sustainable and resilient infrastructure* (pp. 685–705). Routledge.

Cañavera-Herrera, J. S., Tang, J., Nochta, T., & Schooling, J. M. (2022). On the relation between 'resilience'and 'smartness': A critical review. *International Journal of Disaster Risk Reduction, 75*, 102970.

Caragliu, A., Del Bo, C., & Nijkamp, P. (2011). Smart cities in Europe. *Journal of Urban Technology, 18*(2), 65–82.

Choubin, B., Borji, M., Mosavi, A., Sajedi-Hosseini, F., Singh, V. P., & Shamshirband, S. (2019). Snow avalanche hazard prediction using machine learning methods. *Journal of Hydrology, 577*, 123929.

Chourabi, H., Nam, T., Walker, S., Gil-Garcia, J. R., Mellouli, S., Nahon, K., et al. (2012). Understanding smart cities: An integrative framework. In *45th Hawaii international conference on system sciences*. https://doi.org/10.1109/HICSS.2012.615

City of Copenhagen. (2011). *Copenhagen Climate Adaptation Plan*. Retrieved from City of Copenhagen's Official Website: https://international.kk.dk/

Coaffee, J. (2008). Risk, resilience, and environmentally sustainable cities. *Energy Policy, 36*, 4633–4638.

Edgington, D. W. (2010). *Reconstructing Kobe: The geography of crisis and opportunity*. UBC Press.

Erdelj, M., Król, M., & Natalizio, E. (2017). Wireless sensor networks and multi-UAV systems for natural disaster management. *Computer Networks, 124*, 72–86. https://doi.org/10.1016/j.comnet.2017.06.009

Fang, H. (2015, June). Managing data lakes in big data era: What's a data lake and why has it became popular in data management ecosystem. In *2015 IEEE international conference on cyber technology in automation, control, and intelligent systems (CYBER)* (pp. 820–824). IEEE.

Godschalk, D. R. (2003). Urban Hazard mitigation: Creating resilient cities. *Natural Hazards Review, 4*, 136–143. https://doi.org/10.1061/(ASCE)1527-6988(2003)4:3(136)

Gungor, V. C., Sahin, D., Kocak, T., Ergut, S., Buccella, C., Cecati, C., & Hancke, G. P. (2013). Smart grid technologies: Communication technologies and standards. *IEEE Transactions on Industrial Informatics, 7*(4), 529–539. https://doi.org/10.1109/TII.2011.2166794

Hollands, R. G. (2008). Will the real smart city please stand up? Intelligent, progressive or entrepreneurial? *City, 12*(3), 303–320. https://doi.org/10.1080/13604810802479126

Ichimura, Y., Dalaklis, D., Kitada, M., & Christodoulou, A. (2022). Shipping in the era of digitalization: Mapping the future strategic plans of major maritime commercial actors. *Digital Business, 2*(1), 100022. https://doi.org/10.1016/j.digbus.2022.100022

Kitchin, R. (2014). The real-time city? Big data and smart urbanism. *GeoJournal, 79*(1), 1–14. https://doi.org/10.1007/s10708-013-9516-8

Kitchin, R. (2016). The ethics of smart cities and urban science. *Philosophical Transactions of the Royal Society A: Mathematical, Physical and Engineering Sciences, 374*(2083), 20160115. https://doi.org/10.1098/rsta.2016.0115

Li, M., Porter, A. L., & Suominen, A. (2018). Insights into relationships between disruptive technology/innovation and emerging technology: A bibliometric perspective. *Technological Forecasting and Social Change, 129*, 285–296.

Li, L., Gao, S., Wang, Y., & Liu, X. (2019). Artificial intelligence in disaster management: A review. *International Journal of Disaster Risk Reduction, 35*, 101096.

Li, Q., Zhang, H. G., & Hong, X. P. (2020). Knowledge structure of technology licensing based on co-keywords network: A review and future directions. *International Review of Economics & Finance, 66*, 154–165.

Meier, P. (2015). *Digital humanitarians: How big data is changing the face of humanitarian response*. CRC Press.

Meijer, A., & Bolívar, M. P. R. (2016). Governing the smart city: A review of the literature on smart urban governance. *International Review of Administrative Sciences, 82*(2), 392–408. https://doi.org/10.1177/0020852314564308

Munawar, H. S., Mojtahedi, M., Hammad, A. W., Kouzani, A., & Mahmud, M. P. (2022). Disruptive technologies as a solution for disaster risk management: A review. *Science of the Total Environment, 806*, 151351.

Nabian, M. A., & Meidani, H. (2018). Deep learning for accelerated seismic reliability analysis of transportation networks. *Computer-Aided Civil and Infrastructure Engineering, 33*(6), 443–458.

Neirotti, P., De Marco, A., Cagliano, A. C., Mangano, G., & Scorrano, F. (2014). Current trends in smart city initiatives: Some stylised facts. *Cities, 38*, 25–36. https://doi.org/10.1016/j.cities.2013.12.010

Ovesný, M., Křížek, P., Borkovec, J., Švindrych, Z., & Hagen, G. M. (2014). ThunderSTORM: A comprehensive ImageJ plug-in for PALM and STORM data analysis and super-resolution imaging. *Bioinformatics, 30*(16), 2389–2390.

Reilly, J., Dashti, S., Ervasti, M., Bray, J. D., Glaser, S. D., & Bayen, A. M. (2013). Mobile phones as seismologic sensors: Automating data extraction for the iShake system. *IEEE Transactions on Automation Science and Engineering, 10*(2), 242–251.

Samarakkody, A., Amaratunga, D., & Haigh, R. (2023). Technological innovations for enhancing disaster resilience in smart cities: A comprehensive urban scholar's analysis. *Sustainability, 15*(15), 12036.

San Francisco (2022). *San Francisco's strategic approach to disaster preparedness*. Retrieved from City and County of San Francisco's Official Website: https://www.sf.gov/resource/2022/san-francisco-emergency-plans

Seng, K. P., Ang, L. M., & Ngharamike, E. (2022). Artificial intelligence Internet of Things: A new paradigm of distributed sensor networks. *International Journal of Distributed Sensor Networks, 18*(3), 15501477211062835.

Shah, S. A., Seker, D. Z., Rathore, M. M., Hameed, S., Yahia, S. B., & Draheim, D. (2019). Towards disaster resilient smart cities: Can internet of things and big data analytics be the game changers? *IEEE Access, 7*, 91885–91903.

Shahat, E., Hyun, C. T., & Yeom, C. (2020). Conceptualizing smart disaster governance: An integrative conceptual framework. *Sustainability, 12*(22), 9536.

Sharifi, A., & Allam, Z. (2023). *Contributions of smart technologies to disaster resilience*. Deakin University. Chapter retrieved from https://hdl.handle.net/10779/DRO/DU:22092611.v1

Sharma, K., Anand, D., Sabharwal, M., Tiwari, P. K., Cheikhrouhou, O., & Frikha, T. (2021). A disaster management framework using internet of things-based interconnected devices. *Mathematical Problems in Engineering, 2021*, 1–21.

Shi, W., Cao, J., Zhang, Q., Li, Y., & Xu, L. (2016). Edge computing: Vision and challenges. *IEEE Internet of Things Journal, 3*(5), 637–646. https://doi.org/10.1109/JIOT.2016.2579198

Syifa, M., Park, S. J., Achmad, A. R., Lee, C. W., & Eom, J. (2019). Flood mapping using remote sensing imagery and artificial intelligence techniques: A case study in Brumadinho, Brazil. *Journal of coastal research, 90*(SI), 197–204.

Sun, W., Bocchini, P., & Davison, B. D. (2020). Applications of artificial intelligence for disaster management. *Natural Hazards, 103*(3), 2631–2689.
Townsend, A. M. (2013). *Smart cities: Big data, civic hackers, and the quest for a new utopia.* W.W. Norton & Company.
Vanolo, A. (2014). Smartmentality: The smart city as disciplinary strategy. *Urban Studies, 51*(5), 883–898. https://doi.org/10.1177/0042098013494427
Vorbach, S., Wipfler, H., & Schimpf, S. (2017). Business model innovation vs. business model inertia: The role of disruptive technologies. BHM berg-und Hüttenmännische. *Monatshefte, 162*(9), 382–385.
Xie, X., Fu, G., Xue, Y., Zhao, Z., Chen, P., Lu, B., & Jiang, S. (2019). Risk prediction and factors risk analysis based on IFOA-GRNN and apriori algorithms: Application of artificial intelligence in accident prevention. *Process Safety and Environmental Protection, 122*, 169–184.
Xiong, K., Sharifi, A., & He, B. J. (2022). Resilient-smart cities: Theoretical insights. In *Resilient smart cities: Theoretical and empirical insights* (pp. 93–118). Springer.
Zanella, A., Bui, N., Castellani, A., Vangelista, L., & Zorzi, M. (2014). Internet of Things for smart cities. *IEEE Internet of Things Journal, 1*(1), 22–32. https://doi.org/10.1109/JIOT.2014.2306328
Zhang, Y., Geng, P., Sivaparthipan, C. B., & Muthu, B. A. (2021). Big data and artificial intelligence based early risk warning system of fire hazard for smart cities. *Sustainable Energy Technologies and Assessments, 45*, 100986.

Dr. Apurvaa Trivedi is currently working as an Associate Professor at Himalayan School of Management Studies, Swami Rama Himalayan University. He has 15+ years of blended experience in the field of IT, HR, and Management Education. He is actively involved in training and development activities and is a Management Entrepreneurship and Professional Skills Council certified trainer. He is a life member of the Indian Society for Training and Development (ISTD) and the National Institute of Personnel Management (NIPM). He is the Vice-Chairperson of state chapters of ISTD and NIPM. He is also involved in research and consultancy and has successfully completed many management development programs. His research interests include technology in human resource management, innovation and design thinking, women empowerment, disaster risk reduction, and reflexivity and its application in research. He has published his research work in national and international peer-reviewed journals and frequently presents his work at various national and international conferences.

Dr. Hardik Shah is currently working as an Associate Professor and Chairperson—Human Resource Management (HRM) and Management Development Program (MDP) at the Institute of Management, Nirma University. He is a passionate faculty, trainer, and researcher having around 20+ years of experience in the HRM and organizational development (OD) domain. Prior to joining Nirma University, he worked with IMT, Ghaziabad (IMT-G), Adani Institute of Infrastructure Management, Academy of HRD, AIMS, and UPES. He successfully facilitated and steered several training/consulting assignments in different organizations, including IIT-Roorkee, Vidhya Knowledge Park, UGC-HRD (Nainital), ISTD, AHRD, Johnson Controls Hitachi Air-Conditioning, IOCL, BPCL, Aga Khan Foundation, Nuvoco Vistas Corp, Mahindra Comviva, NIAM, CEL, Apollo Tyres, Reckitt Benkiser (RB), EXL Service Inc., IFB, NIIT, NHPC, ISP, GSRTC, NTPC, SAIL, UPCL, HPCL, GUVNL, EISS, Hindustan Zink, GETRI, Cadila Pharma, Eureka Forbes, etc.

He has authored 5 books and over 35+ research papers. He is on the editorial board of *AI Practitioner*, the only journal worldwide, dedicated to "appreciative inquiry." He is also reviewer for the *Journal of Public Administration and Policy Research* (JPAPR), *Journal of Management Development* (JMD), and *European Journal of Training and Development* (EJTD). His current areas of interest include competency mapping, project teams, managerial capability building, psychometric testing/assessment, and appreciative inquiry.

Index

S. Majumdar et al. (eds.), *Smart Cities*, S.M.A.R.T. Environments,
https://doi.org/10.1007/978-3-031-59846-3

SPRINGER NATURE

GPSR Compliance

The European Union's (EU) General Product Safety Regulation (GPSR) is a set of rules that requires consumer products to be safe and our obligations to ensure this.

If you have any concerns about our products, you can contact us on ProductSafety@springernature.com

In case Publisher is established outside the EU, the EU authorized representative is:

Springer Nature Customer Service Center GmbH
Europaplatz 3
69115 Heidelberg, Germany

The manufacturer's authorised representative in the EU is Springer Nature Customer Service Centre GmbH, Europaplatz 3, 69115 Heidelberg, Germany. If you have any concerns regarding our products, please contact ProductSafety@springernature.com

Printed and bound by CPI Group (UK) Ltd, Croydon, CR0 4YY
01/10/2026
02225907-0015